Principles of Osteoimmunology

Peter Pietschmann
Editor

Principles of Osteoimmunology

Molecular Mechanisms and Clinical Applications

Second Edition

Springer

Editor
Peter Pietschmann
Department of Pathophysiology and Allergy Research
Center of Pathophysiology, Infectiology and Immunology
Medical University of Vienna
Vienna
Austria

ISBN 978-3-319-34236-8 ISBN 978-3-319-34238-2 (eBook)
DOI 10.1007/978-3-319-34238-2

Library of Congress Control Number: 2016949412

Printed on acid-free paper

This Springer imprint is published by Springer Nature
The registered company is Springer International Publishing AG Switzerland

Preface

Osteoimmunology is a rapidly developing research field on the crosstalk between bone and the immune system. Examples of such immune-bone interactions are pathogenic mechanisms of bone diseases that are caused by or related to altered immune reactions. The English term "osteoimmunology" was first used in 2000 by Arron and Choi in a comment in *Nature* (430: 535). Nevertheless, the concept that osteoclasts, multinucleated bone-resorbing cells, develop from the monocyte-macrophage lineage dates back to the 1920s. In the 1980s, proinflammatory cytokines such as interleukin-1 or TNF-alpha were shown to stimulate bone degradation. The discovery of the RANK/RANKL/osteoprotegerin system and the development of antibody-based targeted therapies for osteoporosis and other bone diseases have significantly increased the momentum of osteoimmunology.

The purpose of this book is to give an introduction to the field of osteoimmunology to scientists and clinicians working in immunology, pathophysiology and osteology. The book is organized into 12 chapters. The first chapters give an introduction to cell and molecular biology of bone and the immune system, including methodological issues such as automated cell detection and bone markers. Dedicated chapters also describe effects of vitamin D on the immune system and immunological aspects of biomechanics. The next chapters deal with molecular mechanisms, clinical aspects and the treatment of osteoimmune diseases such as osteoporosis and rheumatoid arthritis. A special chapter is dedicated to osteoporosis in organ transplant patients. The final chapter describes the osteoimmunology of periodontal diseases.

I am very thankful to all authors who contributed to this book for their valuable time, expertise and effort. Moreover, I would like to acknowledge the great help and dedication of the staff from Springer Nature, in particular Dr. Silvia Herold, Dr. Amrei Strehl, Mag. Angelika Heller and Prakash Jagannathan. Special thanks is given also to Birgit Schwarz for her continuous support of this book project.

I am convinced that our readers will enjoy the book as much as I enjoyed editing it.

Vienna, Austria
March 2016

Peter Pietschmann

Contents

Contributors

Ursula Azizi-Semrad, MD Department of Pathophysiology and Allergy Research, Center for Pathophysiology, Immunology and Infectiology, Medical University of Vienna, Vienna, Austria

Kristina Bertl, DMD, PhD, MSc Department of Periodontology, Faculty of Odontology, University of Malmö, Malmö, Sweden

Gerold Ebenbichler, MD Department of Physical Medicine and Rehabilitation, Medical University of Vienna, Vienna, Austria

Rupert Ecker, PhD TissueGnostics GmbH, Vienna, Austria

Isabella Ellinger, PhD Department of Pathophysiology and Allergy Research, Center for Pathophysiology, Infectiology and Immunology, Medical University of Vienna, Vienna, Austria

Jessica Furst, MD Department of Medicine, Columbia University Medical Center, New York, NY, USA

Andreas Heindl, PhD Department of Pathophysiology and Allergy Research, Center for Pathophysiology, Infectiology and Immunology, Medical University of Vienna, Vienna, Austria

Seewald Solutions, Weißkirchen an der Traun, Austria

Division of Molecular Pathology and Centre for Evolution and Cancer, The Institute of Cancer Research, London, UK

Lorenz C. Hofbauer, MD Division of Endocrinology, Diabetes and Metabolic Bone Diseases, Department of Medicine III, Dresden Technical University Medical Center, Dresden, Germany

Center for Healthy Aging, Dresden Technical University Medical Center, Dresden, Germany

Katharina Kerschan-Schindl, MD Department of Physical Medicine and Rehabilitation, Medical University of Vienna, Vienna, Austria

Roland Kocijan, MD Medical Department II, St. Vincent Hospital Vienna, Academic Teaching Hospital of the Medical University Vienna, Vienna, Austria

Peter Mikosch, MD Department of Internal Medicine, State Hospital Oberpullendorf, Oberpullendorf, Austria

Robert Nica TissueGnostics GmbH, Vienna, Austria

Barbara Obermayer-Pietsch, MD Division of Endocrinology and Metabolism, Medical University of Graz, Graz, Austria

Peter Pietschmann, MD Department of Pathophysiology and Allergy Research, Center of Pathophysiology, Infectiology and Immunology, Medical University Vienna, Vienna, Austria

Martina Rauner, PhD Division of Endocrinology, Diabetes and Metabolic Bone Diseases, Department of Medicine III, Dresden Technical University Medical Center, Dresden, Germany

Martin Schepelmann, PhD Department of Pathophysiology and Allergy Research, Center for Pathophysiology, Infectiology and Immunology, Medical University of Vienna, Vienna, Austria

School of Pharmacy and Pharmaceutical Sciences, Cardiff University, Cardiff, UK

Georg Schett, MD Department of Internal Medicine 3, University of Erlangen-Nuremberg, Erlangen, Germany

Verena Schwetz, MD, PhD Division of Endocrinology and Nuclear Medicine, Medical University of Graz, Graz, Austria

Alexander K. Seewald, PhD Seewald Solutions, Weißkirchen an der Traun, Austria

Elizabeth Shane, MD Department of Medicine, Columbia University Medical Center, New York, NY, USA

Wolfgang Sipos, DVM Clinical Department for Farm Animals, University of Veterinary Medicine, Vienna, Austria

Andreas Stavropoulos, DDS, PhD Department of Periodontology, Faculty of Odontology, University of Malmö, Malmö, Sweden

Theresia Thalhammer, PhD Department of Pathophysiology and Allergy Research, Center for Pathophysiology, Infectiology and Immunology, Medical University of Vienna, Vienna, Austria

Douglas H.N. White, M.D., BSc, MBChB, FRACP Rheumatology Department, Waikato Hospital, Hamilton, New Zealand

Martin Willheim, MD Department of Laboratory Medicine, Wilhelminenspital Vienna, Vienna, Austria

Maria Winzer, MD Division of Endocrinology, Diabetes and Bone Diseases, Medical Faculty of the Technische Universität Dresden, Dresden, Germany

Division of Rheumatology, Department of Medicine III, Medical Faculty of the Technische Universität Dresden, Dresden, Germany

Basics of Bone Biology

1

Martina Rauner and Lorenz C. Hofbauer

1.1 Introduction to Bone

1.1.1 Bone Function and Structure

Bone is the major constituent of the skeleton which is a hallmark of all higher vertebrates. Besides the protection of internal organs and the support of body structures, the most important functions of bone are to serve as an attachment site for ligaments, tendons, and muscles allowing locomotion and provide a cavity for hematopoiesis in the bone marrow (Mendez-Ferrer et al. 2010; Zaidi 2007). Moreover, bone has a central role in mineral homeostasis as it functions as a reservoir for inorganic ions that can be mobilized rapidly on metabolic demand.

Although bone is often considered an inert, static material, it is a highly organized, living tissue that undergoes constant remodeling. Different cell lineages have emerged to serve distinct skeletal functions. While cells from the hematopoietic lineage, such as osteoclasts, break down bone tissue to remove old and damaged bone or release calcium to maintain calcium homeostasis, cells from the mesenchymal lineage, including chondroblasts, fibroblasts, and osteoblasts, construct and later remodel bone tissue (Jiang et al. 2002). Osteoblasts produce the organic components

M. Rauner, PhD (✉)
Division of Endocrinology, Diabetes and Metabolic Bone Diseases,
Department of Medicine III, Dresden Technical University Medical Center,
Fetscherstr. 74, Dresden 01307, Germany
e-mail: martina.rauner@uniklinikum-dresden.de

L.C. Hofbauer, MD
Division of Endocrinology, Diabetes and Metabolic Bone Diseases,
Department of Medicine III, Dresden Technical University Medical Center,
Fetscherstr. 74, Dresden 01307, Germany

Center for Healthy Aging, Dresden Technical University Medical Center,
Fetscherstr. 74, Dresden 01307, Germany

P. Pietschmann (ed.), *Principles of Osteoimmunology*,
DOI 10.1007/978-3-319-34238-2_1

of the extracellular matrix, which mainly includes type I collagen (approximately 95 %), but also noncollagenous proteins (i.e., osteocalcin, osteopontin, osteonectin, bone sialoprotein) and proteoglycans. The inorganic matrix predominantly contains calcium and phosphorus, appearing as hydroxyapatite crystals ($Ca_{10}(PO_4)_6(OH)_2$), and is deposited into the collagenous matrix. This complex organization confers rigidity and strength to the skeleton while maintaining a high degree of elasticity.

Two types of osseous tissues are found in all bones: cortical or compact bone and trabecular or cancellous bone, sometimes also referred to as spongy bone (Fig. 1.1). Cortical bone is mainly found in the shafts of long bones (diaphyses) and is made of numerous overlapping cylindrical units termed Haversian systems or osteons. The central Haversian canal, containing the blood vessel and nerves, is surrounded by densely packed collagen fibrils which are formed to concentric lamellae. Osteocytes, terminally differentiated osteoblasts, are located between concentric lamellae and are connected to each other via canaliculi, allowing the exchange of nutrients and metabolic waste and the sensation of mechanical stress. Volkmann's canals are

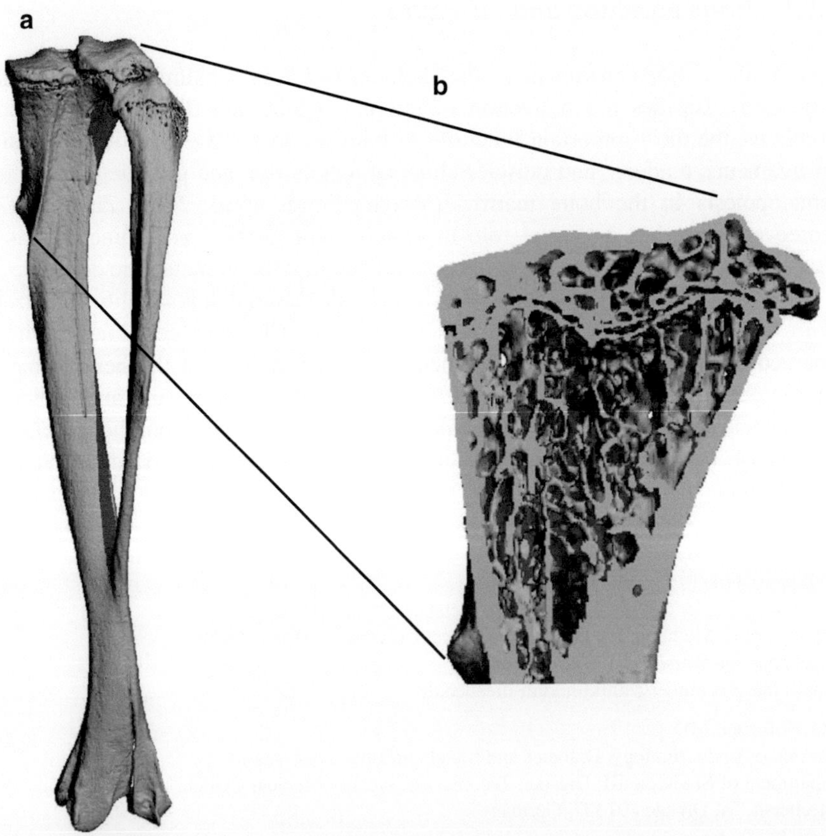

Fig. 1.1 Illustration of compact and cancellous bone. (**a**) Whole tibial structure measured by micro-computed tomography. Twelve-micron isotropic spatial resolution. (**b**) Lateral view of the tibia. The trabecular meshwork structure is clearly visible at the epiphysis. Cortical bone is found at the shaft (*Source*: Peter Varga and Phillipe Zysset)

responsible for the conjunction of blood vessels from the inner and outer bone surfaces to the vessels of the Haversian canals. The dense organization of cortical bone thus provides maximum strength and load-bearing capacity by being highly resistance to bending and torsion. Cancellous bone, on the other side, is predominantly found at the ends of long bones (epiphyses) as well as in flat bones and in vertebral bodies where force may be applied at variable angles. It is composed of a meshwork of trabeculae, thereby reducing skeletal weight without compromising strength. This particular construction also establishes a vast surface area. Considering that bone remodeling only takes place at bone surfaces, cancellous bone is quick to render metabolic activities but also disproportionally susceptible to damage when net bone loss occurs.

1.1.2 Ossification Processes

Ossification occurs either intramembranous or endochondral. During skeletal development, flat bones (e.g., calvariae) and some irregular bones are formed by intramembranous ossification where bony tissue directly forms from the connective tissue without an intermediate cartilage stage (Blair et al. 2008). Within this process, mesenchymal stem cells (MSCs) condense into highly vascularized sheets of primitive connective tissue at sites of eventual bone formation. Certain MSCs group together and differentiate into osteoblasts that deposit extracellular matrix (osteoid) which subsequently is mineralized forming the bone matrix. These small aggregates of bone tissue, termed bone spicules, continuously expand with new MSCs lining on the surface, differentiating into osteoblasts, and secreting extracellular matrix. Once they become embedded by the secreted mineralized matrix, osteoblasts terminally differentiate into osteocytes. As the bone spicules grow and interconnect with others, a trabecular network of woven bone – also referred to as primary spongiosa – is formed. Although woven bone forms quickly with the collagen fibers being randomly organized, it is structurally weak. Thus, it is soon replaced by a more solid lamellar bone, which is composed of a highly organized collagen structure (Frost and Jee 1994). Several collagen fibers align in the same layer, and several such concentric layers stacked in alternating orientations finally constitute a bone unit called osteon (Parfitt 1988). This highly sophisticated organization confers strength and resistance to torsion forces to lamellar bone. However, the complex architecture and orderly deposition of collagen fibers require more time and restrict the formation of osteoid to 1–2 μm per day. Besides the creation of woven bone in fetal bone development, it may also occur in adults after fractures or in patients with Paget's disease (Parfitt 1994).

In contrast to flat and irregular bones, bones of the vertebral column, pelvis, and extremities develop by endochondral ossification. Thereby, hyaline cartilage devoid of blood vessels is first formed and then replaced by bone matrix starting at the primary ossification center. During embryonic development, chondrocytes congregate to a cartilaginous model that alleges the shape of the future bone, and after the local enlargement of chondrocytes (hypertrophy), endochondral bone formation is initiated in the middle of the shaft at the primary ossification center. The perichondrium, which surrounds the cartilage model, becomes invaded with blood vessels and then is called periosteum (Stanka et al. 1991; Streeten and Brandi 1990; Trueta and Buhr 1963). The periosteum contains layers of MSCs that differentiate into osteoblasts during development, when the bone increases its width (appositional growth), or after fractures, when new bone formation is required. In addition to its important function to supply nutrients via the blood vessels, the periosteum contains nociceptor nerve endings that allow the sensation of pain (Fortier and Nixon 1997; Grubb 2004; Jimenez-Andrade et al. 2010).

The growth plates are characterized by the orderly proliferation and maturation of chondrocytes in longitudinal columns, forming stratified zones of reserve, proliferative, maturing, and hypertrophic cartilage (Poole et al. 1991). Hypertrophic chondrocytes secrete large amounts of a specialized extracellular matrix rich in collagen type X and alkaline phosphatase, which becomes calcified. After the calcification of the collagenous matrix, hypertrophic chondrocytes start producing matrix metalloproteinase 13, which is crucial for the subsequent degradation of the cartilage matrix, and undergo apoptosis (Stickens et al. 2004). By doing so, transverse septa of cartilage matrix surrounding them are broken down, leaving vertical septa largely intact, but allowing the entry of capillaries and invading cells of the ossification front. These cells mainly include cells of the mesenchymal (osteoblast precursors and stromal cells) and hematopoietic lineages (osteoclast precursors and other hematopoietic lineages that constitute the bone marrow). After osteoblast precursor cells have migrated to the surface of remnant cartilage spicules, they differentiate into fully mature osteoblasts and deposit a predominantly type I collagen-containing extracellular matrix (osteoid), which subsequently becomes mineralized into the mature bone matrix. The ossification continues toward the ends of the bones, where the further elongation of long bones occurs in the growth plates of the metaphysis. Finally, the trabecular bone in the diaphysis is broken down by osteoclasts to open up the medullary cavity.

The same procedure is true for the secondary ossification center, located in the epiphysis, except that the trabecular bone is retained (Alini et al. 1996). The length of bones increases until the early twenties through a process similar to endochondral ossification (Riggs et al. 1999). The cartilage in the epiphyseal plate remains proliferating and is continuously replaced by bone matrix until the skeleton has reached maturity and the epiphyseal plate has become almost completely ossified. The articular cartilage remains uncalcified and covers the ends of the long bones. Due to its incredibly low coefficient of friction, coupled with its ability to bear very large compressive loads, articular cartilage is ideally suited for placement in joints, such as the knee and hip.

1.2 Bone Remodeling

During a person's lifetime, continuously changing functional demands require permanent adaptation of the bone structure and microarchitecture. Wolff has observed this principle of functional adaptation already over 100 years ago (Wolff 1892). The process of where "form follows function" occurs in conditions of disuse (as during immobility, space flights, or long-term bed rest), overloading (weight gain), and growth and after fracture healing and consists of two activities, namely, bone formation and bone resorption (Sommerfeldt and Rubin 2001; Frost 1990 #133). While these processes are locally separated in modeling (Frost 1990), bone remodeling is characterized by the spatial and temporal coupling of bone formation by osteoblasts and bone resorption by osteoclasts (Rodan and Martin 1981). The so-called basic multicellular unit (BMU) is covered by a canopy of cells that creates a bone remodeling compartment (BRC). While the nature of the canopy cells remains under debate, evidence in humans suggests that it is bone lining cells with tissue-specific macrophages (osteomacs) interspersed, generating a unique microenvironment to facilitate coupled osteoclastic bone resorption and osteoblastic synthesis (Andersen et al. 2009; Chang et al. 2008; Sims and Martin 2014). Interestingly, the action of BMUs slightly differs in cortical (endocortical as well as intracortical surfaces) and trabecular bone. While in cortical bone, the BMU forms a cylindrical tunnel of about 2,000 μm long and 200 μm wide and the BMU burrows through the bone with a speed of 20–40 μm/day, the remodeling process in trabecular bone is mainly a surface event reaching a depth of approximately 50 μm. With a speed of 25 μm/day, active remodeling sites of BMUs in trabecular bone cover areas of varying sizes ranging from about 100–1,000 μm^2. In general, approximately 5–25 % of bone surface are undergoing bone remodeling (Parfitt 1994; Raisz 1988), thereby, restoring microdamages and ensuring mechanical integrity as well as regulating the release of calcium and phosphorus, while maintaining the global bone morphology.

An active BMU performs one bone remodeling cycle that occurs over several weeks and includes four main processes: activation, resorption, reversal, and formation (Parfitt 1988). While the process of bone resorption is usually accomplished within 2–3 weeks, the new synthesis of bone requires around 2–3 months. The remodeling cycle is initiated by the detection of signals that induce the activation of the quiescent bone surface, which is covered with bone lining cells. These signals may be provided through osteocytes that sense mechanical strain or are affected by structural damage, which severs the processes of osteocytes in their canaliculi and leads to osteocyte apoptosis (Aguirre et al. 2006; Bonewald 2007; Hazenberg et al. 2006; Verborgt et al. 2002). Alternatively, hormone actions (e.g., estrogen or parathyroid hormone (PTH)) due to more systemic changes in homeostasis or effects of corticosteroids on bone cells may negatively alter osteocyte biology. Current research points toward an intricate communication between osteocytes, which senses bone damage deep within the osteon or hemiosteons, and lining cells on the bone surface, which receive signals through the long processes of osteocytes and communicate the health status of the bone to the marrow environment to initiate the establishment of a BRC (Hauge et al. 2001). Osteocyte apoptosis may also contribute to the recruitment of osteoclast precursor cells by diminishing the osteocytic secretion of factors that usually inhibit osteoclast formation such as transforming growth factor-β (TGF-β) (Heino et al. 2002). In vivo evidence indicates that osteocyte apoptosis precedes osteoclast formation as osteocyte apoptosis occurs within 3 days of immobilization and is followed within 2 weeks by osteoclastogenesis (Aguirre et al. 2006). Although the process of osteoclast precursor attraction is not fully understood yet, osteoblast-secreted products including monocyte chemoattractant protein-1 (MCP-1), the osteoclast differentiating factor receptor activator of NF-kB ligand (RANKL), as well as the osteoclast-attractant sphingosine-1-phosphate (S1P) may play an important role (Ishii et al. 2009, 2010; Keller et al. 2014; Li et al. 2007; Nakashima et al. 2011; Xiong et al. 2011). After osteoclast precursor cells are recruited to the activated surface, they fuse to form mature, bone resorbing osteoclasts (Vaananen and

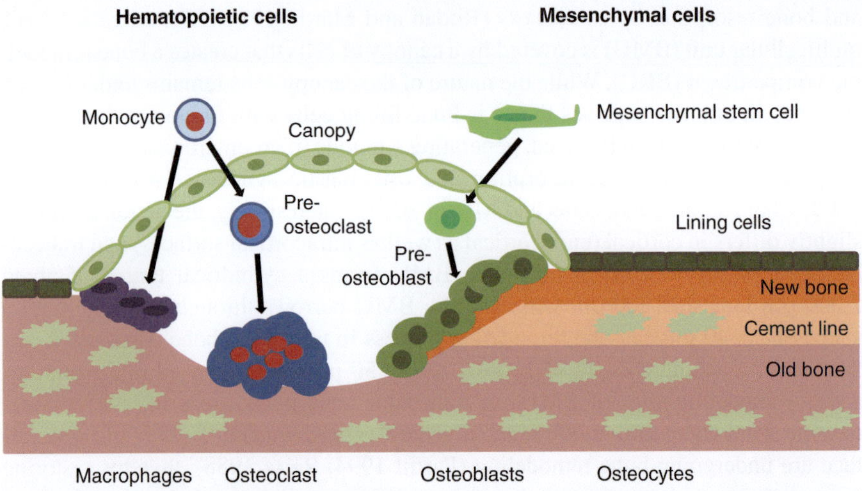

Fig. 1.2 Bone remodeling. Monocytes from the hematopoietic lineage differentiate into osteoclasts, which resorb old and damaged bone tissue. Macrophages, which also originate from the hematopoietic lineage, contribute to the initiation of bone remodeling and attract osteoblast precursors that mature to bone-forming osteoblasts at the bone surface. After filling the resorption lacunae, osteoblasts become embedded by the bone matrix and turn into osteocytes. Quiescent lining cells remain at the bone surface

Horton 1995). The osteoclasts attach to the surface and form a ruffled boarder at the bone/osteoclast surface that is completely surrounded by a sealing zone. Thereby, osteoclasts create an isolated microenvironment to dissolve the inorganic matrix by producing an acidic microenvironment and degrade the organic matrix with specific enzymes (Teitelbaum 2000). As bone resorption subsides and a resorption pit with a demineralized collagen matrix remains, osteoclasts disappear and mononuclear cells of undetermined lineage remove the collagen remnants and prepare the surface for bone formation. This phase is called reversal. Currently, there is a debate about whether the reversal cell is of hematopoietic or mesenchymal origin. Recent evidence suggests that this cell type may be a resident macrophage of the bone termed osteomacs (Pettit et al. 2008). These cells are positive for the macrophage markers F4/80+ and CD68, but negative for the osteoclast marker tartrate-resistant acid phosphatase (TRAP), and are found throughout in the periosteum and endosteum. Moreover, these cells have been shown to produce MMPs, which are required for matrix degradation, as well as TGF-β and ephrin B2, which may promote osteoblast recruitment, differentiation, and/or activation of bone lining cells (Chang et al. 2008; Compagni et al. 2003). Thus, these cells would be ideal couplers of bone resorption and formation. However, further research is needed to clarify the nature of the reversal cells. After the reversal phase, the bone remodeling cycle is finished with the synthesis and deposition of bone matrix by osteoblasts until an equal amount of bone was reproduced. Also in this case, the mechanisms that terminate bone formation are not known, but may be mediated by signals from osteocytes that have become embedded in the mature bone matrix. Finally, bone lining cells build a canopy covering the surface keeping the material dormant until the next cycle (Fig. 1.2).

1.3 Key Players of Bone Remodeling

1.3.1 Cells of the Osteoblast Lineage: Osteoblasts, Osteocytes, and Bone Lining Cells

1.3.1.1 Functions of Osteoblast Lineage Cells

Osteoblasts are derived from MSCs, and their primary function is to synthesize the organic collagenous matrix and promote its mineralization by producing bone matrix proteins including osteocalcin, osteopontin, and bone sialoprotein and providing optimal environmental conditions for crystal formation (Ducy et al. 2000). Due to their active protein machinery, osteoblasts have a prominent Golgi apparatus and endoplasmatic reticulum. As mentioned earlier, osteoblasts are also the main producers of RANKL and its decoy receptor osteoprotegerin (OPG) and are therefore critically involved in regulating osteoclastogenesis (see also Sect. 1.4.2). Fully differentiated osteoblasts that are surrounded by mineralized bone tissue are called osteocytes and act as mechanosensors in bone tissue (Bonewald 2011; Paic et al. 2009). They are the most numerous cells within the bone tissue and scattered evenly through the matrix. With their flattened morphology and long processes, they form a sensory network which allows the detection of abnormal strain situations such as generated by microcracks (Hirao et al. 2007; Martin and Seeman 2008). By communicating these signals to bone lining cells (the second terminally differentiated osteoblast cell type) or secrete factors that recruit osteoclasts, osteocytes initiate the repair of damaged bone.

Other emerging roles of osteoblast lineage cells include the maintenance of hematopoietic stem cell (HSC) niches and HSC homing as well as acting as nonprofessional antigen-presenting cells in conditions of inflammation (Fleming et al. 2008; Jung et al. 2007; Mendez-Ferrer et al. 2010; Ruiz et al. 2003; Schrum et al. 2003; Skjodt et al. 1989). While the capacity to stimulate effector cells of the immune system may only be relevant under pathophysiological conditions, the osteoblast-driven maintenance of the stem cell niche is of critical importance for the homeostasis of hematopoiesis. Experiments in mice have shown that the number of long-term repopulating HSCs increases or decreases in parallel with in vivo osteoblast stimulation by PTH or osteoblast ablation using a mouse genetic approach (Visnjic et al. 2004). Although the underlying signaling events are not fully understood yet, several mechanisms such as the selective expression of signaling molecules (i.e., jagged, G-protein Gsα), adhesion molecules (i.e., integrins, N-cadherin), and components of the ECM (i.e., proteoglycans) may determine the long-term repopulating ability of HSCs and their ability to home into the bone marrow.

Finally, one of the most important discoveries in the past years was the appreciation of bone as an endocrine organ (Karsenty and Ferron 2012). Next to other factors that have been described to possess hormonal features (e.g., fibroblast growth factor 23, an osteocyte-produced hormone that regulates phosphate homeostasis (Feng et al. 2006)), several novel endocrine functions of osteocalcin have been discovered. First, it has been shown to control glucose metabolism by stimulating insulin production in beta cells in the pancreas and by directing fat cells to release adiponectin, both leading to enhanced insulin sensitivity (Lee et al. 2007 and Ferron et al. 2010). In addition, osteocalcin increases energy expenditure in skeletal muscle and brown adipose tissue, further contributing to the protective effect of osteocalcin on obesity and insulin resistance (Ferron et al. 2008; Lee et al. 2007; Puigserver et al. 1998). Next to the control of energy balance, osteocalcin has also been linked to the regulation of reproduction. By stimulating the biosynthesis of testosterone in Leydig cells, osteocalcin has also been shown to regulate male fertility (Oury et al. 2011). Thus, the osteoblast does not only serve as a matrix-producing cell but controls several critical organ functions.

1.3.1.2 Osteoblast Differentiation and Involved Signaling Pathways

MSCs give rise to a variety of cells including osteoblasts, adipocytes, chondrocytes, and myoblasts (Pittenger et al. 1999). Several steps of commitment are undertaken by the MSC to generate progeny with more limited differentiation capacities until the differentiated end-stage cell is able to express distinct functional markers and morphological traits. Typical osteoblast markers include alkaline phosphatase (ALP) and type I collagen, as well as various noncollagenous proteins such as osteocalcin, osteopontin, or bone sialoprotein. However, cells of the osteoblastic lineage also selectively express proteins at distinct differentiation stages such as type I collagen in matrix-producing osteoblasts or osteocalcin and sclerostin in fully mature osteoblasts or osteocytes.

Osteoblasts express receptors for various hormones including PTH, 1,25-dihydroxyvitamin D_3, estrogen, glucocorticoids, and leptin, which are involved in the regulation of osteoblast differentiation (see Sect. 1.4.1). Furthermore, osteoblasts are regulated by multiple local factors including bone morphogenetic proteins (2, 4, 6, and 7) (Shore et al. 2006; Storm and Kingsley 1999; Wu et al. 2003; Wutzl et al. 2010), growth factors (transforming growth factor-β, epidermal growth factor, insulin-like growth factor) (Canalis 2009), Sonic and Indian hedgehogs (Guan et al. 2009; Maeda et al. 2007), as well as members of the Wnt family in a paracrine and autocrine fashion (Baron and Kneissel 2013). Because the Wnt signaling pathway is of such critical importance for bone mass maintenance, it will be discussed here in more detail.

Wnt signaling is highly conserved throughout evolution among a variety of species and plays an important role in regulating cellular processes such as proliferation, differentiation, cell survival, and motility (van Amerongen and Nusse 2009). Wnt signaling further plays a key role in embryonic development and maintenance of tissue homeostasis, including bone (Baron and Kneissel 2013). Wnt proteins are cysteine-rich glycoproteins that act on target cells by binding to the seven-span transmembrane receptor protein Frizzled (FZD) and low-density lipoprotein receptor-related proteins 5 and 6 (LRP 5/6). In bone, various components of this pathway have been shown to positively or negatively regulate osteoblast differentiation (Baron and Kneissel 2013). First evidence that the Wnt/β-catenin pathway is involved in bone mass homeostasis has been provided by observations of mutations in the LRP5 gene, in which gain-of-function mutations led to a high bone mass phenotype in humans and mice and loss-of-function mutations led to low bone mass phenotypes (Boyden et al. 2002; Gong et al. 2001; Van Wesenbeeck et al. 2003). Since then, the discovery of several other members of the Wnt pathway has been implicated in participating in the regulation of bone remodeling (summarized in Baron and Kneissel 2013).

Wnt signaling comprises several pathways that are usually divided into the canonical or β-catenin-dependent pathway and noncanonical or β-catenin-independent pathways. Both canonical and noncanonical pathways regulate bone metabolism. The canonical pathway is activated upon binding of Wnt proteins to a receptor complex consisting of Frizzled (FZD) and its co-receptor, lipoprotein receptor-related protein (LRP)-5 or LRP-6. This interaction results in the stabilization and nuclear translocation of β-catenin, where it induces the transcription of osteoblast proteins (Clevers 2006). Noncanonical signaling pathways on the other hand are activated by distinct receptors such as ROR2 and RYK and comprise various downstream pathways such as the Ca^{2+}-dependent pathways, protein kinase C, as well as Rho/Rac small GTPase and Jun N-terminal kinase (JNK) (Baron and Kneissel 2013).

Wnt signaling is regulated at various levels such as through the presence or absence of multiple Wnt proteins, co-receptors, intracellular signaling molecules, and transcription factors. Furthermore, it is tightly regulated by a series of extracellular inhibitors including members of the secreted frizzled-related protein (sFRP) family and Wnt inhibitory factor that bind to Wnt ligands, as well as dickkopfs (Dkks) and sclerostin, both binding to LRP5/6 (Semenov et al. 2005; Tian et al. 2003). In both cases, these interactions lead to the blockade of Wnt ligands binding to FZD receptors. Many Wnt inhibitors have been proposed as therapeutic targets for increasing bone mass by applying neutralizing antibodies, whereas sclerostin may be of particular interest due to its specific expression in osteocytes (Keller and Kneissel 2005; Paszty et al. 2010). Romosozumab and blosozumab, antibodies that block sclerostin actions in humans, are currently in phase 3 and phase 2 clinical trials, respectively, to test their efficacy to prevent fractures in postmenopausal women (McClung et al. 2014; Recker et al. 2015).

At a transcriptional level, osteoblast differentiation is induced by the master transcription factor runx2 (runt-related transcription factor 2, also called core binding factor 1, Cbfa1) and several signaling pathways converge to increase runx2 expression. Runx2-deficient mice have no osteoblasts and thus only contain a cartilage-like skeleton (Harada et al. 1999; Miller et al. 2002). Intriguingly, although runx2 supports osteogenic differentiation, it inhibits osteoblast maturation into osteocytes, keeping osteoblasts in an immature state (Lian et al. 2006). Runx2 expression is induced by BMPs, TGFβ1, Indian hedgehog, and members of the Wnt pathway and is tightly regulated by various posttranslational modifications as well as co-repressors, such as Twist and menin-1, and co-activations, such as TAZ. Even though runx2 is regarded as the master transcription factor for osteoblasts, also other transcription factors including osterix (also called specificity protein 7, sp7) (Kim et al. 2006b), β-catenin (Krishnan et al. 2006), dlx3 and dlx5 (distal-less homeobox) (Harris et al. 2003), msx2 (homeobox factor) (Liu et al. 1999; Satokata et al. 2000), ATF4 (activating transcription factor 4) (Tozum et al. 2004), as well as NFATc1 (nuclear factor of activated T cells c1) (Koga et al. 2005) also participate in the regulation of osteoblast differentiation. Many of these factors control osteoblast differentiation at very specific locations such as dlx proteins in the skull.

After osteoblasts have fully matured and deposited a mineralized matrix surrounding them, they become osteocytes, which serve different functions than matrix deposition. As mentioned above, osteocytes are evenly located throughout the bone tissue and produce a dense network by connecting each other via gap junctions on their processes. In this respect, connexion-43 seems to play a critical role in the formation of hemichannels which allow an extensive communication between two osteocytes (Plotkin et al. 2002, 2008). Mice rendered osteocyte depletion exhibit enhanced bone fragility, intracortical porosity, and microfractures, indicating their crucial function to maintain bone integrity (Tatsumi et al. 2007). Also several other studies have shown that loss of osteocyte viability is related to bone loss (Aguirre et al. 2006; Teti and Zallone 2009; Weinstein et al. 2000). Besides mechanosensation and mechanotransduction, osteocytes control phosphate metabolism by producing FGF-23 and express several mineralization inhibitors including fetuin-A, dentin matrix protein-1, phex, and the Wnt inhibitor sclerostin, which allows them to control the amount and quality of the bone matrix (Bonewald 2011; Coen et al. 2009; Liu et al. 2009; Poole et al. 2005). Dentin matrix protein-1-deficient mice, for example, show impaired osteocyte maturation, increased fibroblast growth factor-23 expression, and severe abnormalities of bone mineralization (Feng et al. 2006). Of note, also glucocorticoids have the potential to increase the expression of mineralization inhibitors, thereby compromising bone quality and bone strength (Yao et al. 2008).

1.3.2 Cells of the Osteoclast Lineage: Osteomacs and Osteoclasts

Tissue-resident macrophages, also referred to as osteomacs, and osteoclasts both derive from the hematopoietic monocytic lineage. The concept of osteomacs has only recently been developed due to thorough observations of periosteal and endosteal tissues and the bone remodeling compartment (Pettit et al. 2008). Pettit et al. determined that osteomacs constitute about one sixth of the total cells within osteal tissues and span a network along bone surfaces with their stellate morphology. Due to their abundance and wide-spread location, it is likely that osteomacs contribute to immune surveillance in the bone marrow compartment and react quickly to inflammatory stimuli. Osteomacs are distinguishable from osteoclasts by the expression of the murine macrophage marker F4/80, which is not present on osteoclasts, and by being negative for osteoclast-specific markers such as TRAP (Chang et al. 2008). As they are also located at the bone remodeling compartment, osteomacs have been proposed to participate in the reversal phase of bone remodeling and closely interact with osteoblasts through the production of osteoblast-stimulating factors such as bone morphogenetic protein-2 or transforming growth factor-β.

Osteoclasts are tissue-specific giant polykaryons (up to 100 μm in diameter) derived from the monocyte/macrophage hematopoietic lineage and are the only cells capable of breaking down large amounts of mineralized bone, dentine, and calcified cartilage (Teitelbaum 2000). Bone resorption is a crucial step in bone remodeling which is necessary for healthy bone homeostasis, thereby repairing microdamages and adapting to new mechanical loads and altered metabolic conditions. Bone remodeling starts with the retraction of bone lining cells uncovering bone tissue and attracting mononuclear precursors to the bone surface. The earliest step in osteoclastogenesis is the determination of the stem cell precursor to the osteoclastic lineage following the induction of PU.1 (Tondravi et al. 1997). Soon thereafter, precursors express the M-CSF receptor, *c-fms*, and after activation with the ligand, proliferation is induced. The next determination step toward a mature osteoclast is the expression of receptor activator of NFkB (RANK). The presence of its ligand, RANKL, is essential for the formation and fusion of multinucleated cells. Mice lacking either RANKL or RANK have no osteoclasts and severe osteopetrosis (for more details, see Sect. 1.4.2) (Anderson et al. 1997; Dougall et al. 1999; Kong et al. 1999b; Lacey et al. 1998; Yasuda et al. 1998). RANK signaling activates several transcription factors that are essential for osteoclastogenesis including activated protein-1, NFkB, or nuclear factor of activated T cells (NFAT). In the osteoclast, most signals converge to induce the activity of NFATc1. This is also proven genetically, as embryonic precursors lacking NFATc1 fail to become osteoclasts (Takayanagi et al. 2002). Importantly, NFATc1 is indispensible and sufficient for osteoclastogenesis, as its overexpression yields osteoclasts even in the absence of RANK signaling (Matsuo et al. 2004).

Although several downstream effectors of RANK signaling induce NFATc1 expression, the mechanisms that induce NFATc1 in a calcium-dependent way have only recently been identified. Therein, immunoreceptor tyrosine-based activation motif (ITAM)-containing adaptor molecules such as DAP (DNAX-activating protein) 12 and Fc common receptor-γ chain (FcR-γ) have been shown to be indispensable for osteoclastogenesis as mice deficient for both receptors are severely osteopetrotic (Koga et al. 2004; Mocsai et al. 2004). The activation of phospholipase-Cγ, Syk, and Tec kinases has been shown to be required for the activation of calcineurin-dependent calcium (Faccio et al. 2005; Mocsai et al. 2004; Wada et al. 2005). Paired immunoglobulin-like receptor-A (PIR-A) and osteoclast-associated receptor (OSCAR) have been found to associate with FcR-γ (Kim et al. 2002), whereas triggering receptor expressed on myeloid cells-2 (TREM-2) and signal-regulatory protein-β1 (SIRP-β1) bind to DAP12. These signals are considered to act as co-stimulatory signals for RANKL in osteoclast precursors, since those signals alone are not able to induce osteoclastogenesis (Koga et al. 2004).

Mature osteoclasts express several specific proteins including TRAP, cathepsin K, calcitonin receptor (CTR), and integrin receptors (Teitelbaum 2000; Teitelbaum and Ross 2003). Via integrins, osteoclasts attach very tightly to the matrix (sealing zone), thereby creating an isolated lacuna (Howship's lacuna) able to maintain an acidic environment necessary for matrix dissolution (Mimura et al. 1994; Miyauchi et al. 1991). At least four integrin receptors are expressed in osteoclasts, including $\alpha_v\beta_3$, $\alpha_v\beta_5$, $\alpha_2\beta_1$, and $\alpha_v\beta_1$ binding to various extracellular matrix proteins such as vitronectin, collagen, osteopontin, and BSP. After attachment, intracellular rearrangements lead to the polarization of the cell borders, whereas the sealing zone is adjacent to the basolateral domain and the ruffled border, respectively. At the opposite side of the ruffled border emerges the functional secretory domain. The ruffled border and the functional secretory domain are connected to each other via microtubules on which exocytotic vesicle traffic has been observed, suggesting the secretion of resorbed material into the extracellular space (Vaananen and Horton 1995). In addition to the development of distinct membrane domains, the cytoskeleton undergoes organizational changes creating a dense actin-ring in osteoclasts preparing for resorption (Silver et al. 1988). This process has been shown to be greatly dependent on Rho-GTPases, which require the mevalonate pathway for isoprenylation and activation (Chellaiah 2006). Of note, bisphosphonates have been shown to block osteoclast activity by inhibiting farnesyl diphosphate synthase, a critical enzyme in the mevalonate pathway.

The resorption of bone matrix takes place in the resorption lacuna. The ruffled border is formed by the fusion of cytoplasmic acidic vacuoles, thereby releasing acid into the resorption lacuna and initiating rapid dissolution of the hydroxyapatite crystals (Blair et al. 1989; Teti et al. 1989). Furthermore, ATPases, located in the ruffled border, additionally transport protons into the Howship's lacuna (Li et al. 1999; Mattsson et al. 1994). The protons are supplied by the reaction of water and carbon dioxide catalyzed by the enzyme carbonic anhydrase II resulting in the formation of protons and HCO_3^-. Whereas H^+ is pumped into the resorption lacuna, HCO_3^- is transported into the extracellular space via HCO_3/Cl exchanger. The imported chloride ions are also pumped into the resorption lacuna to form hydrochloric acid with a pH as low as 4, which is capable of dissolving the mineralized matrix (Silver et al. 1988). The organic matrix is degraded by various enzymes, including TRAP, cathepsin K, and matrix MMP-9. Cathepsin K is a lysosomal cysteine proteinase capable of degrading type I collagen (Gelb et al. 1996). Although osteoclasts form in cathepsin K-deficient mice that build a ruffled boarder and are able to mobilize bone mineral, they are unable to efficiently degrade the collagen matrix and thus resorb bone (Saftig et al. 1998). Furthermore, active osteoclasts express high levels of matrix metalloproteinases such as TRAP and MMP-9 (Okada et al. 1995; Wucherpfennig et al. 1994). Using electron microscopy, Okada and colleagues were able to show that MMP-9 degraded collagen into fragments, suggesting the involvement of MMP-9 in the resorption process. Stronger evidence is provided by mice lacking MMP-9, which are severely osteopetrotic and have difficulties in the endochondral ossification process, as the collagen matrix is only insufficiently being broken down (Engsig et al. 2000).

After the resorption of bone tissue, osteoclasts die by apoptosis and are quickly removed by phagocytes (Teitelbaum and Ross 2003). At present, little is known about the molecular mechanisms that terminate osteoclast resorption and initiate osteoclast apoptosis in vivo. Nevertheless, targeting osteoclasts for apoptosis, such as by using bisphosphonates or denosumab, a monoclonal antibody targeting RANKL, has been the predominant approach to prevent bone destruction in conditions of bone loss such as postmenopausal osteoporosis and therapy-induced or cancer-related bone loss (Rachner et al. 2011).

1.4 Regulation of Bone Remodeling

1.4.1 Hormones

Bone formation and resorption, as well as the cell machinery that performs those tasks, are under the subtle control of various hormones, whereas the most extensively studied ones are estrogens and androgens, parathyroid hormone (PTH), 1,25-dihydroxyvitamin D_3, and glucocorticoids, due to their common use as anti-inflammatory drugs. These major endocrine regulators will be discussed in more detail. However, it should be noted that bone homeostasis is also regulated by other hormones such as calcitonin (Huebner et al. 2008), leptin (Karsenty and Ducy 2006), and hormones of the anterior pituitary gland (follicle-stimulating hormone, thyroid-stimulating hormone, and adrenocorticotropic hormones) (Imam et al. 2009).

PTH is a peptide hormone and one of the most important regulators of calcium ion homeostasis (Kronenberg 2006; Lanske et al. 1999). PTH is produced and secreted by C cells in the parathyroid gland in response to low blood calcium levels and acts on the kidney, bone, and intestine to maintain blood calcium concentrations. In bone, PTH stimulates the production of interleukin-6 and RANKL by osteoblasts and stromal cells, thereby promoting the differentiation, activation, and survival of osteoclasts (Dai et al. 2006; Greenfield et al. 1993). Thus, PTH and PTHrP (PTH-related protein) promote bone resorption and consequently the release of calcium (Lanske et al. 1999; Pollock et al. 1996). However, it should be noted that an intermittent exposure to PTH has bone anabolic effects mainly increasing osteoblast functions and, thus, is currently the only approved anabolic treatment option in the treatment of postmenopausal osteoporosis (Bilezikian and Kurland 2001).

Calcitriol (1α,25-dihydroxyvitamin D_3), the active hormonal form of vitamin D, is a steroid hormone either ingested from the diet or synthesized in the skin from 7-dehydrocholesterol through exposure to sunlight (Webb and Holick 1988). Its importance for the development and maintenance of the mineralized skeleton was demonstrated in studies using vitamin D receptor or 1α(OH)ase knockout mice (Dardenne et al. 2001; Panda et al. 2004). The mineralization defect was normalized after a high-calcium, high-phosphate, and high-lactose diet (rescue diet) was administered. However, the administration of only $1,25(OH)_2D_3$ to 1α(OH)ase knockout mice was not sufficient to normalize the impaired mineralization if hypocalcemia was not corrected (Panda et al. 2004). Moreover, vitamin D-deficient mice showed an increase in osteoblast number, bone formation, and bone volume as well as increased serum ALP levels. Additionally, osteoclast numbers were decreased due to a decreased production of RANKL and an enhanced production of OPG (Kitazawa et al. 2003).

Besides PTH and calcitriol, which mainly regulate calcium homeostasis, estrogens and androgens are sex steroid with profound effects on bone. In contrast to PTH and 1,25-dihydroxyvitamin D_3, they enhance bone formation and inhibit bone resorption (Carani et al. 1997; Khosla et al. 2001; Leder et al. 2003). Lack of estrogen as well as testosterone inevitably leads to an increased bone turnover rate with a simultaneous increase in osteoclastic bone resorption as well as osteoblastic bone formation

(Eghbali-Fatourechi et al. 2003; Khosla and Riggs 2003; Weitzmann et al. 2002), although the latter of less magnitude. However, the net effect of estrogen deficiency is bone loss as a result of an increased production of RANKL and a decreased production of OPG in osteoblastic cells as well as an increase in the secretion of pro-inflammatory and pro-resorptive cytokines in lymphocytes such as IL-1, IL-6, and tumor necrosis factor-alpha (TNFα) (Jilka et al. 1992, 1995; Tanaka et al. 1993). Although clinical trials have shown that hormone replacement therapy decreased the incidence of major osteoporotic fractures (Cauley et al. 1995; Orwoll et al. 1996), serious side effects including cardiovascular disease and cancer have occurred, and therefore other medications are now used in the treatment of osteoporosis (e.g., selective estrogen receptor modulators such as raloxifene) (Riggs and Hartmann 2003).

Bone cells also contain glucocorticoid receptors that confer responsiveness to endogenously produced and exogenously administered glucocorticoids. While active forms of endogenous glucocorticoids such as cortisol are necessary for bone development, glucocorticoid excess is detrimental to many metabolic systems including bone. Studies in mice lacking 11β-hydroxysteroid dehydrogenase-2, an enzyme that inactivates active glucocorticoids, showed that endogenous glucocorticoids are necessary to prime MSCs to the osteoblastic lineage (Eijken et al. 2005; Hamidouche et al. 2008; Sher et al. 2004; Zhang et al. 2008) and support osteoblastogenesis. This is also recapitulated in human osteoblast cultures, which require physiological amounts of glucocorticoids to differentiate into fully mature, mineralizing osteoblasts (Rauner et al. 2010). Moreover, when the glucocorticoid receptor was specifically deleted in osteoclasts, glucocorticoids enhanced the lifetime of osteoclasts, but at the same time inhibited bone resorption by disrupting the osteoclastic cytoskeleton (Kim et al. 2006a). In contrast to these bone anabolic physiological effects of glucocorticoids, the prolonged exposure to synthetic glucocorticoids, such as required to treat inflammation or organ rejection, results in severe bone loss already within the first months of administration. The pathophysiology of glucocorticoid-induced bone loss includes the transient hyperactivation of osteoclasts due to an increased RANKL/OPG ratio in osteoblasts (Hofbauer et al. 1999, 2009) and a severely inhibited osteoblast function, mediated by the suppression of critical pro-osteoblastic factors such as runx2, Wnt, and BMP signaling, as well as the induction of mineralization inhibitors such as dentin matrix protein-1 or phex (O'Brien et al. 2004; Rauner et al. 2010; Thiele et al. 2012; Wang et al. 2005, 2008; Yao et al. 2008). Animal studies suggest that glucocorticoid-induced osteoporosis may be successfully prevented administering bisphosphonates, PTH, denosumab, or an anti-sclerostin antibody (Hofbauer et al. 2009; Yao et al. 2008, 2015). Analogous to selective estrogen receptor modulators, also selective glucocorticoid receptor modulators have been developed that display an improved benefit/risk ratio, but need to be verified in human studies.

1.4.2 Osteoblast-to-Osteoclast Signaling: RANKL/OPG

Although interactions between osteoblasts and osteoclasts have already been observed in the 1980s by Rodan and colleagues, it took another 15 years to identify

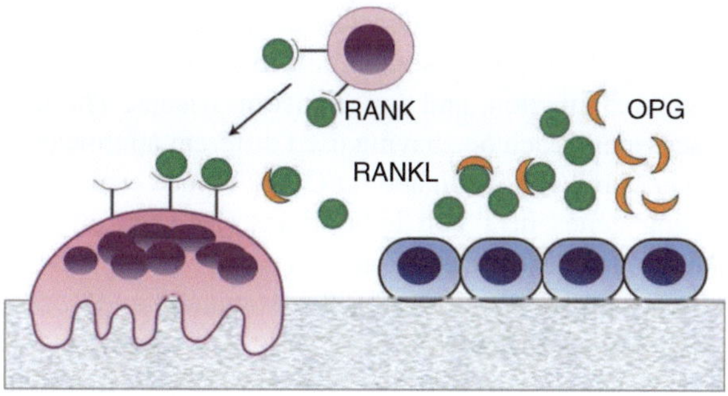

Fig. 1.3 RANK/RANKL/OPG system. RANK is expressed on mononuclear osteoclast precursor cells (*pink*). Upon binding of RANKL (*circles*), produced by osteoblasts (*blue*), osteoclast differentiation is induced. RANKL/RANK signaling is also active in mature osteoclasts (*pink*) to promote resorption activity and prolong survival. OPG (*half circle*), which is also produced by osteoblasts, is a soluble decoy receptor for RANKL and can thereby prevent binding of RANKL to RANK, thus the induction of osteoclastogenesis

the two main negotiators in osteoblast-osteoclast communication, RANKL and OPG (Anderson et al. 1997; Kong et al. 1999b; Lacey et al. 1998; Yasuda et al. 1998) (Fig. 1.3). Today, the high efforts invested in understanding and characterizing the RANKL/RANK/OPG system have led to detailed knowledge of the pathogenesis of metabolic bone diseases and have already contributed to the development of innovative therapeutic drugs that are now already in clinical use (human anti-RANKL antibody, denosumab, Prolia^{TM}) (Cummings et al. 2009; Smith et al. 2009).

As mentioned earlier, bone formation and bone resorption are coupled processes in remodeling. Often, dysregulations favoring osteoclastogenesis are responsible for the development of metabolic bone diseases such as osteoporosis, Paget's disease, rheumatoid arthritis, or osteoarthritis. The discovery of RANKL and its receptors RANK and OPG has finally highlighted the molecular processes in osteoclastogenesis raising the possibility to inhibit the development of osteoclasts, rescuing bone from exorbitant resorption.

In 1997 Simonet et al. discovered a protein which exposed an osteopetrotic phenotype when overexpressed in transgenic mice (Simonet et al. 1997). Investigating even further, they found that this protein was secreted by preosteoblasts/stromal cells and was capable to inhibit osteoclast development and activation. Due to its bone-protective effects, they named it osteoprotegerin. OPG belongs to the TNF receptor superfamily, however, lacking a transmembrane and cytoplasmic domain. OPG is expressed on a variety of tissues, including the lung, heart, kidney, liver, stomach, intestine, brain, spinal cord, thyroid gland, and bone, indicating multiple possible functions. The most prominent role of OPG has been assigned to bone protection; however, recent investigations have also proposed important functions of OPG in endothelial cell survival (Holen et al. 2005; Malyankar et al. 2000) and vascular calcification (Al-Fakhri et al. 2005; Rasmussen et al. 2006; Bucay et al. 1998).

After the identification of OPG followed the discovery of RANKL, which does not only have a huge repertoire of names (TRANCE, TNF-related activation-induced cytokine; ODF, osteoclast differentiating factor; OPGL, osteoprotegerin ligand; TNFSF11, TNF superfamily member 11) but also many faces regarding its structure, function, and appearance in tissues. The names originated from the four discoverers, each one having used different approaches to identify the protein. They either searched for a ligand for OPG (Yasuda et al. 1998), screened for apoptosis-regulating genes in T cell hybridomas (Kong et al. 1999b), or found RANKL to induce osteoclastogenesis (Lacey et al. 1998) and enhance the life span of dendritic cells (Anderson et al. 1997). Kartsogiannis and colleagues detected RANKL protein and mRNA expression in a variety of tissues, including the bone, brain, heart, kidney, liver, lung, intestine, skeletal muscle, mammary tissue, placenta, spleen, thymus, and testis (Kartsogiannis et al. 1999). This extensive distribution of RANKL throughout the body already indicates its multiple functions, whereas the most important one is dedicated to the regulation of bone remodeling. RANKL knockout mice reveal a severe osteopetrotic phenotype due to the absence of osteoclasts. Furthermore, defects in tooth eruption, lymph node genesis, mammary gland, and lymphocyte development were reported as well as disturbances in T cell/dendritic cell interactions and thermoregulation (Kong et al. 1999a; Martin and Gillespie 2001). RANKL is a member of the TNF superfamily and is mainly expressed in preosteoblasts/stromal cells as well as activated T cells. It exists in three isoforms: RANKL1 and RANKL2 are type II transmembrane proteins, whereas RANKL2 encodes for a shorter intracellular domain. RANKL3 is a soluble protein, supposed to be cleaved by TACE (TNFα-converting enzyme, a metalloprotease) from the transmembrane form. Other isoforms of RANKL are likely to exist, for Kong and colleagues mentioned a primary secreted form of RANKL in activated T cells (Kong et al. 1999a) and other groups found diverse posttranslational modifications on the N-terminus (Dossing and Stern 2005).

The third participant in the bone remodeling regulatory system is RANK and belongs to the TNFR superfamily alike OPG (Anderson et al. 1997). RANK represents a type I transmembrane protein and is expressed in tissues as ubiquitously as RANKL, though most commonly found in osteoclasts and dendritic cells (Hofbauer and Heufelder 2001). RANK-deficient mice show similar phenotypes to those of RANKL knockout mice, including lack of tooth eruption, osteopetrosis, and missing lymph nodes (Dougall et al. 1999). The RANK signaling cascade is initiated when RANKL binds to the extracellular domain of RANK which passes the signal along to TRAF6 (TNF receptor-associated factor 6). TRAF6 has various downstream mediators including the transcription factors NFkB, NFATc1 (nuclear factor of activated T cells), and AP-1 (activator protein-1) as well as the cascades of mitogen-activated protein kinases (MAPK) such as p38 stress kinase, JNK (c-Jun N-terminal kinase), and ERK (extracellular signal-regulated kinase) (Koga et al. 2004; Rauner et al. 2007).

The expression of OPG and RANKL is highly inducible by various systemic and local factors. Among others, estrogen, bone morphogenetic protein-2, INF-γ, and TGF-β positively regulate OPG, whereas PTH, 1,25(OH)$_2$ vitamin D$_3$, glucocorticoids, prostaglandin E$_2$, IL-6, IL-8, and IL-11 enhance the expression of RANKL (summarized in Khosla (2001) and Leibbrandt and Penninger (2009)).

1.4.3 Osteoclast-to-Osteoblast Coupling Factors

Even though a great deal of research has mainly focused on the communication of osteoblasts to osteoclasts through the RANKL/OPG system in the past 15 years, recent observations in patients taking antiresorptive drugs (i.e., bisphosphonates or denosumab) or patients with osteoclast-poor osteopetrosis have suggested that bone resorption may stimulate bone formation as these patients do not only present with suppressed bone resorption but also with reduced bone formation. Besides factors that are released from the bone matrix and may also promote osteoblast function, such as TGF-β or IGF-I and IFG-II, several novel factors have been identified that are secreted by osteoclasts and act on osteoblasts to regulate their function. For example, mature osteoclasts have been shown to produce cardiotrophin-1 (a member of the IL-6 cytokine superfamily), collagen triple helix repeat containing 1, Wnt1, and Wnt10b, all of which stimulate osteoblast differentiation and function in vitro and in vivo (Ota et al. 2013; Kimura et al. 2008; Pederson et al. 2008; Takeshita et al. 2013; Walker et al. 2008; Weivoda et al. 2015).

One intriguing class of coupling factors is ephrin signaling, which can transmit signaling in a bidirectional manner, signaling either through the receptor (forward signaling) or through the ligand (reverse signaling). Initially, ephrinB2 has been discovered as a membrane-bound ligand on osteoclasts, while its receptor EphB4 is expressed on osteoblasts. Activation of the EphB4 receptor by ephrinB2 has been shown to promote osteoblastogenesis, while signaling through ephrinB2 suppressed osteoclast differentiation through downregulating NFATc1 and c-Fos (Zhao et al. 2006). However, as mice lacking ephrinB2 in myeloid cells displayed a normal bone phenotype, the question remained how the ephrinB2/EphB4 signaling impacts on bone physiology (Zhao et al. 2006). Additional studies have shown that ephrinB2 is also induced by PTH and that it is in fact the osteoblast-derived ephrinB2 that is critical to regulate bone formation, rather than the osteoclast-derived ephrinB2 (Allan et al. 2008; Takyar et al. 2013). Nevertheless, the dual regulation of osteoblast and osteoclast activity by this pathway suggests an important role not only in physiological but also in pathophysiological conditions.

The last class of coupling factors that will be discussed here is the semaphorins. Semaphorins were originally described as axon guidance repellents and use plexins and neuropilins as their primary receptors. Whereas membrane-bound semaphorins usually use plexins and soluble semaphorins often associate with neuropilins (Tamagnone et al. 1999). Meanwhile, semaphorins have been described to function in a variety of different tissues, including bone. In osteoclasts, Sema4D is expressed highest. Targeted deletion of Sema4D in osteoclasts resulted in an increased bone mass, due to an increased number and function of osteoblasts, whereas osteoclast numbers and function were not altered (Negishi-Koga et al. 2011). More detailed studies showed that the binding of Sema4D on osteoclasts to its cognate receptor plexin B1 on osteoblasts suppresses osteoblast activity by attenuating IGF-1 signaling in a RhoA-dependent manner (Negishi-Koga et al. 2011). In addition to Sema4D, also other semaphorins have been implicated in the regulation of bone metabolism, but most of them are not expressed by osteoclasts. Sema3D has been shown to be expressed in osteoblasts and stimulate osteoclastogenesis (Sutton et al. 2008). Moreover, polymorphisms in Sema7a are associated with low bone mineral density and a high risk of vertebral fractures in postmenopausal women (Koh et al. 2006). However, underlying mechanisms remain to be elucidated. More detailed insights have been gained into the regulation of bone mass through Sema3A, which is also produced by osteoblasts. While the initial characterization of Sema3A-deficient mice led to the conclusion that Sema3A suppresses osteoclast function and promotes osteoblast differentiation, a subsequent study using cell type-specific cre lines indicated that Sema3A does not directly act on bone cells to regulate bone homeostasis but does so indirectly by modulating sensory nerve development (Fukuda et al. 2013; Hayashi et al. 2012). However, more studies will be needed to clarify the complex role of semaphorins on the regulation of bone mass, especially in light of their regulation of nervous system, whose interactions with the skeletal system are still largely unexplored.

1.4.4 Cytokines and Chemokines

Bone remodeling is also critically regulated by various cytokines and chemokines, not only in pathophysiological conditions but also within physiological bone remodeling. Many pro-inflammatory cytokines including TNFα and interleukin (IL)-1, IL-6, IL-7, IL-11, IL-15, and IL-17 potentiate bone loss either by increasing osteoclast generation and activation or by inducing RANKL expression by the osteoblasts. On the other hand, IL-4, IL-5, IL-10, IL-12, IL-13, and IL-18 and interferon (IFN)-α, IFN-β, and IFN-γ are inhibitors of osteoclastogenesis by blocking RANKL signaling, either directly or indirectly (reviewed in Lorenzo et al. (2008); Sipos et al. (2008)). Interestingly, IL-1 directly stimulates TRAF6 expression on the osteoclast, thereby potentiating RANK signaling, whereas IFN-γ is known to downregulate TRAF6 by targeting it for proteosomal degradation, thereby aborting osteoclast formation (Takayanagi et al. 2000). TGF-β is described to both directly suppress osteoclastogenesis and induce osteoclastogenesis via suppressor of cytokine signaling 3 (SOCS3) (Lovibond et al. 2003; Ruan et al. 2010).

In contrast to osteoclasts, little is known about the effects of cytokines on osteoblasts. TNFα, IL-1, and IFN-γ were shown to inhibit osteoblast differentiation and block collagen synthesis (Canalis 1986; Centrella et al. 1992; Gilbert et al. 2000; Kuno et al. 1994). IL-6 and the IL-6 receptors were shown to be produced by osteoblasts and stromal cells, but the effects on osteoblastogenesis remain unclear (Franchimont et al. 1997a, b). IL-4 has been reported to be a chemoattractant for osteoblasts and to directly stimulate the proliferation of osteoblasts (Ura et al. 2000). However, it has an inhibitory effect on osteoblast differentiation. Accordingly, IL-4-overexpressing mice exhibited a decrease in bone formation and decreased differentiated osteoblasts on their bone surface (Jilka et al. 1998).

So far, only little is known about the regulation of bone mass by chemokines. However, Binder et al. reported on a critical role of the C-C chemokine receptor 2 (CCR2) in normal and pathological bone mass maintenance by regulating osteoclastogenesis (Binder et al. 2009). In this study, CCR2, the receptor for several monocyte chemoattractant proteins, was found to induce the expression of RANK in osteoclast precursor cells using the NFkB and ERK signaling pathways, thereby making them more susceptible for RANKL-induced osteoclastogenesis. Osteoblast differentiation and activity, on the other hand, were not affected. Thus, chemokines and chemokine receptors as well as cytokines are likely to play an important role in the maintenance of bone mass, but their definitive functions still remain to be determined in more detail.

1.5 Concluding Remarks

Bone is a highly dynamic tissue that undergoes constant remodeling to repair structural damage or adapt to changing functional demands. Osteoblasts, osteocytes, and osteoclasts intensively communicate with each other to coordinate the remodeling process, and their functions are tightly regulated by various systemic and local factors such as hormones or cytokines. Local factors may be produced by bone cells themselves to act in an autocrine manner or by other cell types (e.g., immune cells, vascular cells, adipocytes, neurons) that also participate in the regulatory process. Due to the increasing knowledge of cellular and molecular mechanisms of bone remodeling, efficient therapies have already been developed (bisphosphonates, PTH, denosumab) and will continue to develop (i.e., anti-sclerostin antibodies) to encounter exacerbated bone loss that occurs with aging, estrogen deficiency, or malignant and inflammatory disease.

References

Aguirre JI, Plotkin LI, Stewart SA, Weinstein RS, Parfitt AM, Manolagas SC, Bellido T (2006) Osteocyte apoptosis is induced by weightlessness in mice and precedes osteoclast recruitment and bone loss. J Bone Miner Res 21:605–615

Al-Fakhri N, Hofbauer LC, Preissner KT, Franke FE, Schoppet M (2005) Expression of bone-regulating factors osteoprotegerin (OPG) and receptor activator of NF-kappaB ligand (RANKL) in heterotopic vascular ossification. Thromb Haemost 94:1335–1337

Alini M, Marriott A, Chen T, Abe S, Poole AR (1996) A novel angiogenic molecule produced at the time of chondrocyte hypertrophy during endochondral bone formation. Dev Biol 176: 124–132

Allan EH, Häusler KD, Wei T, Gooi JH, Quinn JM, Crimeen-Irwin B, Pompolo S, Sims NA, Gillespie MT, Onyia JE, Martin TJ (2008) EphrinB2 regulation by PTH and PTHrP revealed by molecular profiling in differentiating osteoblasts. J Bone Miner Res 23:1170–1181

Andersen TL, Sondergaard TE, Skorzynska KE, Dagnaes-Hansen F, Plesner TL, Hauge EM, Plesner T, Delaisse JM (2009) A physical mechanism for coupling bone resorption and formation in adult human bone. Am J Pathol 174:239–247

Anderson DM, Maraskovsky E, Billingsley WL, Dougall WC, Tometsko ME, Roux ER, Teepe MC, DuBose RF, Cosman D, Galibert L (1997) A homologue of the TNF receptor and its ligand enhance T-cell growth and dendritic-cell function. Nature 390:175–179

Baron R, Kneissel M (2013) WNT signaling in bone homeostasis and disease: from human mutations to treatments. Nat Med 19:179–192

Bilezikian JP, Kurland ES (2001) Therapy of male osteoporosis with parathyroid hormone. Calcif Tissue Int 69:248–251

Binder NB, Niederreiter B, Hoffmann O, Stange R, Pap T, Stulnig TM, Mack M, Erben RG, Smolen JS, Redlich K (2009) Estrogen-dependent and C-C chemokine receptor-2-dependent pathways determine osteoclast behavior in osteoporosis. Nat Med 15:417–424

Blair HC, Teitelbaum SL, Ghiselli R, Gluck S (1989) Osteoclastic bone resorption by a polarized vacuolar proton pump. Science 245:855–857

Blair HC, Zaidi M, Huang CL, Sun L (2008) The developmental basis of skeletal cell differentiation and the molecular basis of major skeletal defects. Biol Rev Camb Philos Soc 83:401–415

Bonewald LF (2007) Osteocytes as dynamic multifunctional cells. Ann N Y Acad Sci 1116: 281–290

Bonewald LF (2011) The amazing osteocyte. J Bone Miner Res 26:229–238

Boyden LM, Mao J, Belsky J, Mitzner L, Farhi A, Mitnick MA, Wu D, Insogna K, Lifton RP (2002) High bone density due to a mutation in LDL-receptor-related protein 5. N Engl J Med 346:1513–1521

Bucay N, Sarosi I, Dunstan CR, Morony S, Tarpley J, Capparelli C, Scully S, Tan HL, Xu W, Lacey DL, Boyle WJ, Simonet WS (1998) Osteoprotegerin-deficient mice develop early onset osteoporosis and arterial calcification. Genes Dev 12:1260–1268

Canalis E (1986) Interleukin-1 has independent effects on deoxyribonucleic acid and collagen synthesis in cultures of rat calvariae. Endocrinology 118:74–81

Canalis E (2009) Growth factor control of bone mass. J Cell Biochem 108:769–777

Carani C, Qin K, Simoni M, Faustini-Fustini M, Serpente S, Boyd J, Korach KS, Simpson ER (1997) Effect of testosterone and estradiol in a man with aromatase deficiency. N Engl J Med 337:91–95

Cauley JA, Seeley DG, Ensrud K, Ettinger B, Black D, Cummings SR (1995) Estrogen replacement therapy and fractures in older women Study of Osteoporotic Fractures Research Group. Ann Intern Med 122:9–16

Centrella M, McCarthy TL, Kusmik WF, Canalis E (1992) Isoform-specific regulation of platelet-derived growth factor activity and binding in osteoblast-enriched cultures from fetal rat bone. J Clin Invest 89:1076–1084

Chang MK, Raggatt LJ, Alexander KA, Kuliwaba JS, Fazzalari NL, Schroder K, Maylin ER, Ripoll VM, Hume DA, Pettit AR (2008) Osteal tissue macrophages are intercalated throughout human and mouse bone lining tissues and regulate osteoblast function in vitro and in vivo. J Immunol 181:1232–1244

Chellaiah MA (2006) Regulation of podosomes by integrin alphavbeta3 and Rho GTPase-facilitated phosphoinositide signaling. Eur J Cell Biol 85:311–317

Clevers H (2006) Wnt/beta-catenin signaling in development and disease. Cell 127:469–480

Coen G, Ballanti P, Silvestrini G, Mantella D, Manni M, Di Giulio S, Pisano S, Leopizzi M, Di Lullo G, Bonucci E (2009) Immunohistochemical localization and mRNA expression of matrix Gla protein and fetuin-A in bone biopsies of hemodialysis patients. Virchows Arch 454: 263–271

Compagni A, Logan M, Klein R, Adams RH (2003) Control of skeletal patterning by ephrinB1-EphB interactions. Dev Cell 5:217–230

Cummings SR, San Martin J, McClung MR, Siris ES, Eastell R, Reid IR, Delmas P, Zoog HB, Austin M, Wang A, Kutilek S, Adami S, Zanchetta J, Libanati C, Siddhanti S, Christiansen C (2009) Denosumab for prevention of fractures in postmenopausal women with osteoporosis. N Engl J Med 361:756–765

Dai JC, He P, Chen X, Greenfield EM (2006) TNFalpha and PTH utilize distinct mechanisms to induce IL-6 and RANKL expression with markedly different kinetics. Bone 38:509–520

Dardenne O, Prud'homme J, Arabian A, Glorieux FH, St-Arnaud R (2001) Targeted inactivation of the 25-hydroxyvitamin D(3)-1(alpha)-hydroxylase gene (CYP27B1) creates an animal model of pseudovitamin D-deficiency rickets. Endocrinology 142:3135–3141

Dossing DA, Stern PH (2005) Receptor activator of NF-kappaB ligand protein expression in UMR-106 cells is differentially regulated by parathyroid hormone and calcitriol. J Cell Biochem 95:1029–1041

Dougall WC, Glaccum M, Charrier K, Rohrbach K, Brasel K, De Smedt T, Daro E, Smith J, Tometsko ME, Maliszewski CR, Armstrong A, Shen V, Bain S, Cosman D, Anderson D, Morrissey PJ, Peschon JJ, Schuh J (1999) RANK is essential for osteoclast and lymph node development. Genes Dev 13:2412–2424

Ducy P, Schinke T, Karsenty G (2000) The osteoblast: a sophisticated fibroblast under central surveillance. Science 289:1501–1504

Eghbali-Fatourechi G, Khosla S, Sanyal A, Boyle WJ, Lacey DL, Riggs BL (2003) Role of RANK ligand in mediating increased bone resorption in early postmenopausal women. J Clin Invest 111:1221–1230

Eijken M, Hewison M, Cooper MS, de Jong FH, Chiba H, Stewart PM, Uitterlinden AG, Pols HA, van Leeuwen JP (2005) 11beta-Hydroxysteroid dehydrogenase expression and glucocorticoid synthesis are directed by a molecular switch during osteoblast differentiation. Mol Endocrinol 19:621–631

Engsig MT, Chen QJ, Vu TH, Pedersen AC, Therkidsen B, Lund LR, Henriksen K, Lenhard T, Foged NT, Werb Z, Delaisse JM (2000) Matrix metalloproteinase 9 and vascular endothelial growth factor are essential for osteoclast recruitment into developing long bones. J Cell Biol 151:879–889

Faccio R, Teitelbaum SL, Fujikawa K, Chappel J, Zallone A, Tybulewicz VL, Ross FP, Swat W (2005) Vav3 regulates osteoclast function and bone mass. Nat Med 11:284–290

Feng JQ, Ward LM, Liu S, Lu Y, Xie Y, Yuan B, Yu X, Rauch F, Davis SI, Zhang S, Rios H, Drezner MK, Quarles LD, Bonewald LF, White KE (2006) Loss of DMP1 causes rickets and osteomalacia and identifies a role for osteocytes in mineral metabolism. Nat Genet 38:1310–1315

Ferron M, Hinoi E, Karsenty G, Ducy P (2008) Osteocalcin differentially regulates beta cell and adipocyte gene expression and affects the development of metabolic diseases in wild-type mice. Proc Natl Acad Sci U S A 105:5266–5270

Ferron M, Wei J, Yoshizawa T, Del Fattore A, DePinho RA, Teti A, Ducy P, Karsenty G (2010) Insulin signaling in osteoblasts integrates bone remodeling and energy metabolism. Cell 142:296–308

Fleming HE, Janzen V, Lo Celso C, Guo J, Leahy KM, Kronenberg HM, Scadden DT (2008) Wnt signaling in the niche enforces hematopoietic stem cell quiescence and is necessary to preserve self-renewal in vivo. Cell Stem Cell 2:274–283

Fortier LA, Nixon AJ (1997) Distributional changes in substance P nociceptive fiber patterns in naturally osteoarthritic articulations. J Rheumatol 24:524–530

Franchimont N, Gangji V, Durant D, Canalis E (1997a) Interleukin-6 with its soluble receptor enhances the expression of insulin-like growth factor-I in osteoblasts. Endocrinology 138:5248–5255

Franchimont N, Rydziel S, Canalis E (1997b) Interleukin 6 is autoregulated by transcriptional mechanisms in cultures of rat osteoblastic cells. J Clin Invest 100:1797–1803

Frost HM (1990) Skeletal structural adaptations to mechanical usage (SATMU): 2. Redefining Wolff's law: the remodeling problem. Anat Rec 226:414–422

Frost HM, Jee WS (1994) Perspectives: a vital biomechanical model of the endochondral ossification mechanism. Anat Rec 240:435–446

Fukuda T, Takeda S, Xu R, Ochi H, Sunamura S, Sato T, Shibata S, Yoshida Y, Gu Z, Kimura A, Ma C, Xu C, Bando W, Fujita K, Shinomiya K, Hirai T, Asou Y, Enomoto M, Okano H, Okawa A, Itoh H (2013) Sema3A regulates bone-mass accrual through sensory innervations. Nature 497:490–493

Gelb BD, Shi GP, Chapman HA, Desnick RJ (1996) Pycnodysostosis, a lysosomal disease caused by cathepsin K deficiency. Science 273:1236–1238

Gilbert L, He X, Farmer P, Boden S, Kozlowski M, Rubin J, Nanes MS (2000) Inhibition of osteoblast differentiation by tumor necrosis factor-alpha. Endocrinology 141:3956–3964

Gong Y, Slee RB, Fukai N, Rawadi G, Roman-Roman S, Reginato AM, Wang H, Cundy T, Glorieux FH, Lev D, Zacharin M, Oexle K, Marcelino J, Suwairi W, Heeger S, Sabatakos G, Apte S, Adkins WN, Allgrove J, Arslan-Kirchner M, Batch JA, Beighton P, Black GC, Boles RG, Boon LM, Borrone C, Brunner HG, Carle GF, Dallapiccola B, De Paepe A, Floege B,

Halfhide ML, Hall B, Hennekam RC, Hirose T, Jans A, Juppner H, Kim CA, Keppler-Noreuil K, Kohlschuetter A, LaCombe D, Lambert M, Lemyre E, Letteboer T, Peltonen L, Ramesar RS, Romanengo M, Somer H, Steichen-Gersdorf E, Steinmann B, Sullivan B, Superti-Furga A, Swoboda W, van den Boogaard MJ, Van Hul W, Vikkula M, Votruba M, Zabel B, Garcia T, Baron R, Olsen BR, Warman ML (2001) LDL receptor-related protein 5 (LRP5) affects bone accrual and eye development. Cell 107:513–523

Greenfield EM, Gornik SA, Horowitz MC, Donahue HJ, Shaw SM (1993) Regulation of cytokine expression in osteoblasts by parathyroid hormone: rapid stimulation of interleukin-6 and leukemia inhibitory factor mRNA. J Bone Miner Res 8:1163–1171

Grubb BD (2004) Activation of sensory neurons in the arthritic joint. Novartis Found Symp. 260:28–36; discussion 36–48, 100–104, 277–279.

Guan CC, Yan M, Jiang XQ, Zhang P, Zhang XL, Li J, Ye DX, Zhang FQ (2009) Sonic hedgehog alleviates the inhibitory effects of high glucose on the osteoblastic differentiation of bone marrow stromal cells. Bone 45:1146–1152

Hamidouche Z, Hay E, Vaudin P, Charbord P, Schule R, Marie PJ, Fromigue O (2008) FHL2 mediates dexamethasone-induced mesenchymal cell differentiation into osteoblasts by activating Wnt/beta-catenin signaling-dependent Runx2 expression. Faseb J 22:3813–3822

Harada H, Tagashira S, Fujiwara M, Ogawa S, Katsumata T, Yamaguchi A, Komori T, Nakatsuka M (1999) Cbfa1 isoforms exert functional differences in osteoblast differentiation. J Biol Chem 274:6972–6978

Harris SE, Guo D, Harris MA, Krishnaswamy A, Lichtler A (2003) Transcriptional regulation of BMP-2 activated genes in osteoblasts using gene expression microarray analysis: role of Dlx2 and Dlx5 transcription factors. Front Biosci 8:s1249–s1265

Hauge EM, Qvesel D, Eriksen EF, Mosekilde L, Melsen F (2001) Cancellous bone remodeling occurs in specialized compartments lined by cells expressing osteoblastic markers. J Bone Miner Res 16:1575–1582

Hayashi M, Nakashima T, Taniguchi M, Kodama T, Kumanogoh A, Takayanagi H (2012) Osteoprotection by semaphorin 3A. Nature 485:69–74

Hazenberg JG, Freeley M, Foran E, Lee TC, Taylor D (2006) Microdamage: a cell transducing mechanism based on ruptured osteocyte processes. J Biomech 39:2096–2103

Heino TJ, Hentunen TA, Vaananen HK (2002) Osteocytes inhibit osteoclastic bone resorption through transforming growth factor-beta: enhancement by estrogen. J Cell Biochem 85:185–197

Hirao M, Hashimoto J, Yamasaki N, Ando W, Tsuboi H, Myoui A, Yoshikawa H (2007) Oxygen tension is an important mediator of the transformation of osteoblasts to osteocytes. J Bone Miner Metab 25:266–276

Hofbauer LC, Heufelder AE (2001) Role of receptor activator of nuclear factor-kappaB ligand and osteoprotegerin in bone cell biology. J Mol Med 79:243–253

Hofbauer LC, Gori F, Riggs BL, Lacey DL, Dunstan CR, Spelsberg TC, Khosla S (1999) Stimulation of osteoprotegerin ligand and inhibition of osteoprotegerin production by glucocorticoids in human osteoblastic lineage cells: potential paracrine mechanisms of glucocorticoid-induced osteoporosis. Endocrinology 140:4382–4389

Hofbauer LC, Zeitz U, Schoppet M, Skalicky M, Schuler C, Stolina M, Kostenuik PJ, Erben RG (2009) Prevention of glucocorticoid-induced bone loss in mice by inhibition of RANKL. Arthritis Rheum 60:1427–1437

Holen I, Cross SS, Neville-Webbe HL, Cross NA, Balasubramanian SP, Croucher PI, Evans CA, Lippitt JM, Coleman RE, Eaton CL (2005) Osteoprotegerin (OPG) expression by breast cancer cells in vitro and breast tumours in vivo--a role in tumour cell survival? Breast Cancer Res Treat 92:207–215

Huebner AK, Keller J, Catala-Lehnen P, Perkovic S, Streichert T, Emeson RB, Amling M, Schinke T (2008) The role of calcitonin and alpha-calcitonin gene-related peptide in bone formation. Arch Biochem Biophys 473:210–217

Imam A, Iqbal J, Blair HC, Davies TF, Huang CL, Zallone A, Zaidi M, Sun L (2009) Role of the pituitary-bone axis in skeletal pathophysiology. Curr Opin Endocrinol Diabetes Obes 16:423–429

Ishii M, Egen JG, Klauschen F, Meier-Schellersheim M, Saeki Y, Vacher J, ProiaL RL, Germain RN (2009) Sphingosine-1-phosphate mobilizes osteoclast precursors and regulates bone homeostasis. Nature 458(7237):524–528

Ishii M, Kikuta J, Shimazu Y, Meier-Schellersheim M, Germain RN (2010) Chemorepulsion by blood S1P regulates osteoclast precursor mobilization and bone remodeling in vivo. J Exp Med 207:2793–2798

Jiang Y, Jahagirdar BN, Reinhardt RL, Schwartz RE, Keene CD, Ortiz-Gonzalez XR, Reyes M, Lenvik T, Lund T, Blackstad M, Du J, Aldrich S, Lisberg A, Low WC, Largaespada DA, Verfaillie CM (2002) Pluripotency of mesenchymal stem cells derived from adult marrow. Nature 418:41–49

Jilka RL, Hangoc G, Girasole G, Passeri G, Williams DC, Abrams JS, Boyce B, Broxmeyer H, Manolagas SC (1992) Increased osteoclast development after estrogen loss: mediation by interleukin-6. Science 257:88–91

Jilka RL, Passeri G, Girasole G, Cooper S, Abrams J, Broxmeyer H, Manolagas SC (1995) Estrogen loss upregulates hematopoiesis in the mouse: a mediating role of IL-6. Exp Hematol 23:500–506

Jilka RL, Weinstein RS, Bellido T, Parfitt AM, Manolagas SC (1998) Osteoblast programmed cell death (apoptosis): modulation by growth factors and cytokines. J Bone Miner Res 13:793–802

Jimenez-Andrade JM, Mantyh WG, Bloom AP, Xu H, Ferng AS, Dussor G, Vanderah TW, Mantyh PW (2010) A phenotypically restricted set of primary afferent nerve fibers innervate the bone versus skin: therapeutic opportunity for treating skeletal pain. Bone 46:306–313

Jung Y, Wang J, Song J, Shiozawa Y, Wang J, Havens A, Wang Z, Sun YX, Emerson SG, Krebsbach PH, Taichman RS (2007) Annexin II expressed by osteoblasts and endothelial cells regulates stem cell adhesion, homing, and engraftment following transplantation. Blood 110:82–90

Karsenty G, Ducy P. (2006) The hypothalamic control of bone mass, implication for the treatment of osteoporosis. Ann Endocrinol (Paris) 67:123.

Karsenty G, Ferron M (2012) The contribution of bone to whole-organism physiology. Nature 481:314–320

Kartsogiannis V, Zhou H, Horwood NJ, Thomas RJ, Hards DK, Quinn JM, Niforas P, Ng KW, Martin TJ, Gillespie MT (1999) Localization of RANKL (receptor activator of NF kappa B ligand) mRNA and protein in skeletal and extraskeletal tissues. Bone 25:525–534

Keller H, Kneissel M (2005) SOST is a target gene for PTH in bone. Bone 37:148–158

Keller J, Catala-Lehnen P, Huebner AK, Jeschke A, Heckt T, Lueth A, Krause M, Koehne T, Albers J, Schulze J, Schilling S, Haberland M, Denninger H, Neven M, Hermans-Borgmeyer I, Streichert T, Breer S, Barvencik F, Levkau B, Rathkolb B, Wolf E, Calzada-Wack J, Neff F, Gailus-Durner V, Fuchs H, de Angelis MH, Klutmann S, Tsourdi E, Hofbauer LC, Kleuser B, Chun J, Schinke T, Amling M (2014) Calcitonin controls bone formation by inhibiting the release of sphingosine 1-phosphate from osteoclasts. Nature Commun 5:5215

Khosla S (2001) Minireview: the OPG/RANKL/RANK system. Endocrinology 142:5050–5055

Khosla S, Riggs BL (2003) Androgens, estrogens, and bone turnover in men. J Clin Endocrinol Metab. 88:2352; author reply 2352–2353.

Khosla S, Melton LJ 3rd, Riggs BL (2001) Estrogens and bone health in men. Calcif Tissue Int 69:189–192

Kim N, Takami M, Rho J, Josien R, Choi Y (2002) A novel member of the leukocyte receptor complex regulates osteoclast differentiation. J Exp Med 195:201–209

Kim HJ, Zhao H, Kitaura H, Bhattacharyya S, Brewer JA, Muglia LJ, Ross FP, Teitelbaum SL (2006a) Glucocorticoids suppress bone formation via the osteoclast. J Clin Invest 116: 2152–2160

Kim YJ, Kim HN, Park EK, Lee BH, Ryoo HM, Kim SY, Kim IS, Stein JL, Lian JB, Stein GS, van Wijnen AJ, Choi JY (2006b) The bone-related Zn finger transcription factor Osterix promotes proliferation of mesenchymal cells. Gene 366:145–151

Kimura H, Kwan KM, Zhang Z, Deng JM, Darnay BG, Behringer RR, Nakamura T, de Crombrugghe B, Akiyama H (2008) Cthrc1 is a positive regulator of osteoblastic bone formation. PLoS One 3:e3174

Kitazawa S, Kajimoto K, Kondo T, Kitazawa R (2003) Vitamin D3 supports osteoclastogenesis via functional vitamin D response element of human RANKL gene promoter. J Cell Biochem 89:771–777

Koga T, Inui M, Inoue K, Kim S, Suematsu A, Kobayashi E, Iwata T, Ohnishi H, Matozaki T, Kodama T, Taniguchi T, Takayanagi H, Takai T (2004) Costimulatory signals mediated by the ITAM motif cooperate with RANKL for bone homeostasis. Nature 428:758–763

Koh JM, Oh B, Lee JY, Lee JK, Kimm K, Kim GS, Park BL, Cheong HS, Shin HD, Hong JM, Kim TH, Park EK, Kim SY (2006) Association study of semaphorin 7a (sema7a) polymorphisms with bone mineral density and fracture risk in postmenopausal Korean women. J Hum Genet 51(2):112–117

Koga T, Matsui Y, Asagiri M, Kodama T, de Crombrugghe B, Nakashima K, Takayanagi H (2005) NFAT and Osterix cooperatively regulate bone formation. Nat Med 11:880–885

Kong YY, Feige U, Sarosi I, Bolon B, Tafuri A, Morony S, Capparelli C, Li J, Elliott R, McCabe S, Wong T, Campagnuolo G, Moran E, Bogoch ER, Van G, Nguyen LT, Ohashi PS, Lacey DL, Fish E, Boyle WJ, Penninger JM (1999a) Activated T cells regulate bone loss and joint destruction in adjuvant arthritis through osteoprotegerin ligand. Nature 402:304–309

Kong YY, Yoshida H, Sarosi I, Tan HL, Timms E, Capparelli C, Morony S, Oliveira-dos-Santos AJ, Van G, Itie A, Khoo W, Wakeham A, Dunstan CR, Lacey DL, Mak TW, Boyle WJ, Penninger JM (1999b) OPGL is a key regulator of osteoclastogenesis, lymphocyte development and lymph-node organogenesis. Nature 397:315–323

Krishnan V, Bryant HU, Macdougald OA (2006) Regulation of bone mass by Wnt signaling. J Clin Invest 116:1202–1209

Kronenberg HM (2006) PTHrP and skeletal development. Ann N Y Acad Sci 1068:1–13

Kuno H, Kurian SM, Hendy GN, White J, deLuca HF, Evans CO, Nanes MS (1994) Inhibition of 1,25-dihydroxyvitamin D3 stimulated osteocalcin gene transcription by tumor necrosis factor-alpha: structural determinants within the vitamin D response element. Endocrinology 134:2524–2531

Lacey DL, Timms E, Tan HL, Kelley MJ, Dunstan CR, Burgess T, Elliott R, Colombero A, Elliott G, Scully S, Hsu H, Sullivan J, Hawkins N, Davy E, Capparelli C, Eli A, Qian YX, Kaufman S, Sarosi I, Shalhoub V, Senaldi G, Guo J, Delaney J, Boyle WJ (1998) Osteoprotegerin ligand is a cytokine that regulates osteoclast differentiation and activation. Cell 93:165–176

Lanske B, Amling M, Neff L, Guiducci J, Baron R, Kronenberg HM (1999) Ablation of the PTHrP gene or the PTH/PTHrP receptor gene leads to distinct abnormalities in bone development. J Clin Invest 104:399–407

Leder BZ, LeBlanc KM, Schoenfeld DA, Eastell R, Finkelstein JS (2003) Differential effects of androgens and estrogens on bone turnover in normal men. J Clin Endocrinol Metab 88:204–210

Lee NK, Sowa H, Hinoi E, Ferron M, Ahn JD, Confavreux C, Dacquin R, Mee PJ, McKee MD, Jung DY, Zhang Z, Kim JK, Mauvais-Jarvis F, Ducy P, Karsenty G (2007) Endocrine regulation of energy metabolism by the skeleton. Cell 130:456–469

Leibbrandt A, Penninger JM (2009) RANK(L) as a key target for controlling bone loss. Adv Exp Med Biol 647:130–145

Li YP, Chen W, Liang Y, Li E, Stashenko P (1999) Atp6i-deficient mice exhibit severe osteopetrosis due to loss of osteoclast-mediated extracellular acidification. Nat Genet 23:447–451

Li X, Qin L, Bergenstock M, Bevelock LM, Novack DV, Partridge NC (2007) Parathyroid hormone stimulates osteoblastic expression of MCP-1 to recruit and increase the fusion of pre/osteoclasts. J Biol Chem 282:33098–33106

Lian JB, Stein GS, Javed A, van Wijnen AJ, Stein JL, Montecino M, Hassan MQ, Gaur T, Lengner CJ, Young DW (2006) Networks and hubs for the transcriptional control of osteoblastogenesis. Rev Endocr Metab Disord 7:1–16

Liu YH, Tang Z, Kundu RK, Wu L, Luo W, Zhu D, Sangiorgi F, Snead ML, Maxson RE (1999) Msx2 gene dosage influences the number of proliferative osteogenic cells in growth centers of the developing murine skull: a possible mechanism for MSX2-mediated craniosynostosis in humans. Dev Biol 205:260–274

Liu S, Tang W, Fang J, Ren J, Li H, Xiao Z, Quarles LD (2009) Novel regulators of Fgf23 expression and mineralization in Hyp bone. Mol Endocrinol 23:1505–1518

Lorenzo J, Horowitz M, Choi Y (2008) Osteoimmunology: interactions of the bone and immune system. Endocr Rev 29:403–440

Lovibond AC, Haque SJ, Chambers TJ, Fox SW (2003) TGF-beta-induced SOCS3 expression augments TNF-alpha-induced osteoclast formation. Biochem Biophys Res Commun 309:762–767

Maeda Y, Nakamura E, Nguyen MT, Suva LJ, Swain FL, Razzaque MS, Mackem S, Lanske B (2007) Indian hedgehog produced by postnatal chondrocytes is essential for maintaining a growth plate and trabecular bone. Proc Natl Acad Sci U S A 104:6382–6387

Malyankar UM, Scatena M, Suchland KL, Yun TJ, Clark EA, Giachelli CM (2000) Osteoprotegerin is an alpha vbeta 3-induced NF-kappa B-dependent survival factor for endothelial cells. J Biol Chem 275:20959–20962

Martin TJ, Gillespie MT (2001) Receptor activator of nuclear factor kappa B ligand (RANKL): another link between breast and bone. Trends Endocrinol Metab 12:2–4

Martin TJ, Seeman E (2008) Bone remodelling: its local regulation and the emergence of bone fragility. Best Pract Res Clin Endocrinol Metab 22:701–722

Matsuo K, Galson DL, Zhao C, Peng L, Laplace C, Wang KZ, Bachler MA, Amano H, Aburatani H, Ishikawa H, Wagner EF (2004) Nuclear factor of activated T-cells (NFAT) rescues osteoclastogenesis in precursors lacking c-Fos. J Biol Chem 279:26475–26480

Mattsson JP, Schlesinger PH, Keeling DJ, Teitelbaum SL, Stone DK, Xie XS (1994) Isolation and reconstitution of a vacuolar-type proton pump of osteoclast membranes. J Biol Chem 269: 24979–24982

McClung MR, Grauer A, Boonen S, Bolognese MA, Brown JP, Diez-Perez A, Langdahl BL, Reginster JY, Zanchetta JR, Wasserman SM, Katz L, Maddox J, Yang YC, Libanati C, Bone HG (2014) Romosozumab in postmenopausal women with low bone mineral density. N Engl J Med 370:412–420

Mendez-Ferrer S, Michurina TV, Ferraro F, Mazloom AR, Macarthur BD, Lira SA, Scadden DT, Ma'ayan A, Enikolopov GN, Frenette PS (2010) Mesenchymal and haematopoietic stem cells form a unique bone marrow niche. Nature 466:829–834

Miller J, Horner A, Stacy T, Lowrey C, Lian JB, Stein G, Nuckolls GH, Speck NA (2002) The core-binding factor beta subunit is required for bone formation and hematopoietic maturation. Nat Genet 32:645–649

Mimura H, Cao X, Ross FP, Chiba M, Teitelbaum SL (1994) 1,25-Dihydroxyvitamin D3 transcriptionally activates the beta 3-integrin subunit gene in avian osteoclast precursors. Endocrinology 134:1061–1066

Miyauchi A, Alvarez J, Greenfield EM, Teti A, Grano M, Colucci S, Zambonin-Zallone A, Ross FP, Teitelbaum SL, Cheresh D et al (1991) Recognition of osteopontin and related peptides by an alpha v beta 3 integrin stimulates immediate cell signals in osteoclasts. J Biol Chem 266: 20369–20374

Mocsai A, Humphrey MB, Van Ziffle JA, Hu Y, Burghardt A, Spusta SC, Majumdar S, Lanier LL, Lowell CA, Nakamura MC (2004) The immunomodulatory adapter proteins DAP12 and Fc receptor gamma-chain (FcRgamma) regulate development of functional osteoclasts through the Syk tyrosine kinase. Proc Natl Acad Sci U S A 101:6158–6163

Nakashima T, Hayashi M, Fukunaga T, Kurata K, Oh-Hora M, Feng JQ, Bonewald LF, Kodama T, Wutz A, Wagner EF, Penninger JM, Takayanagi H (2011) Evidence for osteocyte regulation of bone homeostasis through RANKL expression. Nat Med 17:1231–1234

Negishi-Koga T, Shinohara M, Komatsu N, Bito H, Kodama T, Friedel RH, Takayanagi H (2011) Suppression of bone formation by osteoclastic expression of semaphorin 4D. Nat Med 17:1473–1480

O'Brien CA, Jia D, Plotkin LI, Bellido T, Powers CC, Stewart SA, Manolagas SC, Weinstein RS (2004) Glucocorticoids act directly on osteoblasts and osteocytes to induce their apoptosis and reduce bone formation and strength. Endocrinology 145:1835–1841

Okada Y, Naka K, Kawamura K, Matsumoto T, Nakanishi I, Fujimoto N, Sato H, Seiki M (1995) Localization of matrix metalloproteinase 9 (92-kilodalton gelatinase/type IV collagenase = gelatinase B) in osteoclasts: implications for bone resorption. Lab Invest 72:311–322

Orwoll ES, Bauer DC, Vogt TM, Fox KM (1996) Axial bone mass in older women Study of Osteoporotic Fractures Research Group. Ann Intern Med 124:187–196
Ota K, Quint P, Ruan M, Pederson L, Westendorf JJ, Khosla S, Oursler MJ (2013) TGF-β induces Wnt10b in osteoclasts from female mice to enhance coupling to osteoblasts. Endocrinology 154:3745–3752
Oury F, Sumara G, Sumara O, Ferron M, Chang H, Smith CE, Hermo L, Suarez S, Roth BL, Ducy P, Karsenty G (2011) Endocrine regulation of male fertility by the skeleton. Cell 144:796–809
Paic F, Igwe JC, Nori R, Kronenberg MS, Franceschetti T, Harrington P, Kuo L, Shin DG, Rowe DW, Harris SE, Kalajzic I (2009) Identification of differentially expressed genes between osteoblasts and osteocytes. Bone 45:682–692
Panda DK, Miao D, Bolivar I, Li J, Huo R, Hendy GN, Goltzman D (2004) Inactivation of the 25-hydroxyvitamin D 1alpha-hydroxylase and vitamin D receptor demonstrates independent and interdependent effects of calcium and vitamin D on skeletal and mineral homeostasis. J Biol Chem 279:16754–16766
Parfitt AM (1988) Bone remodeling. Henry Ford Hosp Med J 36:143–144
Parfitt AM (1994) Osteonal and hemi-osteonal remodeling: the spatial and temporal framework for signal traffic in adult human bone. J Cell Biochem 55:273–286
Paszty C, Turner CH, Robinson MK (2010) Sclerostin: a gem from the genome leads to bone-building antibodies. J Bone Miner Res 25:1897–1904
Pederson L, Ruan M, Westendorf JJ, Khosla S, Oursler MJ (2008) Regulation of bone formation by osteoclasts involves Wnt/BMP signaling and the chemokine sphingosine-1-phosphate. Proc Natl Acad Sci USA 105(52):20764–20769.
Pettit AR, Chang MK, Hume DA, Raggatt LJ (2008) Osteal macrophages: a new twist on coupling during bone dynamics. Bone 43:976–982
Pittenger MF, Mackay AM, Beck SC, Jaiswal RK, Douglas R, Mosca JD, Moorman MA, Simonetti DW, Craig S, Marshak DR (1999) Multilineage potential of adult human mesenchymal stem cells. Science 284:143–147
Plotkin LI, Manolagas SC, Bellido T (2002) Transduction of cell survival signals by connexin-43 hemichannels. J Biol Chem 277:8648–8657
Plotkin LI, Lezcano V, Thostenson J, Weinstein RS, Manolagas SC, Bellido T (2008) Connexin 43 is required for the anti-apoptotic effect of bisphosphonates on osteocytes and osteoblasts in vivo. J Bone Miner Res 23:1712–1721
Pollock JH, Blaha MJ, Lavish SA, Stevenson S, Greenfield EM (1996) In vivo demonstration that parathyroid hormone and parathyroid hormone-related protein stimulate expression by osteoblasts of interleukin-6 and leukemia inhibitory factor. J Bone Miner Res 11:754–759
Poole CA, Glant TT, Schofield JR (1991) Chondrons from articular cartilage. (IV). Immunolocalization of proteoglycan epitopes in isolated canine tibial chondrons. J Histochem Cytochem 39:1175–1187
Poole KE, van Bezooijen RL, Loveridge N, Hamersma H, Papapoulos SE, Lowik CW, Reeve J (2005) Sclerostin is a delayed secreted product of osteocytes that inhibits bone formation. Faseb J 19:1842–1844
Puigserver P, Wu Z, Park CW, Graves R, Wright M, Spiegelman BM (1998) A cold-inducible coactivator of nuclear receptors linked to adaptive thermogenesis. Cell 92:829–839
Rachner TD, Khosla S, Hofbauer LC (2011) Osteoporosis: now and the future. Lancet 377: 1276–1287
Raisz LG (1988) Local and systemic factors in the pathogenesis of osteoporosis. N Engl J Med 318:818–828
Rasmussen LM, Tarnow L, Hansen TK, Parving HH, Flyvbjerg A (2006) Plasma osteoprotegerin levels are associated with glycaemic status, systolic blood pressure, kidney function and cardiovascular morbidity in type 1 diabetic patients. Eur J Endocrinol 154:75–81
Rauner M, Sipos W, Pietschmann P (2007) Osteoimmunology. Int Arch Allergy Immunol 143:31–48
Rauner M, Goettsch C, Stein N, Thiele S, Bornhaeuser M, De Bosscher K, Haegeman G, Tuckermann J, Hofbauer LC (2010) Dissociation of osteogenic and immunological effects by

the selective glucocorticoid receptor agonist, compound A, in human bone marrow stromal cells. Endocrinology 2011;152:103–112

Recker RR, Benson CT, Matsumoto T, Bolognese MA, Robins DA, Alam J, Chiang AY, Hu L, Krege JH, Sowa H, Mitlak BH, Myers SL (2015) A randomized, double-blind phase 2 clinical trial of blosozumab, a sclerostin antibody, in postmenopausal women with low bone mineral density. J Bone Miner Res 30:216–224

Riggs BL, Hartmann LC (2003) Selective estrogen-receptor modulators—mechanisms of action and application to clinical practice. N Engl J Med 348:618–629

Riggs BL, Khosla S, Melton LJ 3rd (1999) The assembly of the adult skeleton during growth and maturation: implications for senile osteoporosis. J Clin Invest 104:671–672

Rodan GA, Martin TJ (1981) Role of osteoblasts in hormonal control of bone resorption—a hypothesis. Calcif Tissue Int 33:349–351

Ruan M, Pederson L, Bradley EW, Bamberger AM, Oursler MJ (2010) Transforming growth factor-{beta} coordinately induces suppressor of cytokine signaling 3 and leukemia inhibitory factor to suppress osteoclast apoptosis. Endocrinology 151:1713–1722

Ruiz C, Perez E, Vallecillo-Capilla M, Reyes-Botella C (2003) Phagocytosis and allogeneic T cell stimulation by cultured human osteoblast-like cells. Cell Physiol Biochem 13:309–314

Saftig P, Hunziker E, Wehmeyer O, Jones S, Boyde A, Rommerskirch W, Moritz JD, Schu P, von Figura K (1998) Impaired osteoclastic bone resorption leads to osteopetrosis in cathepsin-K-deficient mice. Proc Natl Acad Sci U S A 95:13453–13458

Satokata I, Ma L, Ohshima H, Bei M, Woo I, Nishizawa K, Maeda T, Takano Y, Uchiyama M, Heaney S, Peters H, Tang Z, Maxson R, Maas R (2000) Msx2 deficiency in mice causes pleiotropic defects in bone growth and ectodermal organ formation. Nat Genet 24: 391–395

Schrum LW, Bost KL, Hudson MC, Marriott I (2003) Bacterial infection induces expression of functional MHC class II molecules in murine and human osteoblasts. Bone 33:812–821

Semenov M, Tamai K, He X (2005) SOST is a ligand for LRP5/LRP6 and a Wnt signaling inhibitor. J Biol Chem 280:26770–26775

Sher LB, Woitge HW, Adams DJ, Gronowicz GA, Krozowski Z, Harrison JR, Kream BE (2004) Transgenic expression of 11beta-hydroxysteroid dehydrogenase type 2 in osteoblasts reveals an anabolic role for endogenous glucocorticoids in bone. Endocrinology 145:922–929

Shore EM, Xu M, Feldman GJ, Fenstermacher DA, Cho TJ, Choi IH, Connor JM, Delai P, Glaser DL, LeMerrer M, Morhart R, Rogers JG, Smith R, Triffitt JT, Urtizberea JA, Zasloff M, Brown MA, Kaplan FS (2006) A recurrent mutation in the BMP type I receptor ACVR1 causes inherited and sporadic fibrodysplasia ossificans progressiva. Nat Genet 38:525–527

Silver IA, Murrills RJ, Etherington DJ (1988) Microelectrode studies on the acid microenvironment beneath adherent macrophages and osteoclasts. Exp Cell Res 175:266–276

Simonet WS, Lacey DL, Dunstan CR, Kelley M, Chang MS, Luthy R, Nguyen HQ, Wooden S, Bennett L, Boone T, Shimamoto G, DeRose M, Elliott R, Colombero A, Tan HL, Trail G, Sullivan J, Davy E, Bucay N, Renshaw-Gegg L, Hughes TM, Hill D, Pattison W, Campbell P, Sander S, Van G, Tarpley J, Derby P, Lee R, Boyle WJ (1997) Osteoprotegerin: a novel secreted protein involved in the regulation of bone density. Cell 89:309–319

Sims NA, Martin TJ (2014) Coupling the activities of bone formation and resorption: a multitude of signals within the basic multicellular unit. Bonekey Rep 3:481

Sipos W, Pietschmann P, Rauner M (2008) Strategies for novel therapeutic approaches targeting cytokines and signaling pathways of osteoclasto- and osteoblastogenesis in the fight against immune-mediated bone and joint diseases. Curr Med Chem 15:127–136

Skjodt H, Moller T, Freiesleben SF (1989) Human osteoblast-like cells expressing MHC class II determinants stimulate allogeneic and autologous peripheral blood mononuclear cells and function as antigen-presenting cells. Immunology 68:416–420

Smith MR, Egerdie B, Hernandez Toriz N, Feldman R, Tammela TL, Saad F, Heracek J, Szwedowski M, Ke C, Kupic A, Leder BZ, Goessl C (2009) Denosumab in men receiving androgen-deprivation therapy for prostate cancer. N Engl J Med 361:745–755

Sommerfeldt DW, Rubin CT (2001) Biology of bone and how it orchestrates the form and function of the skeleton. Eur Spine J 10(Suppl 2):S86–S95
Stanka P, Bellack U, Lindner A (1991) On the morphology of the terminal microvasculature during endochondral ossification in rats. Bone Miner 13:93–101
Stickens D, Behonick DJ, Ortega N, Heyer B, Hartenstein B, Yu Y, Fosang AJ, Schorpp-Kistner M, Angel P, Werb Z (2004) Altered endochondral bone development in matrix metalloproteinase 13-deficient mice. Development 131:5883–5895
Storm EE, Kingsley DM (1999) GDF5 coordinates bone and joint formation during digit development. Dev Biol 209:11–27
Streeten EA, Brandi ML (1990) Biology of bone endothelial cells. Bone Miner 10:85–94
Sutton AL, Zhang X, Dowd DR, Kharode YP, Komm BS, Macdonald PN (2008) Semaphorin 3B is a 1,25-Dihydroxyvitamin D3-induced gene in osteoblasts that promotes osteoclastogenesis and induces osteopenia in mice. Mol Endocrinol 22(6):1370–1381
Takayanagi H, Ogasawara K, Hida S, Chiba T, Murata S, Sato K, Takaoka A, Yokochi T, Oda H, Tanaka K, Nakamura K, Taniguchi T (2000) T-cell-mediated regulation of osteoclastogenesis by signalling cross-talk between RANKL and IFN-gamma. Nature 408:600–605
Takayanagi H, Kim S, Koga T, Nishina H, Isshiki M, Yoshida H, Saiura A, Isobe M, Yokochi T, Inoue J, Wagner EF, Mak TW, Kodama T, Taniguchi T (2002) Induction and activation of the transcription factor NFATc1 (NFAT2) integrate RANKL signaling in terminal differentiation of osteoclasts. Dev Cell 3:889–901
Takeshita S, Fumoto T, Matsuoka K, Park KA, Aburatani H, Kato S, Ito M, Ikeda K (2013) Osteoclast-secreted CTHRC1 in the coupling of bone resorption to formation. J Clin Invest 123:3914–3924
Takyar FM, Tonna S, Ho PW, Crimeen-Irwin B, Baker EK, Martin TJ, Sims NA (2013) EphrinB2/EphB4 inhibition in the osteoblast lineage modifies the anabolic response to parathyroid hormone. J Bone Miner Res 28:912–925
Tamagnone L, Artigiani S, Chen H, He Z, Ming GI, Song H, Chedotal A, Winberg ML, Goodman CS, Poo M, Tessier-Lavigne M, Comoglio PM (1999) Plexins are a large family of receptors for transmembrane, secreted, and GPI-anchored semaphorins in vertebrates. Cell 99:71–80
Tanaka S, Takahashi N, Udagawa N, Tamura T, Akatsu T, Stanley ER, Kurokawa T, Suda T (1993) Macrophage colony-stimulating factor is indispensable for both proliferation and differentiation of osteoclast progenitors. J Clin Invest 91:257–263
Tatsumi S, Ishii K, Amizuka N, Li M, Kobayashi T, Kohno K, Ito M, Takeshita S, Ikeda K (2007) Targeted ablation of osteocytes induces osteoporosis with defective mechanotransduction. Cell Metab 5:464–475
Teitelbaum SL (2000) Bone resorption by osteoclasts. Science 289:1504–1508
Teitelbaum SL, Ross FP (2003) Genetic regulation of osteoclast development and function. Nat Rev Genet 4:638–649
Teti A, Zallone A (2009) Do osteocytes contribute to bone mineral homeostasis? Osteocytic osteolysis revisited. Bone 44:11–16
Teti A, Blair HC, Schlesinger P, Grano M, Zambonin-Zallone A, Kahn AJ, Teitelbaum SL, Hruska KA (1989) Extracellular protons acidify osteoclasts, reduce cytosolic calcium, and promote expression of cell-matrix attachment structures. J Clin Invest 84:773–780
Thiele S, Ziegler N, Tsourdi E, De Bosscher K, Tuckermann JP, Hofbauer LC, Rauner M (2012) Selective glucocorticoid receptor modulation maintains bone mineral density in mice. J Bone Miner Res 27:2242–2250
Tian E, Zhan F, Walker R, Rasmussen E, Ma Y, Barlogie B, Shaughnessy JD Jr (2003) The role of the Wnt-signaling antagonist DKK1 in the development of osteolytic lesions in multiple myeloma. N Engl J Med 349:2483–2494
Tondravi MM, McKercher SR, Anderson K, Erdmann JM, Quiroz M, Maki R, Teitelbaum SL (1997) Osteopetrosis in mice lacking haematopoietic transcription factor PU.1. Nature 386:81–84
Tozum TF, Oppenlander ME, Koh-Paige AJ, Robins DM, McCauley LK (2004) Effects of sex steroid receptor specificity in the regulation of skeletal metabolism. Calcif Tissue Int 75:60–70

Trueta J, Buhr AJ (1963) The vascular contribution to osteogenesis. V. The vasculature supplying the epiphysial cartilage in rachitic rats. J Bone Joint Surg Br 45:572–581
Ura K, Morimoto I, Watanabe K, Saito K, Yanagihara N, Eto S (2000) Interleukin (IL)-4 and IL-13 inhibit the differentiation of murine osteoblastic MC3T3-E1 cells. Endocr J 47:293–302
Vaananen HK, Horton M (1995) The osteoclast clear zone is a specialized cell-extracellular matrix adhesion structure. J Cell Sci 108(Pt 8):2729–2732
van Amerongen R, Nusse R (2009) Towards an integrated view of Wnt signaling in development. Development 136:3205–3214
Van Wesenbeeck L, Cleiren E, Gram J, Beals RK, Benichou O, Scopelliti D, Key L, Renton T, Bartels C, Gong Y, Warman ML, De Vernejoul MC, Bollerslev J, Van Hul W (2003) Six novel missense mutations in the LDL receptor-related protein 5 (LRP5) gene in different conditions with an increased bone density. Am J Hum Genet 72:763–771
Verborgt O, Tatton NA, Majeska RJ, Schaffler MB (2002) Spatial distribution of Bax and Bcl-2 in osteocytes after bone fatigue: complementary roles in bone remodeling regulation? J Bone Miner Res 17:907–914
Visnjic D, Kalajzic Z, Rowe DW, Katavic V, Lorenzo J, Aguila HL (2004) Hematopoiesis is severely altered in mice with an induced osteoblast deficiency. Blood 103:3258–3264
Wada T, Nakashima T, Oliveira-dos-Santos AJ, Gasser J, Hara H, Schett G, Penninger JM (2005) The molecular scaffold Gab2 is a crucial component of RANK signaling and osteoclastogenesis. Nat Med 11:394–399
Walker E, McGregor N, Poulton I, Pompolo S, Allan E, Quinn J, Gillespie MT, Martin TJ, Sims NA (2008) Cardiotrophin-1 is an osteoclast-derived stimulus of bone formation required for normal bone remodeling. J Bone Miner Res 23:2025–2032
Wang FS, Lin CL, Chen YJ, Wang CJ, Yang KD, Huang YT, Sun YC, Huang HC (2005) Secreted frizzled-related protein 1 modulates glucocorticoid attenuation of osteogenic activities and bone mass. Endocrinology 146:2415–2423
Wang FS, Ko JY, Yeh DW, Ke HC, Wu HL (2008) Modulation of Dickkopf-1 attenuates glucocorticoid induction of osteoblast apoptosis, adipocytic differentiation, and bone mass loss. Endocrinology 149:1793–1801
Webb AR, Holick MF (1988) The role of sunlight in the cutaneous production of vitamin D3. Annu Rev Nutr 8:375–399
Weinstein RS, Nicholas RW, Manolagas SC (2000) Apoptosis of osteocytes in glucocorticoid-induced osteonecrosis of the hip. J Clin Endocrinol Metab 85:2907–2912
Weitzmann MN, Roggia C, Toraldo G, Weitzmann L, Pacifici R (2002) Increased production of IL-7 uncouples bone formation from bone resorption during estrogen deficiency. J Clin Invest 110:1643–1650
Weivoda MM, Ruan M, Pederson L, Hachfeld C, Davey RA, Zajac JD, Westendorf JJ, Khosla S, Oursler MJ (2015) Osteoclast TGF-β Receptor signaling induces Wnt1 secretion and couples bone resorption to bone formation. J Bone Miner Res 2016;31(1):76–85
Wolff J. Das Gesetz der Transformation der Knochen. Berlin, Germany: Verlag von August Hirschwald; 1892
Wu XB, Li Y, Schneider A, Yu W, Rajendren G, Iqbal J, Yamamoto M, Alam M, Brunet LJ, Blair HC, Zaidi M, Abe E (2003) Impaired osteoblastic differentiation, reduced bone formation, and severe osteoporosis in noggin-overexpressing mice. J Clin Invest 112:924–934
Wucherpfennig AL, Li YP, Stetler-Stevenson WG, Rosenberg AE, Stashenko P (1994) Expression of 92 kD type IV collagenase/gelatinase B in human osteoclasts. J Bone Miner Res 9: 549–556
Wutzl A, Rauner M, Seemann R, Millesi W, Krepler P, Pietschmann P, Ewers R (2010) Bone morphogenetic proteins 2, 5, and 6 in combination stimulate osteoblasts but not osteoclasts in vitro. J Orthop Res 28:1431–1439
Xiong J, Onal M, Jilka RL, Weinstein RS, Manolagas SC, O'Brien CA (2011) Matrix-embedded cells control osteoclast formation. Nat Med 17:1235–1241
Yao W, Cheng Z, Pham A, Busse C, Zimmermann EA, Ritchie RO, Lane NE (2008) Glucocorticoid-induced bone loss in mice can be reversed by the actions of parathyroid hormone and risedronate on different pathways for bone formation and mineralization. Arthritis Rheum 58:3485–3497

Yao W, Dai W, Jiang L, Lay EY, Zhong Z, Ritchie RO, Li X, Ke H, Lane NE (2015) Sclerostin-antibody treatment of glucocorticoid-induced osteoporosis maintained bone mass and strength. Osteoporos Int 2016;27(1):283–294.

Yasuda H, Shima N, Nakagawa N, Yamaguchi K, Kinosaki M, Mochizuki S, Tomoyasu A, Yano K, Goto M, Murakami A, Tsuda E, Morinaga T, Higashio K, Udagawa N, Takahashi N, Suda T (1998) Osteoclast differentiation factor is a ligand for osteoprotegerin/osteoclastogenesis-inhibitory factor and is identical to TRANCE/RANKL. Proc Natl Acad Sci U S A 95: 3597–3602

Zaidi M (2007) Skeletal remodeling in health and disease. Nat Med 13:791–801

Zhang W, Yang N, Shi XM (2008) Regulation of mesenchymal stem cell osteogenic differentiation by glucocorticoid-induced leucine zipper (GILZ). J Biol Chem 283:4723–4729

Zhao C, Irie N, Takada Y, Shimoda K, Miyamoto T, Nishiwaki T, Suda T, Matsuo K (2006) Bidirectional ephrinB2-EphB4 signaling controls bone homeostasis. Cell Metab 4:111–121

Toward the Automated Detection and Characterization of Osteoclasts in Microscopic Images

2

Andreas Heindl, Martin Schepelmann, Robert Nica, Rupert Ecker, Peter Pietschmann, Alexander K. Seewald, Theresia Thalhammer, and Isabella Ellinger

2.1 Introduction

Microscopes have been used for a long time to observe biological samples. However, measurements of tissue- and cell-related parameters were conducted by human observers and were consequently ad hoc, not reproducible and restricted to small sample numbers. Since computers have become vastly more powerful, life sciences now routinely take advantage of new opportunities to couple microscopy and in silico methods. Automated image segmentation and analysis of large numbers of digital images allow algorithmic recognition of cell and tissue structures and subsequent numeric measurements of cellular parameters. Nevertheless, these new

A. Heindl, PhD (✉)
Department of Pathophysiology and Allergy Research, Center for Pathophysiology, Infectiology and Immunology, Medical University of Vienna, Vienna, Austria

Seewald Solutions, Weisskirchen an der Traun, Austria

Division of Molecular Pathology and Centre for Evolution and Cancer, The Institute of Cancer Research, 15 Cotswold Road, London SM2 5NG, UK
e-mail: andreas.heindl@icr.ac.uk

M. Schepelmann, PhD
Department of Pathophysiology and Allergy Research, Center for Pathophysiology, Infectiology and Immunology, Medical University of Vienna, Vienna, Austria

School of Pharmacy and Pharmaceutical Sciences, Cardiff University, Cardiff, UK

R. Nica • R. Ecker, PhD
TissueGnostics GmbH, Vienna, Austria

P. Pietschmann, MD • T. Thalhammer, PhD • I. Ellinger, PhD
Department of Pathophysiology and Allergy Research, Center for Pathophysiology, Infectiology and Immunology, Medical University of Vienna, Vienna, Austria

A.K. Seewald, PhD
Seewald Solutions, Weißkirchen an der Traun, Austria

P. Pietschmann (ed.), *Principles of Osteoimmunology*,
DOI 10.1007/978-3-319-34238-2_2

methods also come with technical challenges concerning computational resources like processing capacity, memory and disk space, biological sensor limitations, as well as algorithm development.

Today, state-of-the-art hardware and cloud computing enable high throughput analysis of vast amount of images in reasonable time. In the context of bone research, an application for automated analysis is the in silico quantification of osteoclasts. Cell culture models using either murine or human osteoclasts offer the possibility to study parameters such as osteoclast formation from their mesenchymal precursors, differentiation, maturation, and apoptosis (Marino et al. 2014). Advanced molecular imaging of osteoclasts allows to study pathological processes and to elucidate the effect of osteoclast-targeted therapies for diseases in which excess bone resorption is a crucial pathological process. This includes osteoporosis, rheumatoid arthritis, as well as bone tumors such as giant cells tumors, osteosarcomas, and bone metastases.

Currently, quantification of multinucleated osteoclasts in culture is performed manually (Marino et al. 2014). When studying the influence of endogenous metabolites, hormones or therapeutic agents on osteoclast biology, a lot of potentially valuable and relevant information, such as number of nuclei, osteoclast size, number and properties of precursor cells, as well as abundance of target proteins in each cell class, cannot be assessed reliably by manual evaluation. These parameters, however, could be important to elucidate the effects of natural and synthetic agents on osteoclast biology.

To exemplify a typical question in basic osteoclast research, we were looking at whether the pineal hormone melatonin affects osteoclast formation in vitro. Melatonin has long been sought to exert beneficial effects on bone structure and has been proposed to prevent osteoporosis in premenopausal and menopausal women (reviewed in Maria and Witt-Enderby (2014)). Melatonin was found to favor the differentiation of human adult's mesenchymal stem cell into osteoblasts via binding and signaling through a G-protein-coupled melatonin receptor (Radio et al. 2006; Koyama et al. 2002). Thus, beneficial effects of melatonin on bone structure were mainly attributed to its effects on osteoblasts (Maria and Witt-Enderby 2014; Radio et al. 2006). Although melatonin may prevent bone degradation by inhibiting osteoclast formation directly (Ostrowska et al. 2001), the inhibitory effects on osteoclast are also caused by the action of melatonin on osteoblasts. This was suggested from studies in mice, where melatonin increased the osteoprotegerin expression in osteoblasts, which lead to a reduction in the RANKL-mediated osteoclast formation (Maria and Witt-Enderby 2014; Koyama et al. 2002). In a pilot experiment, we re-investigated the possibility of direct effects of melatonin on the formation of mouse osteoclasts from bone-derived precursor cells specifically by focusing on osteoclast number, cell area, multinuclearity, and downregulation of the macrophage (osteoclast precursor cell) marker protein F4/80 as osteoclast marker. As outlined above, limitations of manual evaluation of osteoclast formation would normally prevent the detection of small-scale correlations between changes of cell-associated parameters and the effects of the hormone melatonin. However, when employing a computerized quantification method, all of the abovementioned limitations can be overcome.

In this chapter, we briefly describe standard procedures for osteoclast culture, markers for osteoclast detection, as well as the applied immunofluorescence labeling protocol. We provide a brief introduction to digital images and slide-based microscopy and discuss software packages for image processing as well as valuable tools for handling the workflow.

We then describe the development and application of an image processing algorithm for the detection and quantification of osteoclasts. Principles of ground-truth data and their evaluation as well as the algorithms used for osteoclast detection are explained. Measurements are introduced, and examples are given on how to apply them in real-life scenarios. The described steps are not programming language specific and can be implemented in the framework of the reader's choice. However, we do not cover implementation details, as these are beyond the scope of this work and should be read in existing books on microscopy and digital image processing, for example (Gonzalez and Woods 2008; Wu et al. 2008; Burger and Burge 2008; Solomon and Breckon 2011). The algorithm, which was developed, has been integrated in the software application StrataQuest (TissueGnostics GmbH, Austria), a context-based qualitative image analysis software package. As a "real-life" basic research example, we then show the results of our pilot study using StrataQuest to quantify the direct effect of melatonin on osteoclast formation in culture which yielded novel interesting results that would not have been available without an automated in silico analysis system.

Many of the image processing problems mentioned in these sections are so-called *ill-defined* problems – e.g., segmentation (Martin et al. 2001; Bakushinskiy and Goncharsky 2012) – meaning that there is no unambiguous gold standard to compare these algorithms to. Furthermore, the question of how to compare a developed algorithm to the performance of human experts still remains an important open research problem. Thus, to illustrate these difficulties, we cover not only typical pitfalls including technical problems but also human intuition and limitations in perception and vision that have an influence on the development and evaluation of image analysis algorithms.

In summary, the purpose of this chapter is to introduce biologists and medical scientists to image processing by using a commercially available automated image analysis program to quantify various parameters in osteoclast cultures and exemplify this through evaluating the effect of melatonin on osteoclast formation. We hope that we can thereby raise the audience's awareness and interest in the possibilities and limitations of this new, powerful technology.

2.2 Techniques

2.2.1 Culture Conditions for Isolated Murine Osteoclasts

The culture conditions for osteoclasts suitable for an automatic detection method were developed from standard protocols (Marino et al. 2014; Akatsu et al. 1992), but different culture parameters were evaluated.

Mice (*Mus musculus*) were culled by neck dislocation following asphyxiation. Tibiae and femora were prepared, the caps of the bones were cut off, and the bones were rinsed with 10 ml pre-warmed (37 °C) minimum essential medium containing antibiotics and antifungals. The cells were diluted to 2×10^6 cells per ml, and osteoclast formation-stimulating additives (25 ng/ml Receptor activator of nuclear factor kappa-B ligand (RANKL), 15 ng/ml Macrophage colony-stimulating factor (M-CSF)) were added. 1 ml of cell suspension was added per well (each containing a sterile glass coverslip) of a 24-well culture plate. Culture was maintained at 95 % relative humidity/5 % CO_2 in an incubator for 7 days. Change of medium was performed every second day.

To investigate the effects of melatonin on osteoclast development from bone marrow precursor cells, we applied different doses of melatonin (1, 0.1, and 0.01 μM melatonin from a stock solution of 10 mg/1 ml Dimethyl sulfoxide (DMSO)) to the osteoclast cultures for the entire cultivation time. The respective controls were treated with the solvent only.

2.2.2 Staining Protocol

At the beginning of the development of an algorithm, characteristic features of the target structure or cell need to be defined by biological experts, preferably in a written document for later reference. In the case of osteoclasts, the biological experts defined two important criteria to identify mature osteoclasts: *criterion 1*, the amount of nuclei per cell (≥3) (Andersson and Marks 1989) and *criterion 2*, a low to undetectable expression of the macrophage-antigen F4/80 (van de Wijngaert et al. 1987). The expression of the latter marker is reduced or lost due to the differentiation from precursor cells to osteoclasts. Therefore, the staining protocol included (1) labeling of the nuclei with blue DAPI, (2) staining of the cells with an antibody directed against F4/80 macrophage marker (eBioscience.com), probed with red Alexa Fluor 568 (Invitrogen molecular probes), and (3) making all cells (osteoclasts and precursors alike) "visible" using one antibody against the membrane-bound calcitonin receptor (Acris) and one antibody against the cytoskeleton component α-tubulin (Sigma Aldrich), both probed with Alexa Fluor 647 (Invitrogen molecular probes; far red, appearing white in the image). In the acquired images (e.g., Fig. 2.1), mature osteoclasts appear white (due to the staining for α-tubulin and reduced expression of F4/80 staining), while precursor cells appear red or pink from the α-tubulin – calcitonin receptor – F4/80 overlay. It should be kept in mind however that differentiation and fusion of cells to generate mature osteoclasts is a continuous process. Therefore, mononucleated cells with rather low F4/80 staining can be present in the culture (visible in Fig. 2.1) as well as multinucleated cells with higher expression levels of F4/80. It is up to the biological experts to define a threshold level of mean F4/80 intensity/cell that discriminates between "true" osteoclasts (≥3 nuclei per cell, mean fluorescence intensity of F4/80 < set threshold) and "true" precursor cells (1–2 nuclei/cell, mean fluorescence intensity of F4/80 > set threshold).

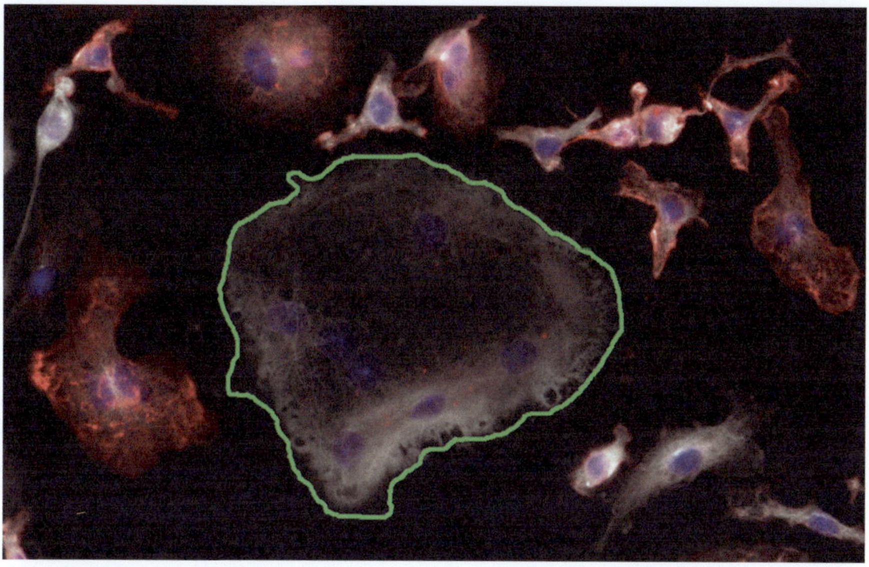

Fig. 2.1 This image shows one immunofluorescence-labeled osteoclast cell in culture surrounded by precursor cells. The osteoclast is marked up in *green*. The perimeter of the target cell is created on a separate layer so that the original content of the image is not modified. This additional layer can later be extracted and processed by computer-based algorithms

Cells were fixed using a 4 % formaldehyde solution, and the remaining aldehyde groups were quenched with 50 mm NH_4Cl. The cells were incubated with blocking/permeabilization buffer (0.5 % Triton X-100+1 % bovine serum albumin in phosphate-buffered saline (PBS)) for 60 min. After this period, the primary antibodies were applied directly in the culture plate at a dilution of 1:1,000 in blocking buffer (parallel approach) for 60 min. After washing with PBS, the secondary antibodies were applied in the dark at a dilution of 1:1,000 in blocking buffer (parallel approach) for 30 min. To stain the nuclei, the cells were incubated for 15 min with DAPI (1 μg/ml) in *aqua bidestilata*. The coverslips were finally mounted with Fluoromount-G (SouthernBiotech) on conventional microscope slides.

Images of the stained cells were acquired using an automated Axio Imager epifluorescence microscope (Zeiss) equipped with TissueFAXS™ hardware and software (TissueGnostics GmbH, Austria) using a 40× Neofluar 1.4 (oil) objective. For special considerations concerning the acquisition process, see the following sections.

2.2.3 Digital Images

Before algorithm development can commence, images have to be acquired and stored. The digitalization of specimens requires a suitable representation of the captured light by an electronic sensor (camera) that can be further processed by the

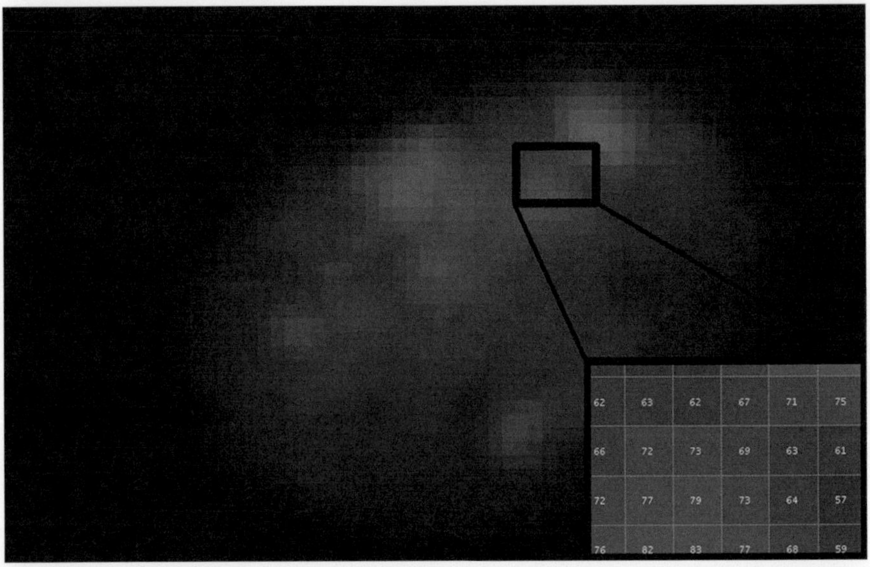

Fig. 2.2 This figure illustrates the digital representation of a captured image derived from fluorescence microscopy. The small insert on the right side shows the gray level value of the selected pixel (8-bit image: a value between 0 and 255)

computer. Exposure time controls the amount of light that is able to hit the sensor elements. Finally, quantification transforms these values to a limited set of intensities that can be further processed by the computer. Typical ranges of 256 ($=2^8=8$ Bit), 4096 ($=2^{12}=12$ Bit), or 65,536 ($=2^{16}=16$ Bit) are used to represent these intensity values. An example of an 8-bit image with a corresponding gray-level matrix is shown in Fig. 2.2.

2.2.4 Automated Slide-Based Microscopy

Sampling of images from large-scale experiments necessitates automation of the acquisition process. A suitable microscope with a fully motorized stage has to be used to automatically acquire images. During acquisition, the stage moves the slides in such a way that the camera captures each region of interest as a set of overlapping field of views (FOVs). For this image acquisition, we used a TissueFAXS™ system (TissueGnostics GmbH, Austria), which offers a convenient workflow to acquire up to eight slides automatically. This was a requirement for further statistical analysis to analyze global effects of various compounds on the growth and formation of osteoclasts in culture. If these cultures contain huge cells, in this case osteoclasts, then stitching (tiling together of the single FOVs to one big image) of acquired regions is essential for further steps, as these cells may have been only partially captured at the border of a FOV. Automatic-stitching algorithms have two main components: finding the best alignment for two neighboring images and finding the best alignment taking into account all neighboring images. TissueFAXS already includes stitching during and after acquisition.

2.2.5 Software for Image Processing

Commercial state-of-the-art solutions like StrataQuest (TissueGnostics GmbH, Austria), in which the osteoclast detection algorithm we have developed was finally incorporated, can automatically detect tissue structures on a digital slide by integrating the detection of objects into detailed context-based quantitative analysis (Stadler et al. 2015; Schmid et al. 2015). StrataQuest allows the analysis of interactions between different types of data like nuclei counts, cell area, protein staining intensities, and the different cell types present via an easy-to-use graphical user interface (GUI). Coupled with TissueFAXS, acquisition and analysis are possible in a homogeneous workflow and CE-marked analysis environment for use in research and in vitro diagnostics. An open source alternative, which requires only basic knowledge of image processing, is Cell Profiler.[1] This application suite offers a framework to build versatile *pipes* (algorithms) for various biological applications. Many predefined pipes for various applications can be downloaded from the online forum and are free of charge. For immunohistochemically stained images, GemIdent[2] (Holmes et al. 2009) is also a versatile tool. A good overview of available software packages is provided in Eliceiri et al. (2012).

However, when designing a new algorithm, a developing framework has to be chosen. Common tools like GNU Image Manipulation Program (GIMP) or Adobe Photoshop are not suitable for algorithm development due to lack of a powerful and fast scripting language. High-performance algorithms made image software tools like OpenCV,[3] ImageJ (Rasband 1997), and Matlab[4] however are very popular: They offer special image-processing toolboxes that are optimized for high throughput while still being relatively easy-to-use for rapid prototyping of image analysis algorithms. For rapid prototyping of image analysis problems, Matlab and ImageJ are good choices. The disadvantage of Matlab and ImageJ is that they are computationally slow compared to OpenCV – so if speed is an issue, algorithms have to be developed or ported to this framework.

In any case, independent of the used framework, generation of ground-truth markups should be started as early as possible so that there is a continuous flow of data. Ground-truth data represent a set of images provided by biological experts that include counts, location, and key features of the objects of interest (Krig 2014). This dataset is then used to evaluate the newly developed algorithms, train machine-learning systems, and optimize parameters throughout the whole development process.

---

[1] http://cellprofiler.org

[2] http://gemident.com

[3] http://opencv.org/

[4] www.mathworks.com/

2.3 Evaluation of Expert Markups and Developed Image-Processing Algorithms

After and during development, the performance of newly implemented algorithms needs to be measured in an unbiased way. For this purpose, visual inspection is the most commonly used method. Unfortunately, this step is error prone and highly dependent on the observer who in many cases will still be a computer scientist rather than a qualified biological researcher. A better method for this evaluation is to let biological domain experts do markups to obtain the so-called ground-truth data from original images (Krig 2014). This can be done using any graphics-editing program like GIMP or Adobe Photoshop. The idea of this markup is to point out those objects of interest (e.g., cells, tissue structures) that should be detected by the algorithm solely by using the provided images without any additional information. The number of markups needed differs from project to project. In case of the osteoclast detection algorithm, about 100 FOVs were manually marked up for osteoclasts. It is essential to draw these markups on a second "layer" (like a "transparency film" on top of the original image) of the target image so that no information from the original image is lost. Alternatively, markups can be stored as simple polygons using popular tools such as Fiji (Schindelin et al. 2015). Either way, manual markups can then be automatically compared to automatically generated masks by the algorithm, ensuring that the newly developed tool and the tissue experts produce comparable results. Supervised machine-learning classifiers like support vector machines, logistic regression, and random forests require ground-truth data as an input to build their decision model.

Before starting to develop an algorithm, the agreement between different human experts has to be confirmed. If this is very low, meaning that there is no consensus between the different human experts, even the best detection algorithm cannot succeed, and therefore development should be postponed, until an agreement between human experts has been achieved. The quality of the ground-truth data can be increased if several human experts provide markups of the same set of images. It is very important that the markups done by humans are performed independently by other experts and by the algorithm developer so that a reliable ground truth can be obtained. A comparison between the experts can be calculated by computing the correlation between the markups. In some cases, like cell recognition covered in this chapter, pixel-level (meaning that the experts actually would have to draw exactly the same lines around the cells) scoring is not very reasonable; therefore, evaluation on a higher level of abstraction, e.g., object-based evaluation, is preferable. This approach counts the number of detected objects and compares it with those that are found in the ground-truth data. Overlaps between detected objects can be used to compute a rough-place agreement beyond the number of detected objects.

In both cases (pixel-/object-based evaluation), having more than one human expert can generate a more objective ground truth (Srivastava et al. 2013), e.g., by performing a majority voting. As the name suggests, majority voting selects those pixels/objects (e.g., osteoclasts) that are marked up by the majority of the human experts.

The next step is the evaluation of the algorithm output. In various scenarios (e.g., segmentation), algorithms have a vast set of parameters. Optimizing them manually is an impossible task, because it would mean evaluating several hundreds of thousands of images by hand. Therefore *exhaustive parameter-optimization* techniques are applied that compare their output with previously created ground-truth data. This technique runs the algorithm with a large set of possible parameter combinations and returns those that have the highest agreement compared to the human experts' ground-truth data. Care must be taken to prevent overfitting, i.e., choosing seemingly optimal parameters, which are only performing well due to chance and only on the images that were used for parameter optimization. Therefore, not all data should be used for this technique – some data should be held back for a final reevaluation of the best parameter settings. This technique is not new and well studied for machine-learning algorithm evaluations: for example, in cross validation, the input data is separated in equally sized sets, employing one for training and one for testing the learned model.

One question that arises is how to compute a score of agreement between the human experts and the algorithm. On pixel level we can calculate the F-score to rank different algorithm output-masks. The F-score (Chinchor and Sundheim 1993) is composed of *precision* and *recall.* Informally, precision represents the percentage of relevant pixels retrieved and is also known as positive prediction rate. Recall on the other hand is also called sensitivity or true positive rate and corresponds to the ratio of correctly classified pixels versus all possible pixels. Formally they are defined as follows:

$$\text{Precision} = \frac{\text{TP}}{\text{TP} + \text{FP}}$$

$$\text{Recall} = \frac{\text{TP}}{\text{TP} + \text{FN}}$$

TP represents the true positive pixel, those that were marked up by both the expert and the algorithm, whereas TN (true negative) stands for pixels that were not marked by either of them. If the pixel was only detected by the algorithm, then it is called a false positive (FP). Similarly, pixels that were not assigned as belonging to the object of interest by the algorithm but were marked up by the expert are called false negative (FN). With these measures, the *balanced F-score* is defined as:

$$F_1 = 2 \cdot \frac{\text{Precision} \cdot \text{Recal}}{\text{Precision} + \text{Recall}}$$

Other common metrics for pixel-level evaluation of segmentation results are the *Jaccard index* (1912) and *Dice coefficient* (1945). Former is defined as the size of the intersection divided by the size of the union of both sets:

$$J(A,B) = \frac{|A \cap B|}{|A \cup B|} = \frac{|A \cap B|}{|A| + |B| - |A \cap B|}$$

J denotes the *Jaccard index*. *A* denotes the set of pixels marked up by the expert and *B* the set of pixels detected by the automated segmentation algorithm.

The *Dice coefficient* is a similarity measure closely related to the *Jaccard index* and often used in ecology. It is less prone to outliers and better suited for heterogeneous data sets (McCune et al. 2002). Formally it is defined as follows:

$$s = \frac{2|A \cap B|}{|A| + |B|}$$

where *s* denotes the *Dice coefficient*, *A* the sets of pixels marked up by the expert, and *B* the automated segmentation algorithm.

The above measures were used in recent state-of-the-art literature to evaluate the performance between expert and machine or between various machine-derived segmentations (Hunter et al. 2013; Esteves et al. 2013; Maska et al. 2014). Besides pixel- and object-based comparison, there are other methods, which may be more suitable for specific scenarios. An extensive survey of segmentation methods was published in Zhang (1996). Segmentation is a well-studied topic in computer vision but still a challenging task. Several evaluation methods have been published (Zhang 2001; Smochina 2010; Benes and Zitova 2015). Recently, even crowdsourcing was employed to evaluate segmentation results of experts with various levels in expertise and automated image processing systems (Irshad et al. 2015).

When applying these discussed points to the example of osteoclast detection, difficulties with their practical implementation become obvious. An example markup of a target cell can be seen in Fig. 2.1 (the green line indicates the manually drawn perimeter of an osteoclast). Since the number of osteoclasts is the desired output of the algorithm, an object-based evaluation is more suitable than a pixel-based one. In this case it is less important whether a specific pixel belongs to the osteoclast or to the background as long as the number, general location, and size of the objects (= osteoclasts) agree with the ground-truth data. As mentioned above, to obtain a general detection algorithm which does not only model the specific knowledge of one expert alone, multiple experts with biological background and experience with osteoclast cultures should deliver ground-truth data. In our setting, two experts provided osteoclast markups, and only those osteoclasts that had been identified by both human experts were considered to be "real." When we evaluated the object-level agreement between these two experts in seven different regions (about 70 images) of osteoclast cultures (Fig. 2.3), the mean agreement (matched) between the two experts was $70 \pm 17\%$. This shows that even though the manual markup of osteoclasts appears simple in theory (i.e., on an image with a clearly isolated osteoclast as shown in Fig. 2.3), in many other "real-world" images (i.e., in images with osteoclasts and precursor cells in close proximity), the opinions of the human experts as to what qualifies as an osteoclast and what does not can be quite incongruent. Reasons why human-based classification may be error prone were published in Baak (1991) and are discussed later in Sect. 2.5. Comparing our algorithm to the experts' matched markup, 86% of osteoclasts identified by both human experts were also classified as osteoclasts by the algorithm. In most experiments however, the algorithm classified more osteoclasts than the matched results of

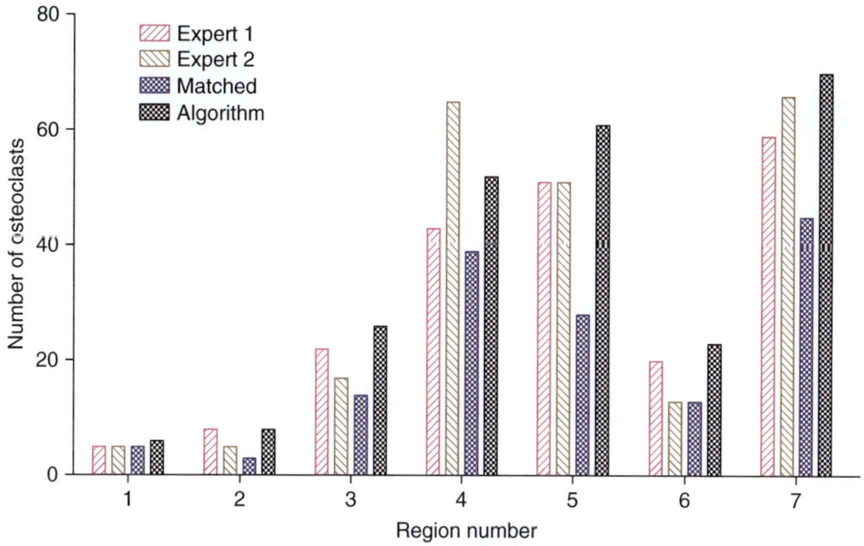

Fig. 2.3 This chart exemplifies the consensus between two human experts. It shows the number of osteoclasts (*y-axis*) detected in seven acquired regions from different cultures (*x-axis*). In each region, the first and the second column represent the osteoclast number detected by each expert (*Expert 1*, *Expert 2*), whereas the third column shows the number of osteoclasts detected congruently by both experts (*matched*)

both human experts (Fig. 2.3). Thus, not only the comparison of the algorithm performance to the matched human experts' ground truth but also the reliability of the ground truth itself will have an impact on the values of precision and recall.

2.4 How to Design an Image-Processing Algorithm Exemplified on Osteoclast Detection in Culture

In this chapter we present the development of an algorithm for osteoclast detection in vitro, starting with the already acquired and stitched images. It follows the general scheme of algorithms for image processing (Fig. 2.4). Most of the existing published image-processing algorithms were derived from this or a similar workflow. Before designing an algorithm, criteria to distinguish between objects of interests (osteoclast) and the remaining objects (precursors) have to be specified. The detailed staining protocol can be found in Sect 2.2. Osteoclasts are defined as multinucleated cells with at least three nuclei (Criterion 1). Additionally, they should not exhibit significant expression levels of F4/80 macrophage marker that identifies only the osteoclast precursor cells (Criterion 2). Such biological criteria should be written down in a document called *customer requirement specification* (CRS) to prevent unexpected results due to misinterpretation by the algorithm developer. The following sections refer to the respective steps of Fig. 2.4: (1) illumination correction, (2) segmentation, (3) post-processing, and (4) labeling).

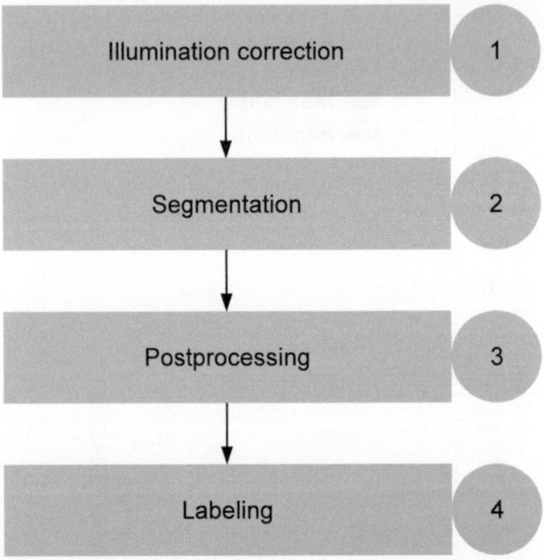

Fig. 2.4 This flowchart demonstrates the general scheme of image-processing algorithms. Four major processing steps must be distinguished

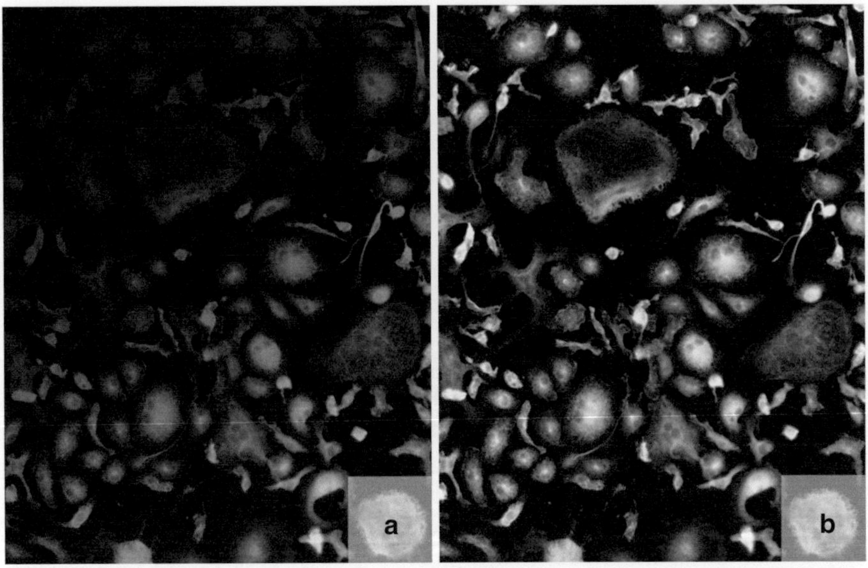

Fig. 2.5 This figure illustrates the effect of a post-acquisition illumination-correction (processing step 1) on images obtained by epifluorescence microscopy. In the original image (**a**), an illumination-gradient is clearly visible from the left upper corner to the right lower corner. This gradient is removed after the application of illumination-correction (**b**)

2.4.1 Illumination Correction

This first step of the detection algorithm is critical, because all further steps will operate on the generated output. Misalignment due to shifts in camera setup or the optical path of the microscope causes uneven illumination (Fig. 2.5a, b) which might not even be visible to the naked eye but can greatly affect the automated

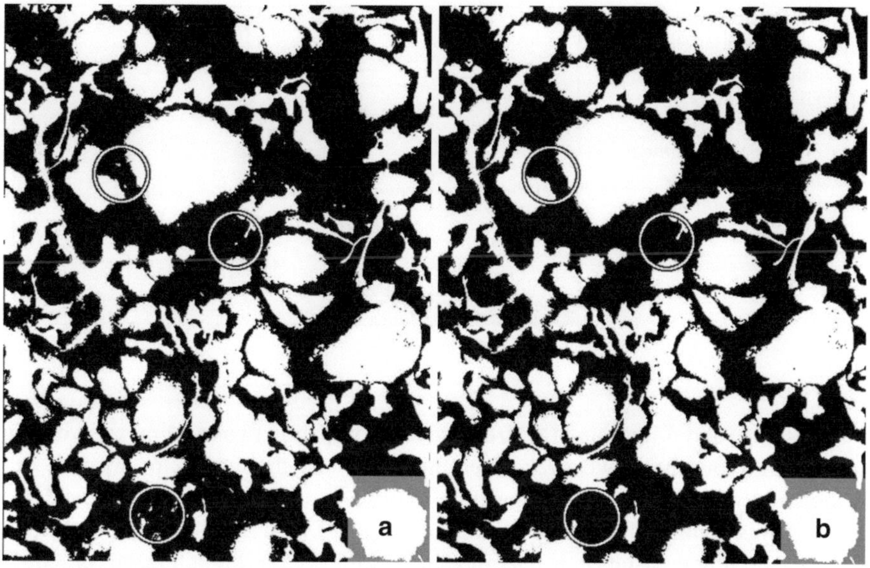

Fig. 2.6 This figure illustrates the results of (**a**) segmentation (processing step 2) and (**b**) post-processing (processing step 3) on images obtained by epifluorescence microscopy. In (**a**), the output image of processing, step 1 (see Fig. 2.5b) has been subjected to automated image segmentation. White structures represent identified objects. This segmentation may introduce artifacts visible as small white blobs that do not belong to actual structures. *Gray circles* indicate example areas where these artifacts are visible. Due to size and shape criteria, they can subsequently be removed resulting in image (**b**)

image analysis. To compensate for this introduced bias, an illumination correction function can be computed (Wu et al. 2008; Wang et al. 2015). It represents a special illumination image that contains the overall pattern of illumination in all future-acquired images. Various publications deal with this process, e.g., (Zhu et al. 2003), Ljosa and Carpenter (2009), and Smith et al. (2015). If the approximation of the illumination function fails, adaptive thresholding can be applied, which is discussed in the next section.

2.4.2 Segmentation

Step 2 splits the pixels in two groups, those belonging to the foreground (cells, including osteoclasts) and those belonging to the background (Fig. 2.6a). This separation can be achieved by applying intelligent thresholding. Examples are the classical triangle algorithm (Zack et al. 1977) or machine-learning-based classifiers with appropriate features (Kapelner et al. 2007). Most of these image segmentation techniques operate on histograms. A histogram is a discrete distribution function of the image's intensity values. It counts the number of gray values that pertain to each of the single categories/bins (intensity values, 0–255 in case of an 8-bit image). The triangle algorithm obtains a threshold to distinguish between background

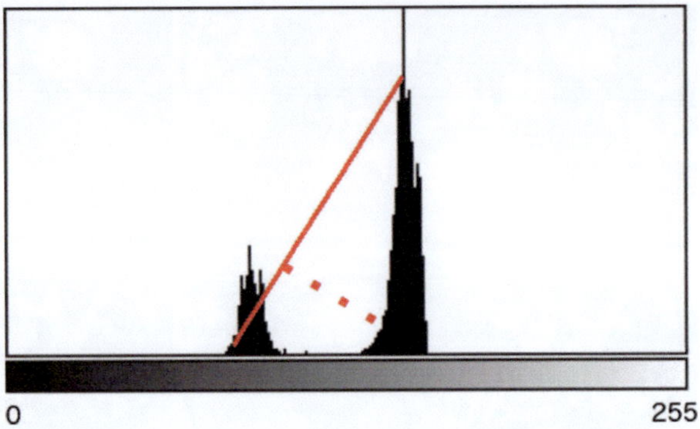

Fig. 2.7 Histogram illustrating the triangle threshold method. The optimal threshold is selected where the maximum distance (*dotted line*) intersects with the base line at the bottom

(< threshold) and foreground (≥ threshold) by computing two local maxima of this histogram. Now these two peaks are connected, and the maximum distance (an orthogonal vector to the line connecting the two peaks) between the line and the histogram is computed (see Fig. 2.7 for an example). The intensity value indicated by the maximum distance is the threshold value. Everything greater or equal (brighter) is considered as foreground. Every value beneath this threshold (darker) is classified as background. The machine-learning-based methods are more complex and require extensive knowledge in learning theory. The interested reader should refer to the publication mentioned above.

If illumination correction (step 1) did not produce an acceptable result, adaptive thresholding can be used to partition images into foreground and background as well. Compared to a single threshold like in the triangle approach, this is a more sophisticated process that chooses different thresholds for each pixel of the image. Because of this method of operation, this is sometimes also called local or dynamic thresholding (Gonzalez and Woods 2008; Stockman and Shapiro 2001; Burger and Burge 2016; Korzynska et al. 2013).

For our osteoclast detection algorithm, we used adaptive thresholding (Liu et al. 2002) and computed local thresholds for each subregion of the image. The threshold is chosen by examining the intensity values of the local neighborhood of each pixel, computing the median intensity. An important parameter of this approach is the neighborhood size which can be determined by thresholding the image with different neighborhood sizes (e.g., 100 pixel, 200 pixel, etc.) and comparing the segmented image with the previously discussed ground-truth markup of the expert. The neighborhood size with the highest concordance is chosen and every subsequent image of the same type processed with the optimized parameter.

2.4.3 Post-processing

Step 3 implements a cleanup step, which removes artifacts (Fig. 2.6a, b) and unwanted cells such as osteoclast precursor cells. This is often achieved by binary operations such as area opening or the computation of morphological features that

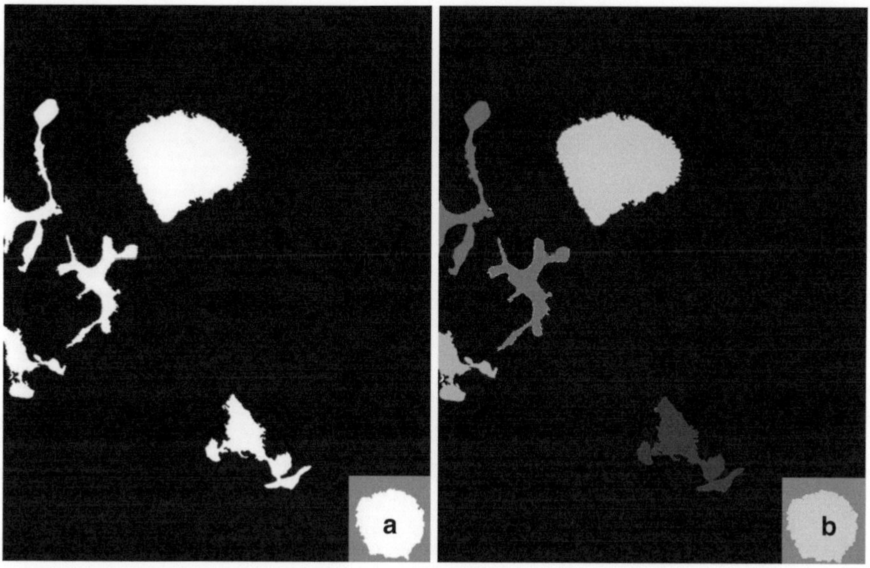

Fig. 2.8 This figure shows the labeling (processing step 4) of images obtained by epifluorescence microscopy. In image (**a**), an enlarged area of the output image of processing step 3 (see Fig. 2.6b) is shown. In image (**b**), the labeling of the segmented image is illustrated. For demonstration purpose, each object is labeled in a different shade of gray. Internally, this would be represented by assigning a unique number to each object

can be used to distinguish between the object of interest and other objects. Area opening removes all objects with an area smaller than a chosen threshold. If this condition is not specific enough, morphological features such as eccentricity, solidity, convex hull, etc., can be used to distinguish between the desired cell and unwanted artifacts. In case of our developed osteoclast detection algorithm, we have also used features from specific staining, i.e., Criterion 1 (osteoclasts: ≥3 nuclei) and Criterion 2 (osteoclast: F4/80 mean intensity/cell <20).

2.4.4 Labeling

Step 4 performs a labeling of the remaining cells (Fig. 2.8a, b) (Samet and Tamminen 1988; He 2012). This is done to directly denominate each single cell on the image so that further measurements can be computed by directly addressing the single cells.

As a side note, the image context is also of great importance during feature calculation. Figure 2.9a, c show two osteoclasts. Reasoning from the example at the top, a feature such as the distance between the nuclei may perfectly identify the target cell. Unfortunately, in the same region, another osteoclast can be found (Fig. 2.9c) where the distance is much larger than in the upper example. So if one were to try and detect osteoclasts only in the DAPI channel, i.e., by looking at the nuclei alone, this might produce false positives or negatives. This is why we

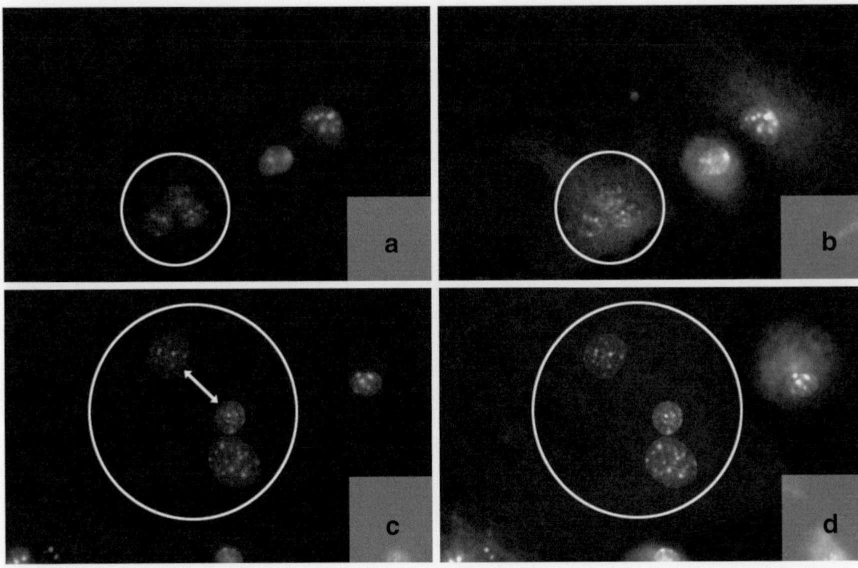

Fig. 2.9 The importance of context is illustrated in this figure. Figures (**a–d**) show two different image details of immunofluorescence-labeled osteoclast cultures. *White circles* indicate osteoclasts. During the staining process, cell nuclei are stained with DAPI (**a**, **c**). Additionally, microtubules and a membrane receptor are immunofluorescence-labeled to visualize cell bodies (**b**, **d**). An important criterion of mature osteoclasts is to have three or more nuclei. An algorithm that operates only on the nuclei (DAPI channel) would probably miss the osteoclast in (**c**) due to the larger distance between the nuclei in contrast to (**a**). Taking the additional staining of cell bodies and borders into account, the same nuclei in (**b**) and (**d**) are now allocated to one cell, and consequently these cells are identified as osteoclasts

introduced the additional immunofluorescence stainings of the microtubules, the calcitonin receptor, and the F4/80 macrophage marker to assign the detected nuclei to whole cells (Fig. 2.9b, d).

Having computed all these features, it is now possible to create statistics and draw conclusions about treatment- or disease-related morphological intensity-based difference or staining intensity-based (i.e., associated protein levels) difference. Our final developed algorithm for image-based osteoclast detection has since been incorporated into the StrataQuest image analysis software package.

2.4.5 Practical Example: Influence of Melatonin on Osteoclast Formation

To exemplify the use of this novel automated system for relevant questions in basic biological bone research, we have investigated the effect of pharmacological (μM) doses of melatonin on osteoclast formation in vitro (Bubenik et al. 1998). Using the automated osteoclast detection algorithm in StrataQuest, a multitude of parameters were measured and computed including (mean) intensity values of labeled proteins

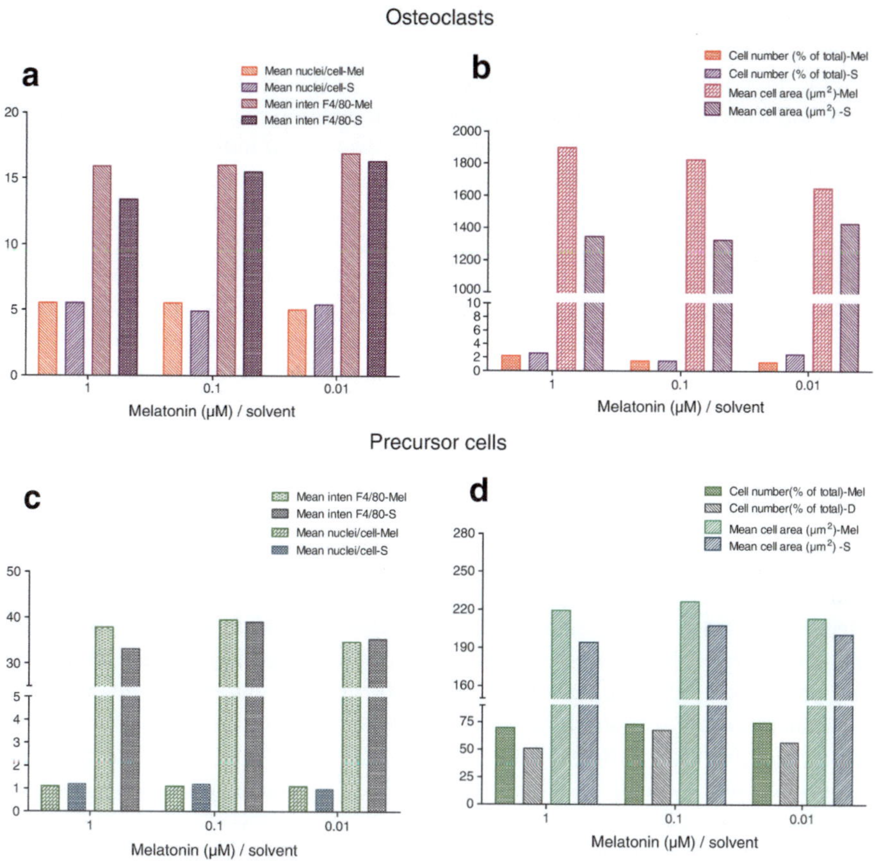

Fig. 2.10 Effect of in melatonin treatment (Mel: 1 μM, 0.1 μM, 0.01 μM) or solvent (S) on cultured osteoclasts (**a**, **b**) and precursor cells (**c**, **d**). (**a**, **c**) Mean number of nuclei/cell and mean intensity levels of F4/80 staining/cell. (**b**, **d**) Relative (% of total) cell number and mean cell area (μm^2). Data from a typical experiment for multinucleated cells with mean intensity levels of F4/80 staining <20 (1.3–2.6 % of cells) and mononucleated cells (precursor cells) with mean intensity levels of F4/80 staining >20 (51–74 % of cells) are shown

per cell (exemplified for F4/80 staining, Fig. 2.10a, c), number of nuclei per cell (Fig. 2.10a, c), total or relative number of individual cells, or mean cell area (Fig. 2.10b, d). Results are displayed for two out of the four discriminated cell populations, namely, the multinucleated osteoclasts with mean F4/80 intensity levels <20 (arbitrary threshold, range 0–255) and the mononucleated precursor cells with mean F4/80 intensity levels >20 (Fig. 2.10 upper and lower graphs, respectively). While, in analogy to published data (Koyama et al. 2002), we could not see any influence of melatonin on osteoclast number (Fig. 2.10b); we found a strong increase ranging from 130 % of control values with 10 nM melatonin to 200 % of control values with 1 μM melatonin in the mean area of osteoclasts after treatment with melatonin, an effect which was not seen for the precursor cells (Fig. 2.10b). In correlation, also the mean number of nuclei/osteoclast increased up to 140 % of control values (Fig. 2.10a). The classical evaluation by humans would have only retrieved the number of osteoclasts/area or well, which in our experiment has not been influenced. The observation that the mean area of and the mean number of nuclei/osteoclast increased (but not of the precursor cells) indicates that melatonin has a stimulating effect on the fusion of osteoclast precursor cells to mature osteoclasts. Whether the morphological alterations are associated with an increased bone-resorbing activity of the osteoclasts needs to be determined in functional assays, e.g., measurement of pit formation on dentine slices (Takahashi et al. 2007). Nevertheless, the results gained by automated evaluation for the first time demonstrate a direct effect of melatonin on bone-resorbing cells, which together with the well-established stimulatory influence on osteoblast activity would greatly influence bone turnover.

2.5 Common Pitfalls

Although high-end microscopy technologies open up new ways to examine tissue and cell culture samples, it also requires detailed knowledge of biology, optics, electronics, and computer science. This includes appropriate sampling of the tissues, optimized cell-culture conditions, furthermore well-chosen fixation, and staining protocols. It also includes the selection of the most appropriate microscopic equipment for acquisition of the specific experiment (Pearson 2007). However, in this section we want to focus on those common problems that occur during image acquisition and evaluation. We discuss pitfalls like uneven illumination, uneven staining, touching clumps of nuclei, and psychological aspects resulting in fallacies due to gestalt laws. For basics about microscopy and optics, please refer to Spector and Goldman (2006).

2.5.1 Imaging-Based Errors

2.5.1.1 Illumination

To perform a quantitative measurement, each step of image acquisition has to be as exact as possible. Noise is often a result of a misaligned light source and increases the error rate in all following algorithm steps. Therefore, having an evenly illuminated image is of great importance. To achieve this goal, calibration slides can be used to align the light source properly before recording microscopy images. Applying an illumination-correction function (as discussed in Sect. 4.1) afterwards is a measure of last resort since it modifies the intensity values of the image and might distort the "real" microscopic picture. If this is done in the channel containing the target protein, the researcher has to consider that she/he may have created an artificial (= false) staining due to the applied correction function. Using these results for further statistical analysis is problematic and may lead to wrong conclusions. The most common optical problem, nonaligned condensers, causes a type of uneven illumination that is almost impossible to repair with image-processing techniques. A priori detection and elimination of illumination gradients before recording images therefore save a lot of effort and avoid unnecessary image manipulation, thus increasing the significance of derived quantitative biological results.

2.5.1.2 Acquisition Parameters

Acquiring the same sample on different instruments or with different settings on the same instrument often results in different-looking images. Depending on the settings and the experience of the user who controls the microscope, image quality may vary. Currently, there is no standard (Yagi and Gilbertson 2005), so the comparability between various acquisition systems is impossible. An example simulating the effect of different acquisition settings on the same FOV is shown in Fig. 2.11. The first image (Fig. 2.11a) of this figure is in focus and has balanced color values as it should be. Figure 2.11b is out of focus; consequently segmentation will prove difficult and unreliable because borders are blurred. To prevent out-of-focus images, techniques like the so-called extended focus (Abrahamsson et al. 2006) can be used: first, a stack of images of the same FOV on five different focal planes is acquired. A subsequent fusing step merges all images per stack and takes only those parts that are in focus on each plane. As a result, the amount of unfocused cells is reduced to a minimum. Figure 2.11c illustrates overexposure. Saturated areas are visible as

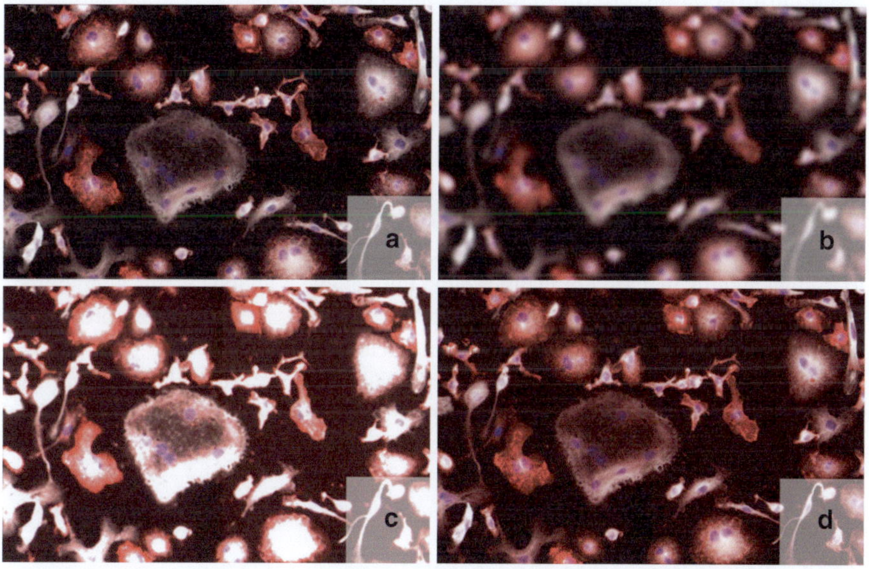

Fig. 2.11 This figure illustrates the effect of acquisition conditions in image processing. Image (**a**) represents an ideal acquisition, showing the cells in focus with balanced color values. In image (**b**), cells are out of focus, and their borders are no longer clearly visible. Image (**c**) is overexposed (exposure time set too long). Saturated areas make further quantitative analysis impossible. Finally, image (**d**) has a significantly increased red value (background) that may mislead the observer. Note: These images have been processed for demonstration purposes

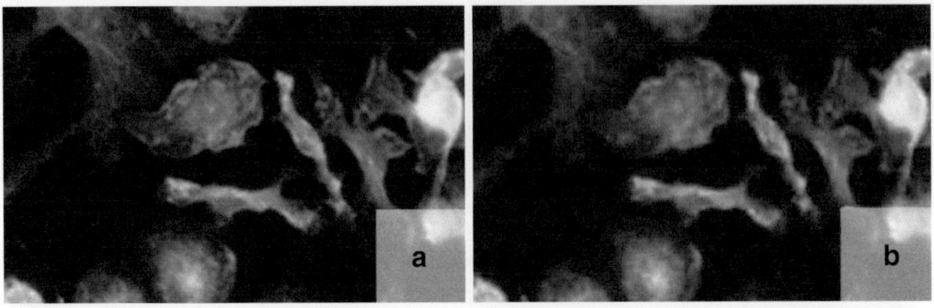

Fig. 2.12 The importance of file formats in image processing is illustrated in this figure. Images (**a**, **b**) represent the same image detail derived from an immunofluorescence-labeled osteoclast culture. While image (**a**) was stored as PNG (with lossless compression), image (**b**) was saved as JPEG. JPEG may reduce image quality dramatically and may introduce rectangular artifacts that could affect machine-learning classification

white spots and prevent useful feature computation. An increased red level is illustrated in Fig. 2.11d. This may mislead the observer to draw incorrect conclusions about the expression of a certain marker.

2.5.1.3 File Formats

Another pitfall can be the file format used to store the image. Lossless formats should always be preferred to prevent quantification artifacts. A good choice for storing high-quality images would be the portable network graphics (PNG) (Fig. 2.12a). This format can be read by a variety of tools and supports lossless compression. Besides PNG, the tagged image file format (TIFF) is popular for storing acquired images. Compared to PNG, it offers a wide range of storage options, which is also the drawback of this format. It cannot be guaranteed that other applications which offer TIFF support are able to read and correctly interpret the chosen TIFF settings. The most commonly used Joint Photographic Experts Group (JPEG) graphics format has to be avoided (Fig. 2.12b). The intention of its compression is to remove details that are not visible to the human eye. Obviously, this alters intensity values and can limit the applicability of machine-learning and image-processing techniques as well as quantification of biological features.

For the osteoclast detection, we selected PNG due to the portability of the format to different operating systems like Windows, Mac OS X, and Linux, since preliminary experiments indicated about 30 % less disk space usage than when using compressed TIFF images.

2.5.2 Errors Related to the Gestalt Laws

The origin of the term Gestalt (the essence of an entity's complex form) goes back to Ernst Mach in 1886 (Mach 1886). Since that time, gestalt laws are extensively examined in psychology. An example of these laws is grouping of objects due to proximity, similarity, closure, good continuation, and connectedness. A summary

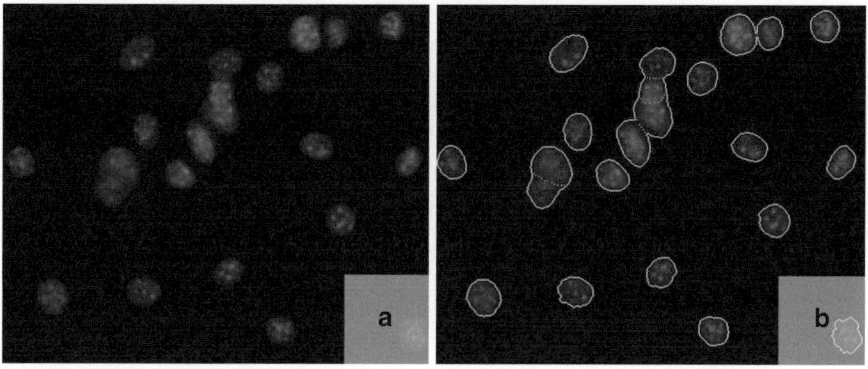

Fig. 2.13 This figure illustrates the limitations of automated image segmentation in comparison to human-object recognition. Figure (**a**) shows an image detail of an osteoclast culture, where cell nuclei are labeled with DAPI. *Clusters* of nuclei, i.e., overlapping nuclei, are visible. The result of a segmentation of these nuclei is indicated in image (**b**). The detected perimeters derived from automated image segmentation are drawn as *solid white lines*, while the *dotted lines* represent the additional separation of the overlapping nuclei, as a human would intuitionally draw them

published in Scientific American describes and illustrates these gestalt laws comprehensively (Rock and Palmer 1990). Especially when ground-truth data is created, gestalt laws play an important role. Human recognition of shapes is still a research field with many open questions. One way to reduce the effect of gestalt laws is to make people aware of it: Training them with examples can increase the output quality of ground-truth data, thus improving detection accuracy of the algorithm. As a result, interdisciplinary projects should be preferred to avoid such imprinted pitfalls.

Up to this day, there are scenarios where image processing cannot compete with human intuition. One frequently occurring example in osteoclast detection is the problem of touching clumps of nuclei (Moffat et al. 2006). Figure 2.13a exemplifies a case where image processing may fail, although nuclei can be intuitively separated by a human (Fig. 2.13b), and different human observers are remarkably consistent in where they separate the given nuclei. Analyses that require identification of a single nucleus fail if no proper segmentation of these clumps is available. There are several approaches to divide clumps in a single nucleus in culture and tissues (Rogojanu et al. 2010; Zhang et al. 2015; Sheeba et al. 2014) like applying watershed algorithms (Sheeba et al. 2014), cutting of nuclei due to angles in the morphological shape (Cloppet and Boucher 2010; Wang et al. 2012), level-set-based processes (Xiong et al. 2006), or dynamic programming that tries to model the human expertise (Nandy et al. 2007). However, each of these approaches has scenarios where they fail, so currently there is no computational algorithm available that can handle all different cases of clumps.

Trying to model this method is tricky due the fact that it is not yet known how the human brain recognizes objects (Liu et al. 2009). Figure 2.14 illustrates an idealized decision-making process of a human. In real life, this process is assumed to be less

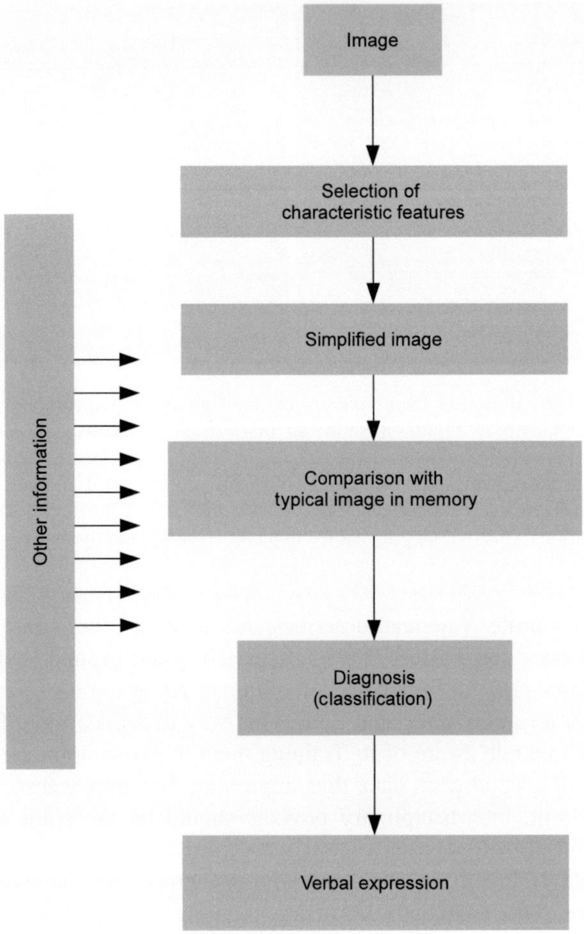

Fig. 2.14 This flowchart shows an assumed human diagnostic process of an idealized decision-making situation (Modified after Baak (1991))

structured and contains trained template recognition, which is believed to be present in the human subconscious and instantly available (Baak 1991; Lennert and Stein 1981) – an example would be recognition of numbers or letters in literate humans. In contrast to the human brain, slight variations in size, staining, orientation, and illumination are not accounted for by the computer and therefore result in poor recognition performance. Currently, image processing uses features (e.g., curvature, intensity changes, homogenous textures, edges, etc.) that seem to be too different and too "weak" to model the human performance of perception.

Nevertheless, visual perception is not the only source of error; verbal expression differs from expert to expert. An example would be the size of osteoclasts. One human expert may assume that a "huge" osteoclast is one with up to 6 nuclei, whereas for another expert, "huge" osteoclast means more than 16 nuclei. Obviously, development of an algorithm requires background knowledge of the target cell structure and texture. Ideally, the computer scientist obtains this information during interviews with the biological expert, and they create a CRS together at the beginning of the project instead of relying on verbal-communicated information only. Quantitative evaluation is given by ground-truth data which should agree with explicit knowledge given in the CRS, and over the course of the project, the CRS should be regularly discussed and possibly adapted. The interpretation of the given features to classify the target cells varies depending on the final knowledge of the algorithm engineer. This was shown for pathologists in Livesey et al. (1978), Pool et al. (1979). However, nowadays when pathological interpretations are quantified by computer-based evaluation, computer scientists take the place of the pathologists by developing tools to support their diagnoses. Clearly, they face the same problems with less medical and biological experience, and additionally they interpret the data based on their technical background which may lead to problems of overgeneralization.

2.5.3 Benefits of Automated Osteoclast Detection

Despite these pitfalls, automated segmentation and analysis of osteoclasts as exemplified in this chapter has significant advantages. Large amount of data is processable – for the results shown in Fig. 2.10, we have processed regions up to 42 mm^2, and the only limits for further scale-up are the acquisition of the images and the computer's processing and memory capacity. Our computer-based evaluation is faster by about two orders of magnitude compared to a trained human expert. Additionally, the result after manual quantification is just the number of osteoclasts, whereas our algorithm yields many more informative measures, such as total cell number, total and relative numbers of osteoclasts and their precursor cells, total area of all cells, total and relative area of osteoclasts and their precursor cells, numbers of nuclei, quantification of associated proteins, and many morphological and statistical features. Furthermore, it is highly unlikely that the human quantification on large regions is reproducible or consistent if compared to another human expert (due to fatigue and different interpretations by different humans). However, reapplying an algorithm to the same set of images always yields the identical result.

Conclusion

Applying image processing and machine-learning techniques to biological and medical images can improve the quality of research and diagnostics dramatically; automated analysis produces consistent quantitative measures. Small differences not visible for the human eye, but possibly linked to disease states, can be detected. Currently applied visual inspection normally produces an overall score rather than measuring each cell. The human mind also cannot keep track of the multiple informative measures of cells or tissue and is generally less able to integrate many weak predictive measures. It should also be emphasized again that machine-based analysis is more efficient after development and can operate 24 h, 7 days a week.

One domain that especially benefits from such systems is bone research. Osteoclast quantification is currently done manually, so that large-scale experiments cannot be conducted. The intra- and inter-variability between experts is normally very high, which is in contrast to an automated system that detects and quantifies these cells, improves the quality of the result, and is always consistent. Applying the algorithm, in addition, enables to measure parameters, which cannot be assessed manually. Indeed, practical application of the novel osteoclast detection system in the software StrataQuest enabled us to identify a direct influence of the indolamine melatonin on in vitro murine osteoclast size and multinuclearity which would not have been accessible without in silico-based image analysis.

By extension, together with automated staining systems, a reproducible, validated, and fully automatic workflow for medium to high throughput evaluation of basic osteoclast biology, but also routine clinical screening, is conceivable. Of course, legal issues have to be considered before such algorithms can successfully be applied to standard diagnosis in hospitals, and the results of automated quantifications are dependent on a number of parameters like correct staining and acquisition. Image processing will never replace human experts completely, because the final diagnosis and interpretation is still up to human expertise, but it can relieve the scientists or physicians from a huge amount of repetitive work and at the same time increase the significance of the obtainable results.

Acknowledgements This project was funded by FFG (Bridge 818094). We also want to express our deepest thanks to Caroline Brünner-Kubarth, Kathrin Burger, Andrea Nußbaumer, Ayse Okay, Magdalena Pilz, Radu Rogojanu, and Roland Stumberger for their contributions and dedication to our project.

References

Abrahamsson, S., Usawa, S., & Gustafsson, M. (2006) A new approach to extended focus for high-speed, high-resolution biological microscopy. In Biomedical Optics 2006 (pp. 60900N-60900N). International Society for Optics and Photonics.

Akatsu T, Tamura T, Takahashi N, Udagawa N, Tanaka S, Sasaki T, Yamaguchi A, Nagata N, Suda T (1992) Preparation and characterization of a mouse osteoclast-like multinucleated cell population. J Bone Miner Res 7(11):1297–1306

Andersson GN, Marks SC Jr (1989) Tartrate-resistant acid ATPase as a cytochemical marker for osteoclasts. J Histochem Cytochem 37(1):115–117

Baak JPA (1991) Manual of quantitative pathology in cancer diagnosis and prognosis. Springer, Berlin/New York, [1991] ©1991

Bakushinskiy A, Goncharsky A (2012) Ill-posed problems: theory and applications. Springer, Netherlands

Benes M, Zitova B (2015) Performance evaluation of image segmentation algorithms on microscopic image data. J Microsc 257(1):65–85

Bubenik GA, Blask DE, Brown GM, Maestroni GJ, Pang SF, Reiter RJ, Viswanathan M, Zisapel N (1998) Prospects of the clinical utilization of melatonin. Biol Signals Recept 7(4):195–219

Burger W, Burge MJ (2008) Digital image processing – an algorithmic introduction using java. Springer, London, XX, 565 pp

Burger W, Burge MJ (2016) Digital image processing: an algorithmic introduction using java. Springer, London

Chinchor N, Sundheim B (1993) MUC-5 evaluation metrics. Association for Computational Linguistics, Baltimore, pp 69–78

Cloppet F, Boucher A (2010) Segmentation of complex nucleus configurations in biological images. Pattern Recogn Lett 31(8):755–761

Dice LR (1945) Measures of the amount of ecologic association between species. Ecology 26(3):297–302

Eliceiri KW, Berthold MR, Goldberg IG, Ibanez L, Manjunath BS, Martone ME, Murphy RF, Peng H, Plant AL, Roysam B, Stuurman N, Swedlow JR, Tomancak P, Carpenter AE (2012) Biological imaging software tools. Nat Methods 9(7):697–710

Esteves T, Oliveira MJ, & Quelhas P (2013) Cancer cell detection and morphology analysis based on local interest point detectors. In Iberian Conference on Pattern Recognition and Image Analysis. Springer Berlin Heidelberg, 624–631

Gonzalez RC, Woods RE (2008) Digital image processing. Prentice Hall, Upper Saddle River

He L, Chao Y, & Suzuki K. (2012) A new two-scan algorithm for labeling connected components in binary images. In Proceedings of the World Congress on Engineering, 2:1141–1146

Holmes S, Kapelner A, Lee PP (2009) An interactive java statistical image segmentation system: GemIdent. J Stat Softw 30(10):i10

Hunter LA, Krafft S, Stingo F, Choi H, Martel MK, Kry SF, Court LE (2013) High quality machine-robust image features: identification in nonsmall cell lung cancer computed tomography images. Med Phys 40(12):121916

Irshad H, Montaser-Kouhsari L, Waltz G, Bucur O, Nowak JA, Dong F, Knoblauch NW, Beck AH (2015) Crowdsourcing image annotation for nucleus detection and segmentation in computational pathology: evaluating experts, automated methods, and the crowd. Pac Symp Biocomput: (p. 294). NIH Public Access.

Jaccard P (1912) The distribution of the flora in the alpine zone. 1. New Phytol 11(2):37–50

Zhu J, Liu B, & Schwartz SC (2003) General illumination correction and its application to face normalization. In Acoustics, Speech, and Signal Processing, 2003. Proceedings. (ICASSP'03). 2003 IEEE International Conference on. IEEE (Vol. 3, pp. III–133)

Kapelner A, Lee PP, & Holmes S (2007) An interactive statistical image segmentation and visualization system. In Medical Information Visualisation-BioMedical Visualisation, 2007. MediVis 2007. International Conference on. IEEE 81–86

Korzynska A, Roszkowiak L, Lopez C, Bosch R, Witkowski L, Lejeune M (2013) Validation of various adaptive threshold methods of segmentation applied to follicular lymphoma digital images stained with 3,3′-Diaminobenzidine&Haematoxylin. Diagn Pathol 8:48

Koyama H, Nakade O, Takada Y, Kaku T, Lau KH (2002) Melatonin at pharmacologic doses increases bone mass by suppressing resorption through down-regulation of the RANKL-mediated osteoclast formation and activation. J Bone Miner Res 17(7):1219–1229

Krig S (2014) Ground truth data, content, metrics, and analysis. Apress, Berkeley, pp 283–311

Lennert K, Stein H (1981) Histopathology of non-Hodgkin's lymphomas: based on the kiel classification. Springer, Berlin

Liu F, Song X, Luo Y, Hu D (2002) Adaptive thresholding based on variational background. Electron Lett 38(18):1017–1018

Liu H, Agam Y, Madsen JR, Kreiman G (2009) Timing, timing, timing: fast decoding of object information from intracranial field potentials in human visual cortex. Neuron 62(2):281–290

Livesey AE, Sutherland FI, Brown RA, Beck JS, Macgillivray JB, Slidders W (1978) Cytological basis of histological typing of diffuse Hodgkin's disease. Demonstration of an implied misnomer in the terminology of the Rye classification. J Clin Pathol 31(6):551–559

Ljosa V, Carpenter AE (2009) Introduction to the quantitative analysis of two-dimensional fluorescence microscopy images for cell-based screening. PLoS Comput Biol 5(12):e1000603

Mach E (1886) Beiträge zur Analyse der Empfindungen. G. Fischer, Jena

Maria S, Witt-Enderby PA (2014) Melatonin effects on bone: potential use for the prevention and treatment for osteopenia, osteoporosis, and periodontal disease and for use in bone-grafting procedures. J Pineal Res 56(2):115–125

Marino S, Logan JG, Mellis D, Capulli M (2014) Generation and culture of osteoclasts. Bonekey Rep 10(3):570

Martin DR, Fowlkes C, Tal D, Malik J (2001) A database of human segmented natural images and its application to evaluating segmentation algorithms and measuring ecological statistics. EECS Department, University of California, Berkeley

Maska M, Ulman V, Svoboda D, Matula P, Matula P, Ederra C, Urbiola A, Espana T, Venkatesan S, Balak DM, Karas P, Bolckova T, Streitova M, Carthel C, Coraluppi S, Harder N, Rohr K, Magnusson KE, Jalden J, Blau HM, Dzyubachyk O, Krizek P, Hagen GM, Pastor-Escuredo D, Jimenez-Carretero D, Ledesma-Carbayo MJ, Munoz-Barrutia A, Meijering E, Kozubek M, Ortiz-de-Solorzano C (2014) A benchmark for comparison of cell tracking algorithms. Bioinformatics 30(11):1609–1617

McCune B, Grace JB, Urban DL (2002) Analysis of ecological communities. MjM Software Design, Gleneden Beach

Moffat J, Grueneberg DA, Yang X, Kim SY, Kloepfer AM, Hinkle G, Piqani B, Eisenhaure TM, Luo B, Grenier JK, Carpenter AE, Foo SY, Stewart SA, Stockwell BR, Hacohen N, Hahn WC, Lander ES, Sabatini DM, Root DE (2006) A lentiviral RNAi library for human and mouse genes applied to an arrayed viral high-content screen. Cell 124(6):1283–1298

Nandy K, Gudla PR, & Lockett SJ (2007) Automatic segmentation of cell nuclei in 2D using dynamic programming. In Proceedings of 2nd Workshop on Microsopic Image Analysis with Applications in Biology

Ostrowska Z, Kos-Kudla B, Swietochowska E, Marek B, Kajdaniuk D, Ciesielska-Kopacz N (2001) Influence of pinealectomy and long-term melatonin administration on GH-IGF-I axis function in male rats. Neuro Endocrinol Lett 22(4):255–262

Pearson H (2007) The good, the bad and the ugly. Nature 447(7141):138–140

Pool CW, Diegenbach PC, Ockeloen BJ (1979) Quantitative succinate-dehydrogenase histochemistry. II. A comparison between visual and quantitative muscle fibre typing. Histochemistry 64(3):263–272

Radio NM, Doctor JS, Witt-Enderby PA (2006) Melatonin enhances alkaline phosphatase activity in differentiating human adult mesenchymal stem cells grown in osteogenic medium via MT2 melatonin receptors and the MEK/ERK (1/2) signaling cascade. J Pineal Res 40(4):332–342

Rasband WS (1997–2015) ImageJ. U. S. National Institutes of Health, Bethesda

Rock I, Palmer S (1990) The legacy of Gestalt psychology. Sci Am 263(6):84–90

Rogojanu R, Bises G, Smochina C, & Manta, V (2010) Segmentation of cell nuclei within complex configurations in images with colon sections. In Intelligent Computer Communication and Processing (ICCP), 2010 IEEE International Conference on. IEEE 243–246

Samet H, Tamminen M (1988) Efficient component labeling of images of arbitrary dimension represented by linear bintrees. IEEE Trans Pattern Anal Mach Intell 10(4):579–586

Schindelin J, Rueden CT, Hiner MC, Eliceiri KW (2015) The ImageJ ecosystem: an open platform for biomedical image analysis. Mol Reprod Dev 82(7-8):518–529

Schmid M, Dufner B, Durk J, Bedal K, Stricker K, Prokoph LA, Koch C, Wege AK, Zirpel H, van Zandbergen G, Ecker R, Boghiu B, Ritter U (2015) An emerging approach for parallel quantification of intracellular protozoan parasites and host cell characterization using TissueFAXS cytometry. PLoS One 10(10):e0139866

Sheeba F, Thamburaj R, Mammen JJ, & Nagar AK (2014) Splitting of Overlapping Cells in Peripheral Blood Smear Images by Concavity Analysis. In International Workshop on Combinatorial Image Analysis. Springer International Publishing 238–249

Smith K, Li Y, Piccinini F, Csucs G, Balazs C, Bevilacqua A, Horvath P (2015) CIDRE: an illumination-correction method for optical microscopy. Nat Methods 12(5):404–406

Smochina C, Manta V, Rogojanu R (2010) New discrepancy measure for evaluation of segmentation quality, In Proc. 11th IASTED International Conference on Computer Graphics and Imaging, Innsbruck, Austria, track 679–053

Solomon C, Breckon T (2011) Fundamentals of digital image processing: a practical approach with examples in matlab. Wiley Publishing, Chichester

Spector DL, Goldman RD (2006) Basic methods in microscopy protocols and concepts from cells: a laboratory manual. Cold Spring Harbor Laboratory Press, Cold Spring Harbor

Srivastava G, Yoder JA, Park J, Kak AC (2013) Using objective ground-truth labels created by multiple annotators for improved video classification: a comparative study. Comput Vis Image Underst 117(10):1384–1399

Stadler M, Walter S, Walzl A, Kramer N, Unger C, Scherzer M, Unterleuthner D, Hengstschlager M, Krupitza G, Dolznig H (2015) Increased complexity in carcinomas: analyzing and modeling the interaction of human cancer cells with their microenvironment. Semin Cancer Biol 35:107–124

Stockman G, Shapiro LG (2001) Computer vision. Prentice Hall PTR, Upper Saddle River

Takahashi N, Udagawa N, Kobayashi Y, Suda T (2007) Generation of osteoclasts in vitro, and assay of osteoclast activity. Methods Mol Med 135(1543-1894 (Print)):285–301

van de Wijngaert FP, Tas MC, Burger EH (1987) Characteristics of osteoclast precursor-like cells grown from mouse bone marrow. Bone Miner 3(2):111–123

Wang H, Zhang H, Ray N (2012) Clump splitting via bottleneck detection and shape classification. Pattern Recogn 45(7):2780–2787

Wang HF, Wang GP, Wang XY, Wang BJ, Zhao XD (2015) Novel background calibration algorithm for image in non-uniform illumination field. Imaging Sci J 63(5):285–289

Wu Q, Merchant F, Castleman K (2008) Microscope image processing. Elsevier Inc., Amsterdam

Xiong G, Zhou X, Ji L, Bradley P, Perrimon N, & Wong S (2006) Segmentation of drosophila RNAi fluorescence images using level sets. In 2006 International Conference on Image Processing. IEEE 73–76

Yagi Y, Gilbertson JR (2005) Digital imaging in pathology: the case for standardization. J Telemed Telecare 11(3):109–116

Zack GW, Rogers WE, Latt SA (1977) Automatic measurement of sister chromatid exchange frequency. J Histochem Cytochem 25(7):741–753

Zhang YJ (1996) A survey on evaluation methods for image segmentation. Pattern Recog 29(8):1335–1346

Zhang YJ (2001) A review of recent evaluation methods for image segmentation. Kuala Lumpur, Malaysia, pp 148–151

Zhang C, Sun C, Su R, Pham TD (2015) Clustered nuclei splitting via curvature information and gray-scale distance transform. J Microsc 259(1):36–52

3 An Introduction to the Immune System

Georg Schett

3.1 Innate and Adaptive Immunity

The term "osteoimmunology" describes the molecular and cellular cross talk between the skeletal and immune systems (Schett et al. 2009). Before referring to these immune–skeletal interactions, we introduce the two main branches of the immune system, the innate immunity and the adaptive immunity. The immune system developed during evolution to protect the organism against invading infectious agents such as bacteria, fungi and viral agents. In principle, the immune system is divided into two major branches that differ from each other in preferentially involved cell types and the mechanisms of action of the immune response. However, innate and adaptive immunities are also closely linked and synergized to ensure best immune defence against microbe.

On the one hand, innate immunity constitutes an old defence system, which has been highly conserved over millions of years among different species. The innate immunity can be characterized as ancient, rapid, nonspecific, invariant and constant, with minimal expansion and limited diversity. Highly conserved molecular patterns such as lipopolysaccharide (LPS) and sugar moieties such as mannans and glycans, which are the so-called pathogen-associated molecular patterns (PAMPs), are recognized by germ line-encoded receptors on a distinct subset of leukocytes. The cells that participate in the innate immunity do not undergo clonal expansion. Therefore, these strategies are suitable for the quick recognition of structurally conserved patterns and an immediate defence against the offending agents. A further characteristic aspect is that there is no immunological memory with innate immune responses.

G. Schett, MD
Department of Internal Medicine 3, University of Erlangen-Nuremberg,
Erlangen 91054, Germany
e-mail: georg.schett@uk-erlangen.de

P. Pietschmann (ed.), *Principles of Osteoimmunology*,
DOI 10.1007/978-3-319-34238-2_3

On the other hand, there is the adaptive immunity that evolved to form a defence that is specifically fitting to the particular pathogen. As a consequence, it requires some time to ensure building up highly specific responses to the infectious agent. Another feature of the adaptive immunity is the development of an immunological memory that ensures rapid and highly specific responses after a second encounter with the same pathogen.

Besides the efficient protection against potential danger caused by various microorganisms, it is very important that the immune response is not directed against the host himself with consequent damage or destruction of the body's own tissues and organs ("horror autotoxicus"). This so-called self-discrimination/non-self-discrimination is the essential feature of the immune system that protects against developing autoimmunity. Here we want to give an overview on the types of cells and pathways of this defence system that creates both innate and adaptive immunity.

3.2 The Innate Immune Response

3.2.1 Effector Cells in Innate Immunity

The distinctive features of innate immunity commonly refer to a broadly distributed variety of myeloid and lymphoid cells that can exert rapid effector functions through a limited repertoire of germ line-encoded receptors. These effector cells can be characterized by the expression of glycoprotein surface molecules, the "cluster of differentiation" (CD) molecules, which are determined by their binding of specific monoclonal antibodies. In particular, the types of cells that are capable of the innate immune response are neutrophils, monocytes, macrophages, natural killer (NK) cells, eosinophils, basophils and mast cells. All these cells except the NK cells are derived from the common myeloid progenitor, the stem cell that also gives rise to megakaryocytes and erythrocytes. As there is a morphological difference between these aforementioned lineages, these cells also differ in their immunological function.

3.2.1.1 Neutrophils

Neutrophils represent the first line of defence as they are attracted by chemotactic factors and accumulate at the site, where a potentially noxious agent is recognized for the first time (Borregaard et al. 2010). As an initial action, neutrophils take up the microorganisms by phagocytosis and destroy the pathogen in cytoplasmic vacuoles by producing oxygen species and lytic enzymes. Besides the quick destruction of the invading agent, neutrophils also produce chemokines and cytokines such as tumor necrosis factor alpha (TNFα) to attract more effector cells such as macrophages to the site of the pathogenic invasion. Therefore, neutrophils are also of key importance for further steps of immune activation. Another task of neutrophils is that they are involved in tissue regeneration after elimination of the pathogen by producing cytokines that promote cell proliferation.

3.2.1.2 Monocytes, Macrophages and Dendritic Cells

Monocytes represent approximately 10 % of leukocytes in human blood (Auffray et al. 2009). Primarily, monocytes function as phagocytes that scavenge toxic components and pathogens. By means of phagocytosis, they are enabled to quickly remove bacteria or virus-infected cells. Possibly dangerous structures are recognized by pathogen recognition receptors such as toll-like receptors that bind particular molecular patterns. These receptors are described below in more detail. Besides their ability to destroy microbes, monocytes have the function to remove apoptotic cells and cellular debris. Thus, monocytes play a crucial role for cellular homeostasis and may prevent the development of autoimmunity. In accordance, it is known that a defective removal of apoptotic cells contributes to the pathogenesis of autoimmune disorders like systemic lupus erythematosus. Monocytes also play a pivotal role in inflammatory conditions, as they are a major source for proinflammatory cytokines such as TNFα. Based on their phagocytic activity and their pattern of cytokines expression, monocytes are further subdivided into three different subsets: The major population are the CD14+CD16-CCR2^{high}CX3CR1^{low} monocytes that represent up to 90 % of human monocytes and are characterized by the production of IL-10 after LPS stimulation in vitro. Other subsets are the CD16+ monocytes, which can be divided into two subclasses: The CD16+CD14+ CD64+CD32+ positive cells function as phagocytes and produce TNFα and IL-1 after LPS stimulation. In contrast, the CD16+CD14^{low} monocytes lack phagocytic activity and do not produce cytokines in response to LPS. Macrophages are derived from monocytes that have left the blood circulation and emigrate to target tissues.

Depending on their localisation in different organs, macrophages are given distinct names: for example, liver-resident macrophages are called Kupffer cells, macrophages in the central nervous system are named microglia, and histiocytes are macrophages embedded in the connective tissue. Macrophages are characterized by surface expression of CD11b, CD14, CD68, F4/80 (mice)/EMR1 (human), lysozyme M and MAC-1/MAC-3. Like neutrophils or monocytes, macrophages also serve as phagocytes. They appear to be mobilized shortly after the recruitment of neutrophils. However, they persist much longer than the neutrophils at sites of chronic inflammation. Macrophages kill bacteria by the production of nitric oxide. Additionally, macrophages play a role in the linking of innate and adaptive immune responses as they produce cytokines like IL-12 or interferon gamma (IFNγ) that direct adaptive immune responses to the TH1 type.

Depending on the co-stimulatory signals, macrophages are polarized into two groups and have broadly been classified as M1 and M2 macrophages (Martinez et al. 2009). M2 macrophages can further be divided into M2a, M2b and M2c macrophages. These two major subsets of macrophages differ in terms of their receptors, cytokine and chemokine expression, as well as effector function. Whereas M1 macrophages are microbicidal and inflammatory, M2 macrophages are immune modulators and are poorly microbicidal. The classically activated M1 macrophages produce cytokines such as IFNγ, IL-6, IL-12 or TNF, which are potent proinflammatory mediators. Alternatively, activated M2 macrophages are induced by IL-4.

IL-10 or IL-13 and act anti-inflammatory by production of IL-10 and transforming growth factor beta (TGF-β), which antagonize proinflammatory cytokines. Thus, macrophage activation can be either proinflammatory or anti-inflammatory.

A further important cross-link to the adaptive immune system is that macrophages serve as antigen-presenting cells (APCs). After phagocytosis by a macrophage, the antigen is processed by proteolysis. Particles of the pathogen are presented on the surface of the macrophage on major histocompatibility complex (MHC) class II molecules, which – different from MHC class I molecules – are only expressed by professional antigen-presenting cells. The presentation of the antigen in a processed form is required to activate T cell-dependent responses in adaptive immunity. This issue is described in more detail below.

The most potent antigen-presenting cells are the dendritic cells (DCs) that can be found in most tissues of the body with a maximum in secondary lymphoid organs. There are two kinds of DCs: The so-called classical, myeloid DCs (mDCs) are enabled to phagocyte antigens and present the processed particle on MHC class II molecules to permit antigen recognition by T cells. The mDCs recognize pathogens commonly via the "toll-like" receptors TLR 2 and TLR 4 and drive the immune response by the production of IL-12 that activates both NK and T cells. Besides these conventional dendritic cells, which are monocyte-derived, a second type of dendritic cells is known, viz. the plasmacytoid DCs (pDCs). These cells are designated according to their morphology and are believed to be derived from lymphoid progenitors. pDCs detect DNA and RNA from viruses by "toll-like" receptors TLR 7 and TLR 9 and respond to pathogen encounter by the secretion of inflammatory chemokines and the interferons IFN-α and IFN-β.

3.2.1.3 Natural Killer Cells

Natural killer (NK) cells are defined morphologically as large granular lymphocytes (Vivier et al. 2011). Comparable to T cells and B cells, they originate from the common lymphoid progenitor. Human NK cells are characterized by the surface expression of CD16 and CD56 and the lack of CD3 or a T cell or B cell receptors. NK cells kill viral-infected cells by cytolysis. The lack of MHC class I molecules on the surface of a virus-infected or a tumour cell, a condition called "missing self", a viral strategy to evade other mechanisms of defence such as apoptosis induction by CD8+ cytotoxic T cells, inactivates mechanisms that prevent NK cells from killing their target. Further, preceding stimulation of NK cells by activating ligand–receptor interaction is required. These receptors bind soluble ligands such as cytokines like IL-1 or IL-2, cell surface molecules or chemokines. Additionally, NK cells are involved in immunoregulatory mechanisms and inflammatory processes as they are able to produce cytokines such as IFNγ or IL-10. NK cells play a crucial role to sustain self-tolerance while efficiently destroying virus-infected or malignant derived cells. Recently it was discovered that NK cells can develop an immunological memory. Therefore, their position is attributed both to the innate immunity and the adaptive immunity.

3.2.2 Pathogen Recognition in Innate Immune Responses by "Toll-Like" Receptors

Toll was first discovered in *Drosophila melanogaster* where it was found to play a key role in embryonic development and the immune response against fungi. Ten human "toll-like" receptors (TLRs) have been identified to date. TLRs recognize pathogen-associated molecular patterns (PAMPs), which mostly are conserved products of microbial metabolism produced by bacteria, but not by the host. Ligands of TLRs are, for example, LPS for TLR-4, peptidoglycans for TLR-2, flagellin for TLR-5 or unmethylated CpG DNA for TLR-9. These receptors are found on phagocytes such as macrophages, neutrophils or dendritic cells and allow to quickly distinguish between self and infectious non-self. Structural properties of all TLRs are the N-terminal leucine-rich repeats and the cytoplasmic toll/IL-1 receptor (TIR) homology domain. Upon ligand recognition, TLRs drive the expression of genes that are involved in host defence, such as antimicrobial proteins, proinflammatory cytokines, chemokines, co-stimulatory and MHC molecules. There are different downstream signal cascades depending on the adaptor molecules that are recruited to the TIR. One important intracellular signalling pathway is the activation of NFkB that promotes the expression of proinflammatory cytokines such as TNFα or IL-1.

3.2.3 Humoral Effectors of the Innate Immunity

3.2.3.1 The Complement System

The complement system consists of over 25 plasma and cell surface proteins and is an important effector mechanism that links both the innate immunity and the adaptive immunity. The main function of the complement system is to mark invading pathogens to have them recognized for phagocytosis by, for example, macrophages, a process called opsonisation. The second aim is to attract other effectors of the innate immunity to the site of pathogen encounter by release of cleavage products that are produced during complement activation and act as chemoattractants. The third way of action is that complement activation can directly promote cell destruction by forming the so-called membrane attack complex (MAC) and inducing lysis of the cell by increased membrane permeability. There are three different ways of complement activation: The classical pathway requires stepwise proteolytic cleavage of complement components and is mediated by antibody–antigen complexes. The alternative way of complement activation is triggered by microbial structures that neutralize inhibitors of spontaneous complement activation and does not require antigen–antibody interaction. The common final path of both modes of activation is the formation of the MAC that promotes removal of viral-infected cells or microbes, respectively. During the formation of the MAC, the complement component C3a is produced which is a potent chemokine and efficiently attracts immune cells to the site of inflammation. The third activation pathway is called the lectin pathway of complement activation because it is started by the contact with mannan-containing microbial cell walls or by complexes of microorganisms and host pentraxins and ficolins.

The mechanisms of complement-mediated cell destruction potently drive local inflammation by recruiting cells of both the innate and adaptive immunities to the place of pathogen encounter. Therefore, the complement system can be seen as a means of cross talk between the two branches of the immune system. This feature can also be applied to different kinds of cytokines that represent a kind of communication between the different effectors of the immune system.

3.2.3.2 Cytokines

Cytokines are glycoprotein or glycopeptide molecules that are produced by a variety of cells including almost all cells, which are involved in the adaptive or innate immunity. Cytokines act as humoral mediators that regulate cellular growth and differentiation, drive inflammatory processes and direct the course of inflammation and defence strategies against pathogens. There are five main groups of cytokines: the interleukins (IL), the interferons (IFN), the tumour necrosis factors (TNF), the colony-stimulating factors (CSF) and the chemokines. Members of all of these five groups are significantly involved in regulation of both the innate immunity and the adaptive immunity.

Interleukins are named after their discovery as they are secreted by leukocytes as a means of communication. There are thirty-seven ILs known to date with a combination of shared and separate functions. Almost all mechanisms of action in both branches of immunity are directed by ILs, as these cytokines are produced by nearly all involved cells such as T cells and B cells, monocytes, macrophages, NK cells and DCs, to only name the major participating subsets. The role of different ILs for the two kinds of T-helper (TH)-mediated cell response is described below in more detail. IFNs are mainly produced as response to viral infection and activate macrophages, NK cells and T lymphocytes. Members of the TNF family are mostly secreted by macrophages. TNFα plays a pivotal role in promoting inflammation and macrophage activation. Chemokines serve as attracting agents for different kinds of immune cells to the site of inflammation or pathogen encounter. There are three different types of chemokines related to the localisation of a cysteine residue in their amino-terminal. Chemokines serve as ligands for the corresponding chemokine receptors, which are expressed by various subsets of effectors in innate and adaptive immunities. In summary, the cytokines regulate a concerted action of the two different branches of immunity to ensure an efficient defence against antigens.

3.3 The Adaptive Immune System

The aim of the adaptive immunity is to generate a response that is perfectly fitting to the particular antigen. This process takes time by necessity but has the advantage of simultaneously establishing an immunological memory. After re-exposure to the antigen, the immunological memory ensures an immediate reaction against the invading agent. B cells and T cells are the key players in adaptive immunity, but they are also dependent on the support of the effectors of innate immune response. After antigen encounter, B cells differentiate and mature, helped by TH cells (T-helper cells), to produce specific antibodies of high affinity. Cytotoxic T cells constitute the defence against intracellular pathogens. Adaptive immunity is dominated by the interplay of B cells and T cells. Therefore these two lymphocyte subsets are described in more detail.

3.3.1 B Cells

3.3.1.1 Early B Cell Development

B cells are named after the organ, in which they develop, the Fabricius' bursa in birds or the bone marrow in mammals. Both B cells and T cells are derived from the common lymphoid progenitor. Depending on co-stimulatory signals, the lymphocytes developing from these stem cells are committed either to the B cell lineage or the T cell lineage. As mentioned, development from the progenitor to the mature B cell takes place in the bone marrow, supported by bone marrow stromal cells that produce cytokines and deliver co-stimulatory signals allowing the survival of B cells. This process is divided into several stages of maturation with the aim to generate B cells with a functional B cell receptor (BCR), which is a membrane-bound immunoglobulin (Ig) molecule. The formation of the Ig is a highly complex process that requires somatic reassembly and hypermutation of the encoding genes. Principally, Ig molecules are composed of a heavy chain and a light chain and have antigen-recognizing variable as well as constant regions that also define the Ig subclass. Antibody diversity is generated through the rearrangement of the V, D and J segments of the Ig heavy and light chain genes. Only one allele in each B cell becomes successfully rearranged, a phenomenon called allelic exclusion. Thereby the "one B cell one specificity concept" is guaranteed, which means that one particular antigen can only be recognized by one specialized B cell.

As a first step of B cell maturation, the stem cell differentiates to the pro-B cell stage that is characterized by the rearrangement of heavy chain genes. If the rearrangement is successfully completed, the cell expresses the pre-B cell receptor (BCR), a membrane-bound μ heavy chain molecule and is then defined as pre-B cell. Signalling via the pre-BCR and the IL-7 receptor is critical for the further development to immature B cells: after the successful rearrangement of the Ig light chain, the B cell expresses the BCR, a membrane-bound IgM molecule, and is called immature B cell. If the B cell fails at any stage of this developmental process, it is removed by the induction of apoptosis. To guarantee immune tolerance, there are two steps of selection that prevent the generation of autoreactive B cells. In the first step (positive selection), only those B cells whose BCR can strongly interact with MHC II molecules receive survival signals. Next, B cells that recognize autoantigens by their BCR are eliminated by apoptosis. This latter process is called negative selection. After completing positive and negative selection, the B cell additionally expresses surface IgD and becomes a mature B cell. Mature B cells leave the bone marrow and migrate to secondary lymphoid organs such as lymph nodes or the mucosa-associated lymphatic tissue (MALT) where they wait for antigen encounter.

3.3.1.2 B Cell Development to Antibody-Forming Plasma Cells and B Cell Memory in T Cell-Dependent Responses

B cells are able to recognize native structures of antigens by their BCR (Kurosaki et al. 2010). In T cell-dependent immune responses to protein antigens, the B cell requires co-stimulatory signals from a CD4+–TH cell that already had contact with the same antigen. After recognition by the BCR, the antigen is processed intracellularly and presented on a MHC class II molecule to the specialized, so-called cognate T cell. Additional co-stimulatory ligand–receptor interactions between the B cell and the TH cell are required to promote further B cell differentiation, such as the interaction between B7 on the surface of the B cell with CD28/CTLA4 on the T cell. Another important co-stimulatory signal is delivered by the interaction of CD40 on the B cell with CD40L/CD154 on the T cell. This interplay between the B cell and T cell takes place in the so-called germinal centres (GC). GCs are the area of lymphoid follicles in lymph nodes and the spleen where the B cells clonally proliferate and mature to antibody-secreting plasma cells or memory B cells. These developing B cells, called germinal centre B cells, are supported by follicular dendritic cells that present antigen–antibody complexes on their surface, which stimulate B cell growth and differentiation. During the clonal expansion, genes encoding for the BCR undergo an extremely high rate of somatic mutation to generate a BCR that interacts highly selectively with its target. Thereby a B cell clone that recognizes the antigen with the highest specificity is selected. The less-specific B cells lack further stimulatory signals and die by apoptosis. The apoptotic cells are removed by macrophages in the GC. Next the selected clone divides several times and differentiates further into plasmablasts, plasma cells and memory B cells.

Plasmablasts produce antibodies to a greater extent than B cells but are still able to produce antigens and to interact with T cells. In contrast, plasma cells are only capable of antibody production (Oracki et al. 2010). Plasma cells secrete an enormous mass of Ig molecules and are the body's "antibody-producing machinery". Contrary to their precursors, plasma cells do not express MHC class II or surface Ig molecules anymore. The class of the secreted Ig molecules depends on the co-stimulatory signals (i.e. cytokines) the B cell received during differentiation by the T cell. This means that there might be a class switch from IgM antibodies that are primarily produced in the antibody response, to another Ig subclass. The lifespan of the plasma cell can be limited to a few days, but there are also plasma cells that migrate to the bone marrow and survive for a lifetime. These long-lived plasma cells ensure immediate antibody production after re-encounter with the same pathogen.

Another developmental pathway of the selected B cell clone is the differentiation to a memory B cell. Human memory B cells can be identified by the expression of CD27, whereas CD27 is not found on their murine equivalent. They usually express IgG, IgA or IgE on their surface, although some also display IgM. They generally lack IgD. Memory B cells migrate actively between the blood and secondary lymphoid organs and have a lifespan up to decades. As a hallmark, memory B cells are able to directly differentiate to high-affinity plasma cells after antigen re-encounter without the preceding process of affinity maturation in a GC. In contrast to the primary response, class-switched antibodies are produced in the secondary response to a pathogen. The ability to immediately respond to a pathogen is due to the already hypermutated Ig V region. The generation of an immunological memory is characteristic of T cell-dependent responses, although it is believed that it may also arise in response to T cell-independent (TI) antigens.

3.3.1.3 T Cell-Independent Immune Responses

Antigens that activate B cells without any help from T cell are called T cell-independent (TI) antigens. As a common feature, these antigens have highly repetitive structures that activate B cells by BCR cross-link. For example, such structures are found in bacterial cell wall components. There are two kinds of TI antigens: TI type 1 (TI-1) antigens such as LPS also activate immature B cells, whereas TI-2 antigens such as bacterial polysaccharides of gram-negative bacteria like *pneumococci* or *meningococci* only activate mature B cells. A specialized B cell subset, the splenic marginal zone (MZ) B cells, is capable of TI-2 responses. These B cells are named after their localisation in the splenic marginal zone, which lies between the red pulp and the white pulp and represents a filter station of the blood. However, humans lack a histologically distinguishable MZ in the spleen, but have MZ B cells, which are recirculating in contrast to their murine equivalents. MZ B cells are characterized as IgM+CD21^{high}CD23^{low}. After antigen recognition, they differentiate into IgM-secreting plasmablasts within hours and ensure the first-line defence against pathogens. The MZ B cell-derived plasmablasts produce IgM in great number, and in the later phase of the response, antibodies of the IgG subclass are also secreted. The loss of MZ B cells in splenectomized individuals explains why they are at increased risk of developing septic complications after infection with gram-negative bacteria.

3.3.1.4 Ig Subclasses

Ig molecules have a variety of functions: The simplest form of action is that antibodies neutralize the antigen and prevent it from causing damage to the body. Further, Igs contribute to the recognition of pathogens and their elimination by initializing various cellular and humoral effector mechanisms. Antibody-tagged microbes can be recognized and eliminated by phagocytes. Another feature is that cell surface-bound antibodies activate the classical pathway of the complement system.

There are five different Ig subclasses: IgM, IgG, IgA, IgE and IgD. The IgG subclass is further subdivided into IgG1 to IgG4. The Ig molecule is structurally similar to a “Y”. It integrates four separate polypeptides, two light (L) and two heavy (H) chains that are covalently joined by disulphide bonds. The protease papain cleaves the Ig molecule into two antigen-binding fragments (Fab) and the crystalline fragment (Fc). The Fab consists of the light chain and the part of the heavy chain that carries the antigen-binding domain. The Fc fragment consists of the constant C region of the heavy chains that determines the antibody subclass. L chains are composed of a variable (V) region, joining (J) region and a constant (C) region. There are two classes of L chains in mammals, k and λ, determined by the V and J regions. There are five different kinds of H chains μ, γ, α, ε and δ

corresponding to the Ig subclasses IgM, IgG, IgA, IgE and IgD. H chains are similar to a combination of a V, a diversity (D), a J and a C region. The V domain of the L chain and the H chain work together as antigen recognition site; the constant domains form the Fc fragment and represent the complement-binding region. Further, the Fc fragments of pathogen-bound antibodies are ligands for the corresponding Fc receptors on phagocytes and facilitate antigen recognition. This process of marking the pathogen with antibodies is called opsonisation.

IgM molecules are found as membrane-bound monomers on B cells. In their secreted form, they occur as high molecular pentamers that are connected by a J chain. IgM is the first Ig subclass to be produced in adaptive immune responses. In addition, the pentameric IgM is the most potent activator of the complement system. Antibodies of the IgG subclass represent the majority of serum Ig with a proportion of 75 %. IgG is the predominant antibody subclass produced during memory responses, but it is also secreted to a small extent in the primary adaptive immune response. As the only antibody subclass, IgG antibodies are able to cross the placental barrier and give passive immunity to the foetus. IgA antibodies occur as monomers in the blood. However, these antibodies are mainly secreted in a dimeric form into different body spaces. Therefore, IgA play a pivotal role for the mucosal immune system in the gut and the lung as these antibodies form a protective layer against invading pathogens on the mucosal surfaces. IgA is also found in the breast milk contributing to the passive immunity of the newborn. Antibodies of the IgE subclass are associated with atopic conditions such as allergic asthma bronchiale and are crucial for the defence against eukaryotic parasites such as helminths. This subclass has the lowest serum concentration. IgD is predominantly expressed on mature B cells and is only found in low levels in the serum.

3.3.2 T Cells

3.3.2.1 T Cell Subsets and Development

The second key player in adaptive immunity is the T cell, named after the thymus, the place where they differentiate. Depending on the composition of their T cell receptor (TCR), T cells can be subdivided into two classes, the αβ T cells and the γδ T cells. The TCR is composed of an αβ heterodimer or a γδ heterodimer that is non-covalently associated with the CD3 complex. CD3 is a pan-T cell marker and is importantly involved in TCR signal transduction. Additional accessory molecules are CD4 and CD8 that determine the interaction with MHC I or MHC II molecules and the functional specificity (Germain et al. 2002). The αβ T cells represent the bulk of T cells in lymphoid organs, and they are restricted to the response to antigens that are either presented on MHC I or II molecules. In contrast, the γδ T cells constitute only 2 % of T cells. γδ T cells are generally not MHC restricted and are mostly found in the gut mucosa where they seem to be involved in microbial surveillance. Further, T cells can be subdivided according to their antigenic specificity and their effector functions into CD4+ TH cells and CD8+ cytotoxic T cells. Similar to the antibody response driven by B cells, memory is also established in the T cell-mediated adaptive immunity. Memory T cells are both of the CD4+ and the CD8+ subclass and rapidly achieve effector functions after re-encounter with the pathogen.

The T cell progenitors are derived from the common lymphoid progenitor in the bone marrow and enter the thymus to differentiate to the mature T cells, supported by antigen-presenting cells of the thymic microenvironment. T cell development can be divided into three steps: the first stage is the CD4-CD8 double-negative stadium, the second is the CD4+CD8+ double-positive stadium and the last is the CD4+ or CD8+ single-positive stadium. The fate of CD4 or CD8 expression is driven by the property of the TCR to better interact with MHC II or I molecules, respectively. Comparable to B cells, T cells are positively and negatively selected during their developmental process to guarantee immunological tolerance and prevent autoimmunity.

3.3.2.2 CD4+ Effector T Cells

TH cells, also known as effector T cells, are important for steering the adaptive immune response. Based on their functional specificity, TH cells can further be subdivided to different populations, the TH1, TH2, Treg and TH17 cells. TH1 cells produce the signature cytokine IFNγ along with proinflammatory cytokines as TNFα and TNFβ, to stimulate innate and T cell-mediated immune responses. The most important function of TH1 cells is to promote cellular immunity that is mediated by CD8+ T cells that possess direct cytolytic activity to eliminate obligate intracellular pathogens. IL-12 drives the differentiation to the TH1 specificity. Self-reactive TH1 cells are related to autoimmune disorders such as insulin-dependent diabetes mellitus or rheumatoid arthritis. TH2 cells are characterized by the production of IL-4, IL-5, IL-9, Il-10 and IL-13 (Mosmann et al. 1989). The TH2 cell-mediated response promotes the humoral immunity and is crucial for the defence against extracellular pathogens. IL-4 essentially directs the TH2 cell response. An aberrant TH2 cell response is implicated in atopic conditions such as allergic asthma bronchiale.

A newly defined subset of TH cells are the IL-17-producing TH17 cells that play a critical role in the induction and propagation of autoimmune conditions such as multiple sclerosis, rheumatoid arthritis, psoriasis or inflammatory bowel disease (Korn et al. 2009). TH17 cell development is driven by TGFβ, IL-6, IL-21 and IL-22. Signal transducer and activator of transcription 3 (STAT3) and retinoic acid receptor-related orphan receptors alpha (RORα) are also critical for TH17 cell differentiation. CD4+CD25+FoxP3+ regulatory Treg cells can either differentiate in the thymus or are derived from naive T cells after stimulation with TGFβ. Treg cells are characterized by the expression of the transcription factor forkhead box P3 (FoxP3), which mainly regulates T cell function and directs the expression of surface molecules and cytokines. In contrast to the other TH cell subsets, Treg cells are immunosuppressive. Importantly, Treg cells sustain self-tolerance and prevent autoreactivity. It is suggested that Treg cells set a limit in immune responses against infections and regenerate homeostasis after pathogen elimination. Disrupted Treg function is associated with a variety of autoimmune disorders.

3.3.2.3 Cytotoxic CD8 + T Cells

Cytotoxic CD8+ T cells recognize antigens that are presented on MHC class I molecules. MHC I molecules are found on nearly all cells of the body and are loaded with peptides that are produced from cytosolic proteins by proteasomal degradation. Therefore, CD8+ T cells are capable of defending against intracellular pathogens such as viruses or obligate intracellular bacteria such as *Listeria monocytogenes*. The majority of CD8+ cytotoxic T cells are resident in lymphoid tissues. Before activation, co-stimulatory signals by an antigen-presenting cell (APC) are required. Cytotoxic T cells possess a variety of effector functions. On the one hand, CD8+ T cells produce cytokines such as TNFα or IFNγ, which are crucial for the defence against microbes. On the other hand, CD8+ T cells have potent cytotoxic activity: the target cell is killed by either apoptosis induction through FAS–FAS ligand interaction or by the granule exocytosis pathway. FAS–FAS ligand interaction initiates apoptosis by the release of cytochrome c out of mitochondria and consecutive caspase activation. Caspases are cysteine proteases that drive apoptosis in a multistep enzymatic process. The granules contain the pore-forming protein perforin, which increases cell permeability by damaging the membrane and facilitates the entry of other granule enzymes. One of these is the granzymes, a family of serine proteases that contribute to apoptosis induction by caspase activation. In addition to the defence against microbial pathogens, CD8+ T cells are also crucial for the elimination of tumour cells. Similar to the antibody response, immunological memory is also established during cellular immune responses. Autoreactive CD8+ T cells are involved in the pathogenesis of various autoimmune disorders such as insulin-dependent diabetes mellitus, systemic lupus erythematosus or antineutrophil cytoplasmic antibody (ANCA)-associated vasculitides. Further, CD8+ T cells contribute to allograft rejection and graft-versus-host disease in transplantation medicine.

3.4 Conclusion

Bone is a tissue with several pivotal functions for the organism: it stabilizes the body's structure, stores calcium and harbours the haematopoietic system. There are many relations between the skeletal and the immune systems: Osteoclasts that resorb bone and play a crucial role in inflammatory conditions are derived from monocytes or dendritic cells. Bone-forming osteoblasts also act as regulators of the haematopoietic niche that gives rise to all cells of both the myeloid lineage and the lymphoid lineage. Further, it is well known that multiple factors that regulate immune cells, including cytokines, chemokines and growth factors, also control osteoblast and osteoclast activity. It is well established that immune cells and cytokines contribute to the loss of bone mass observed in osteoporosis or in inflammatory diseases such as rheumatoid arthritis and inflammatory bowel disease. Cytokines that are secreted by activated immune cells in inflammatory conditions drive increased bone turnover and skeletal pathology. Hence, an improved understanding of osteoimmunology might be the key to develop new therapies for inflammatory disease, involving both bone and the immune system.

References

Auffray C, Sieweke MH et al (2009) Blood monocytes: development, heterogeneity, and relationship with dendritic cells. Annu Rev Immunol 27:669–692

Borregaard N (2010) Neutrophils, from marrow to microbes. Immunity 33:657–670

Germain RN (2002) T-cell development and the CD4-CD8 lineage decision. Nat Rev Immunol 2:309–322

Korn T, Bettelli E et al (2009) IL-17 and Th17 cells. Annu Rev Immunol 27:485–517

Kurosaki T (2010) B-lymphocyte biology. Immunol Rev 237:5–9

Martinez FO, Helming L et al (2009) Alternative activation of macrophages: an immunologic functional perspective. Annu Rev Immunol 27:451–483

Mosmann TR, Coffman RL (1989) TH1 and TH2 cells: different patterns of lymphokine secretion lead to different functional properties. Annu Rev Immunol 7:145–173

Oracki SA, Walker JA et al (2010) Plasma cell development and survival. Immunol Rev 237:140–159

Schett G (2009) Osteoimmunology in rheumatic diseases. Arthritis Res Ther 11:210

Vivier E, Raulet DH et al (2011) Innate or adaptive immunity? The example of natural killer cells. Science 331:44–49

Effects of Vitamin D in the Immune System

4

Ursula Azizi-Semrad, Peter Pietschmann, and Martin Willheim

The term "vitamin" for vitamin D reflects the original finding that plants contain a substance able to restore bone metabolism in patients suffering from rickets. Later on, endogenous production of a related molecule in animal and man was recognized, dependent on sufficient sun exposure of the skin and increasing its activity 10^5 times after further renal metabolism.

This was the discovery of the "hormone" vitamin D_3. In the last century, it was discovered that cells outside the renal system were also capable of generating the most active metabolite of vitamin D_3, namely, $1\alpha,25(OH)_2D_3$, and that this was independent of the hormonal regulatory pathways responsible for homeostasis in bone metabolism. Today, we are aware of a large number of different cells and tissues able to generate and metabolize $1\alpha,25(OH)_2D_3$, and, as a counterpart, many cell types have been identified as targets for this molecule. In many ways, the properties of this part of the vitamin D family classify it as a mediator with some characteristics of cytokines – except for its steroid nature – with paracrine, autocrine, and, only under specific conditions, endocrine functions. Its integral role in the innate and adaptive host defense system is well established now, and in part, there is interference as well as independence between the osteological and the immunological branch of the family.

U. Azizi-Semrad, MD (✉) • P. Pietschmann, MD
Department of Pathophysiology and Allergy Research, Center for Pathophysiology, Immunology and Infectiology, Medical University of Vienna, Währinger Gürtel 18-20, Vienna 1090, Austria
e-mail: ursula.azizi-semrad@meduniwien.ac.at; peter.pietschmann@meduniwien.ac.at

M. Willheim, MD
Department of Laboratory Medicine, Wilhelminenspital Vienna, Montleartstraße 37, Vienna 1160, Austria
e-mail: martin.willheim@wienkav.at

P. Pietschmann (ed.), *Principles of Osteoimmunology*,
DOI 10.1007/978-3-319-34238-2_4

4.1 Vitamin D

In this section, a brief summary of the metabolism and activity of vitamin D is presented; for more detailed information, the reader is referred elsewhere (Norman 2008; Holick 2008; Bikle et al 2013).

The term vitamin D describes a group of secosteroids originally believed to act as essential fat-soluble hormones on bone and mineral homeostasis. Most prominent members of the vitamin D group are vitamin D_2, ergocalciferol, and vitamin D_3, cholecalciferol. While vitamin D_2 derives from plant ergosterol, vitamin D_3 can be produced in the human skin from cholesterol or additionally be provided by nutritional intake of animal products. Since it was originally believed that vitamin D is essential for health but could not be produced by the organism itself, it was classified among fat-soluble vitamins. All vitamin D hormones are provitamins, gaining their biological activity by rupture of their ring b, and are then called "secosteroids."

4.1.1 Vitamin D Metabolism

In the human organism, vitamin D_3 has the greatest significance. Vitamin D_3 is produced from cholesterol in the skin by radiant sun energy: ultraviolet B mediates the conversion from 7-dehydrocholesterol to the instable previtamin D_3, which is stabilized by spontaneous, heat-dependent isomerization to the vitamin D_3. The further classical metabolizing pathway involves two hydroxylation steps taking place in the liver and in the kidneys. Firstly, the parent vitamin D_3 is hydroxylated to $25(OH)D_3$ by hepatocytes, a reaction catalyzed by the 25-hydroxylase (CYP2R1) (Baeke et al. 2010a). A very important event in the regulation of bone and mineral homeostasis is the conversion from $25(OH)D_3$ to the biologically active form of vitamin D_3, $1\alpha,25(OH)_2D_3$ or $24R,25(OH)_2D_3$. This step is mediated by the enzyme 1α-hydroxylase (CYP27B1) produced by the kidney renal proximal tubulus cells and is under tight control by the endocrine system (Peterlik et al. 2009). Most data in publications and within this chapter refer to effects of the active metabolite $1,25(OH)_2D_3$. In accordance with other authors, we use the terms "vitamin D_3" or "calcitriol" as synonyms for $1,25(OH)_2D_3$.

Vitamin D_3 and metabolites are transported in the blood by the vitamin D-binding protein (DBP), a member of the albumin family. DBP not only serves as transporter but also as an important vitamin D_3 reservoir, since only free amounts of $1,25(OH)_2D_3$ are crucial for determination of biologic activity.

Inactivation of $1,25(OH)_2D_3$ is promoted by a further hydroxylation step. $1,25(OH)_2D_3$ itself promotes its own degradation by the induction of the 24-hydroxylase (CYP24A1), an enzyme expressed in most tissues, this mechanism probably serving as an internal feedback loop. Metabolites are secreted into the bile and are partly reabsorbed via the enterohepatic circulation (Bringhurst et al. 2008; Norman 2008).

4.1.2 Vitamin D Signal Transduction

In target tissues, biological functions of vitamin D are realized via binding to the vitamin D receptors (VDR). As to date, two different receptor types are known. The classical pathway comprises the nuclear VDR, a classical nuclear steroid receptor mediating biological activities within hours. In addition, rapid actions of vitamin D taking only seconds to minutes are known. These are mediated by VDR receptors located in the cell membrane. Throughout this chapter, if not mentioned otherwise, we will name the nuclear VDR simply as VDR, since this receptor type is not only crucial for bone and calcium homeostasis but also for immune function regulated by vitamin D (Norman 2008).

4.1.3 The VDR

The VDR is a classical nuclear hormone receptor belonging to the steroid receptor superfamily, such as nuclear receptors for estradiol, glucocorticoids and vitamin A metabolites. After intracellular binding of its key ligand $1,25(OH)_2D_3$, the VDR mainly forms heterodimers with the retinoid X receptor (RXR). The DNA-binding domain of the VDR consists of two zinc-finger domains and the ligand-VDR-RXR complex can bind to vitamin D-responsive elements within promoters of vitamin D-responsive genes in order to regulate mRNA transcription, and, consequently, physiological actions of vitamin D are also modulated by co-activators or corepressors.

In contrast to the nuclear VDR, membrane receptors are responsible for rapid actions, such as regulation of calcium absorption in the intestine known as "transcaltachia." These receptors are located in flask-shaped plasma membrane invaginations called "caveolae." Second messenger systems other than the mechanism involved in signal transduction for the nuclear receptors participate in these processes: Signal transduction via phospholipase C, protein kinase C, G-protein, phosphoinositide 3-kinase, and voltage-gated calcium or chloride channels is assumed. The existence of a crosstalk between the nuclear and membrane vitamin D receptor pathways via the RAF/MAP kinases network has also been reported (Norman 2008).

A summary of classic vitamin D actions is given in Fig. 4.1.

4.1.4 Classic Effects of Vitamin D

The effects of vitamin D regarding bone and calcium homeostasis are well established (Bikle et al. 2013). The regulation of serum calcium and serum phosphate levels is under tight control. Since the first step in hydroxylation in the liver is poorly regulated and almost the total amount of the parent vitamin D_3 is converted

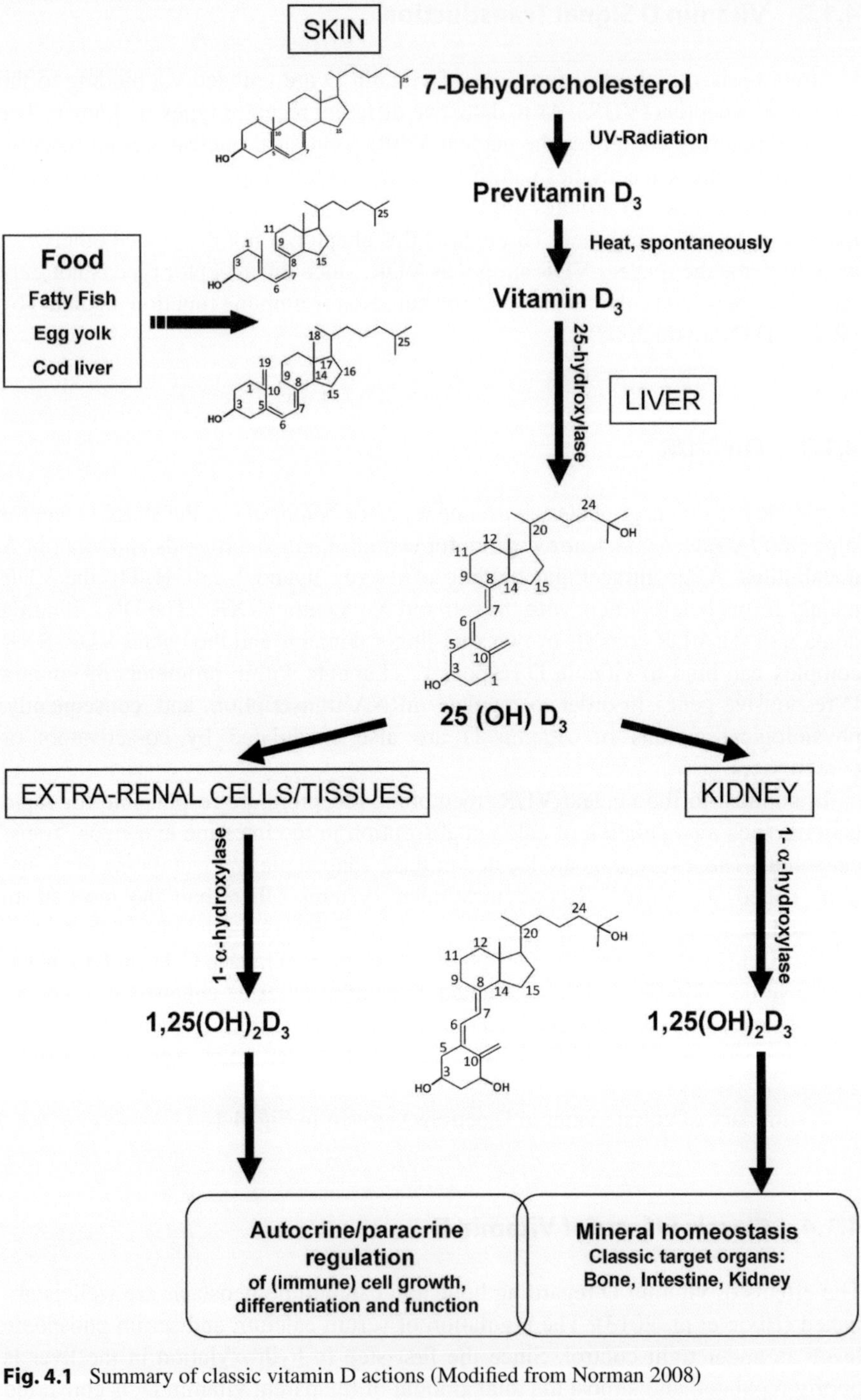

Fig. 4.1 Summary of classic vitamin D actions (Modified from Norman 2008)

to 25(OH)D_3, it is seen as a good indicator for the evaluation of the vitamin D_3 status of an individual (Peterlik and Cross 2006; Baeke et al. 2010a). The real key player in this tightly regulated process is the 1α-hydroxylase, the enzyme catalyzing the second hydroxylation step. Expression of 1α-hydroxylase is strongly influenced by serum calcium and phosphate levels, parathyroid hormone (PTH), and 1,25$(OH)_2D_3$ itself (Baeke et al. 2010a). Decrease in serum calcium levels stimulates PTH secretion from the parathyroid glands. PTH triggers phosphate secretion in the kidneys, thus resulting in decreased serum phosphate levels. Decrease in serum phosphate stimulates conversion from 25(OH)D_3 to 1,25$(OH)_2D_3$ in the kidneys. 1,25$(OH)_2D_3$ elevates the serum calcium levels by facilitating calcium absorption in the intestine and decreases renal calcium excretion. Furthermore, the maturation of osteoclasts allows calcium mobilization from bone to blood (Herold 2008; Norman 2008). 1,25$(OH)_2D_3$ itself induces its own degradation by induction of the 24-hydroxylase (CYP24A1), thus probably serving as an internal feedback loop (Norman 2008).

4.1.5 Additional Effects

Effects of vitamin D are not limited to its significant role in bone and calcium homeostasis. Many tissue and cell types express the VDR (as shown in Table 4.1) (Holick 2008). Furthermore, the key enzyme 1α-hydroxylase is expressed by various cell types and is controlled differently from the renal enzyme (Holick 2008; Peterlik and Cross 2006). There is strong evidence that 1,25$(OH)_2D_3$ operates not only in an endocrine but also in a paracrine fashion, thus being important for the regulation of normal cell differentiation and proliferation in many biological systems. Vitamin D_3 is known to enhance cell differentiation and inhibit cell proliferation. Therefore, adequate vitamin D_3 levels can be considered crucial for normal immune function and normal cell regulation, including inhibition of cancerogenesis (Peterlik et al. 2009).

4.2 Vitamin D and the Immune Cells

For detailed information on the immune system, the reader is referred to the section "Immunology" (see the chapter by Lang und Schett) in this book. The following only refers to basic immunology which is essential for understanding.

Intersection of vitamin D with the immune system has been discovered more than 25 years ago. Clinical observations indicated that patients suffering from the granulomatous disease sarcoidosis showed elevated serum calcium levels, and it could be demonstrated that excessive local 1,25 $(OH)_2D_3$ production in macrophages situated in granulomas was responsible for this (Adams et al. 1983). Furthermore, it was shown that the VDR receptor is expressed by almost all cells of

Table 4.1 Extrarenal 1α-hydroxylase and VDR expression in various cell types and tissues

Extrarenal tissues or cells expressing 1α-hydroxylase	VDR expression in non calcemic tissues/cells
Monocytes/macrophages	Thymus
Osteoblasts	Lymphocytes
Endothelial cells	Monocytes
Synovial cells	Epidermis
Pancreas	Hair follicles
Prostatic gland	Dermis
Ovary	Melanocytes
Uterus	Myocytes
Breast	Cardiac muscle
Large intestine	Pituitary gland
	Pancreas
	Prostatic gland
	Gonads
	Placenta
	Breast
	Stomach
	Brain

Modified from Peterlik and Cross (2005) and Holick (2008)

the immune system (see Table 4.1). While antigen-presenting cells like dendritic cells, macrophages, and monocytes show a constitutive expression of the VDR, resting T and B lymphocytes are able to upregulate VDR expression upon activation (Provvedini et al. 1983; Provvedini and Manolagas 1989; Holick 2008) (also see Table 4.1). Moreover, the key enzymes of $1{,}25(OH)_2D_3$ production (1α-hydroxylase) and degradation (24-hydroxylase) are also expressed by immune cells. Macrophages and dendritic cells show expression of the 1α-hydroxylase identical to the renal form, but activity of this key enzyme in immune cells is differently regulated compared to the tight endocrine control mechanisms of calcium and bone homeostasis. Local production of $1{,}25(OH)D_3$ is strongly dependent on the amount of $25(OH)D_3$ available and predominantly under control of immune signals such as IFN γ, LPS (lipopolysaccharide), and viral infections (Baeke et al. 2010a; van Etten and Mathieu 2005). Furthermore, 1α-hydroxylase activity of immune cells seems to lack a direct negative feedback control, which explains the occurrence of hypercalcemia in granulomatous diseases such as sarcoidosis and tuberculosis, where massive local production of $1{,}25(OH)_2D_3$ is mediated by macrophages situated in the granulomas (Adams et al. 1983). In situ production of $1{,}25(OH)_2D_3$ makes high local levels possible, far exceeding those of serum and allowing a paracrine modulation of immune function virtually independent of systemic serum levels (Baeke et al. 2010a).

4.2.1 Innate Immunity

Historically, there were early signs that vitamin D deficiency could affect the immune function. Already in the seventeenth to nineteenth century, physicians observed that children with rickets were more likely to develop pneumonia or to be infected with tuberculosis. In the nineteenth century, "sunlight" was a well-accepted therapy option for tuberculosis as well as vitamin-D-enriched diets, e.g., cod liver oil (Chesney 2010; Holick 2008). Today, it is a well-supported fact that vitamin D_3 is an enhancer and stimulator of the innate immune response and that vitamin D_3 deficiency influences innate immune functions (Chesney 2010; Mora et al. 2008).

Major components of innate immune responses are monocytes and macrophages. Vitamin D_3 seems to have a pro-differentiating effect on monocytes gaining a macrophage like phenotype. Furthermore, the expression of Fc receptors is modulated, and antigen processing, chemotaxis, phagocytic capacity, tumor cell cytotoxicity, as well as antimicrobial activity is enhanced. Production of IL-1β and TNF α, induced by a variety of stimulators, is decreased in vitamin D_3-treated mature monocytes, although a positive effect of vitamin D on IL-1β and TNF α production has been described for immature monocytic cells (Chesney 2010; Baeke et al. 2010a; Adams et al. 2007b; Boltz-Nitulescu et al. 1995; Zarrabeitia et al. 1992; Almerighi et al. 2009).

Macrophages are able to rapidly recognize certain "danger" signals on microbes, the highly conserved pathogen-associated molecular patterns (PAMPS) by means of the so-called pattern recognition receptors. The most prominent subgroup of pattern recognition receptors are the Toll-like-receptors (TLR) with many of their signaling pathways leading to activation of the transcription factor NFkappaB, consequently enhancing production of pro-inflammatory cytokines such as IL-1 and TNFα (Akira et al. 2006; Chesney 2010, also see Chapter Immunology; Adams et al. 2007b).

Binding of PAMPS to the TLR2/1 on macrophages induces not only expression of VDR, a process probably mediated by IL-15 (Baeke et al. 2010a; Krutzik et al. 2008), but also the 1α-hydroxylase. This enzyme catalyzes local conversion from available $25(OH)D_3$ to $1,25(OH)_2D_3$. After binding of $1,25(OH)_2D_3$ to the VDR, production of fundamental antimicrobial peptides (AMPs), such as cathelicidin with its active form LL-37, is induced. Cathelicidin, localized in autosomes, seems to be crucial in the defense against *Mycobacterium tuberculosis*, since it induces autophagy and therefore promotes killing of intracellular pathogens (Liu et al. 2006). Liu et al. (2006) provided evidence for human TLR induction of cathelicidin on a DNA level. Furthermore, they observed that addition of sera from African-Americans, known to have lower vitamin D_3 serum levels due to skin pigmentation and to be more susceptible to tuberculosis, to monocyte cultures resulted in weaker TLR-dependent induction of cathelicidin mRNA when compared to the addition of sera obtained from a Caucasian cohort. After supplementation of $25(OH)D_3$, a recovery of TLR-induced cathelicidin production could be detected. In a subsequent study, the authors could show a rise of cathelicidin production in vivo induced by TLR in monocytes upon vitamin D_3 supplementation and therefore provide evidence that vitamin D_3 is crucial for maintenance of localized innate immune functions (Adams et al. 2009; Hewison 2010).

On the monocyte surface, expression of CD14, the TLR2 co-receptor, is strongly boosted by VDR (Miller and Gallo 2010; Spittler et al. 1997; Do et al. 2008). Interestingly, at the same time surface expression of TLR2 as well as TLR4 on monocyte and macrophage is diminished by vitamin D_3 (Sadeghi et al. 2006). It is hypothesized that TLR downregulation leading to hyporesponsiveness to PAMPS not only prevents cellular overstimulation and oxidative stress but also limits an inflammatory response (Baeke et al. 2010a). It is postulated that elevated TLR expression is associated with several chronic inflammatory diseases. One study conducted with patients suffering from Behçet disease showed higher expression of TLR 2 and 4 on monocytes compared to healthy controls and remarkably, this alteration was even found to be inversely associated with $25(OH)D_3$ serum levels (Do et al. 2008).

Vitamin D effects on cathelicidin production and surface receptor expression are summarized in Fig. 4.2.

Vitamin D_3 signaling is also essential for induction of defensin β 4, an antimicrobial peptide used in defense against intracellular invaders. Transcription of the defensin β 4 gene may be initiated via convergence of the VDR and IL-1β pathways (Chesney 2010; Liu et al. 2009).

Data on regulation of production of the inducible nitric oxide synthase (iNOS) by vitamin D_3 are conflicting (Baeke et al. 2010a). Enhancement (Rockett et al. 1998) as well as suppression of production (Chang et al. 2004; Pedersen et al. 2007)

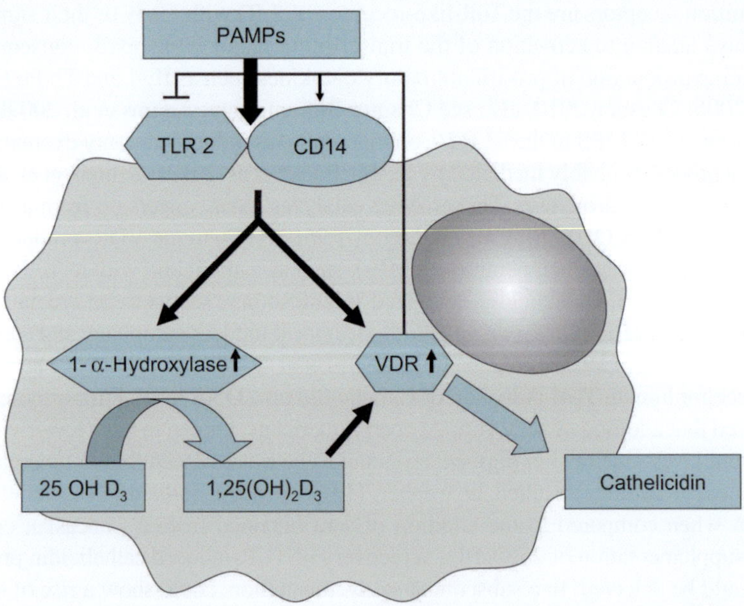

Fig. 4.2 Regulation of innate immune reaction by vitamin D_3 following infection (e.g., *Mycobacterium tuberculosis*); for further explanation, see text. *PAMPS* pathogen-associated molecular patterns, *TLR* Toll-like-receptors, *VDR* vitamin D receptor (Modified from Chesney (2010) and Miller and Gallo (2010))

by vitamin D_3 has been described. NOS is an enzyme important for the generation of the reactive oxygen species nitric oxide (NO). At least in rodents, NO participates in the oxidative burst, an important mechanism of monocytes and macrophages in microbe killing (Baeke et al. 2010a). NO production in human monocytes or macrophages seems to be far less pronounced, and relevance of iNOS for the oxidative burst in humans has still to be elucidated (Chang et al. 2004). Nevertheless, Chang et al. (2004) could show that PBMC from tuberculosis patients showing high vitamin D_3 serum levels released less NO. When mouse macrophages were exposed to vitamin D_3 at increasing concentrations, a dose-dependent decline in iNOS expression was observed. The authors hypothesize that this may protect cells from oxidative injury, since NO metabolites and LDH (lactate dehydrogenase) were diminished at the same time.

4.2.2 Antigen-Presenting Cells (APC): Dendritic Cells and Monocytes/Macrophages

The dendritic cell (DC) is regarded as the most potent APC and therefore essential for priming of naive T cells. Exposing DC to vitamin D_3 in vitro inhibits their differentiation and maturation process, and their antigen-presenting capacity is severely hampered. Expression of MHC II, co-stimulatory molecules (CD40, 80, 86), and other DC maturation surface markers such as CD1a and CD83 is diminished, while CD14, a monocytic marker and co-receptor of TLR2, is persistent (Baeke et al. 2010a; van Etten and Mathieu 2005). IL-12 production, the key cytokine for the initiation of the Th1 pathway, is suppressed, as well as the expression of IL-23, a cytokine essential for Th17 differentiation. Increased production of the immunosuppressive cytokine IL-10, also necessary for regulatory T cell generation, and CCL22, a chemokine able to attract CCR4+ Foxp3+CD4+CD25+ regulatory T cells (Penna et al. 2007; dÁmbrosio 2006), can be observed. Furthermore, upregulation of the inhibitory immunoglobulin-like transcript 3 (ILT-3), an inhibitory receptor that wield negative control on APC activation, is observed even though not crucial for Treg induction (Penna et al. 2005; Unger et al. 2009). Additionally, DC treated with vitamin D_3 are capable of antigen-specific Treg induction and express high levels of programmed death-1 ligand, a member of the co-stimulatory B7 family also negatively affecting T cell response and probably important in tolerance induction (Unger et al. 2009). In summary, exposure of differentiating myeloid (but not plasmacytoid) DC to vitamin D_3 results in the production of tolerogenic DC with reduced capacity to prime and activate T cells (Adorini et al. 2004; Penna et al. 2007), exerting their immunosuppressive effect not only by suppression of the Th1 and Th17 pathway but also by induction of T cells with regulatory properties (Tregs) (Adorini et al. 2004; Penna et al. 2005; Baeke et al. 2010a). While expression of the 1α-hydroxlase and therefore local production of $1,25(OH)_2D_3$ increase with the differentiation stage of DCs, VDR expression is diminished in maturing DC, thus reducing the autocrine effect. As a consequence, initiation of the T cell response is permitted, and high local concentration of vitamin D_3 might prevent further maturation of other DC and excessive stimulation of T cells in a paracrine fashion (Hewison et al. 2003; Hewison 2010).

Regarding monocytes, the immunomodulatory effects of vitamin D_3 on antigen presentation and T cell stimulation are considered similar to the DC: expression of MHC II, co-stimulatory molecules (CD40, 80, 86), is reduced (Baeke et al. 2010a; van Etten and Mathieu 2005; Spittler et al. 1997), and the production of pro-inflammatory cytokines such as IL-1, IL-6, IL-8, and IL-12, as well as of M-CSF (Baeke et al. 2010a; van Etten and Mathieu 2005; Boltz-Nitulescu et al. 1995), is diminished. The influence of vitamin D_3 on TNF α production – subsequently triggered by a variety of different stimulators – seems to be dependent on the differentiation status of monocytic cells with increased production in immature cells, such as cell lines and bone marrow cells, and a decrease in more mature cells, such as peripheral blood monocytes (Hakim and Bar-Shavit 2003). Vitamin D_3 also affects Fc-receptor expression on monocytes, leading to downregulation of FcγRI/CD64, FcγRII/CD32, FcγRIII/CD16, and FcεRII/CD23, and stimulates the expression of FcαR/CD89 (Boltz-Nitulescu et al. 1995). Furthermore, vitamin D_3 exerts several effects on monocytes, reflecting their differentiaton to macrophages (Kreutz and Andreesen 1990).

In several cell types, the 24-hydroxylase, responsible for degradation of the active $1{,}25(OH)_2D_3$, is inducible by $1{,}25(OH)_2D_3$, thus providing negative feedback control. In monocytes, expression of the 24-hydroxylase seems to depend on their maturation and activation. Undifferentiated/resting monocytes show inducible expression of the 24-hydroxylase upon exposure to vitamin $D_{3,}$ while activated macrophages are resistant to $1{,}25(OH)_2D_3$ linked to 24-hydroxylase induction. This could be explained by the observation that infection-induced INFγ production activates STAT 1 α (signal transducers and activators of transcription 1 α) which negatively regulates the expression of the 24-hydroxylase (van Etten and Mathieu 2005; Chesney 2010). Upregulation of 1α-hydroxylase with subsequent (local) production of $1{,}25(OH)_2D_3$, accompanied by the lack of its negative feedback 24-hydroxylase, eventually may even result in systemic hypercalcemia as observed in sarcoidosis.

4.2.3 T Cells

CD4+T cells have important modulating functions in adaptive immunity (also see chapter Immunology). In the 1980s, Mosmann and Coffman initially described two different effector T cell subsets, the Th1 and Th2 cells (Mosmann and Coffman 1989). It was believed that after antigen presentation to a naive T cell, immune responses either take the Th1 or the Th2 pathway, dependent on the respective pathogen involved and the local cytokine environment present. Several years ago, the classical Th1/Th2 dichotomy has been expanded by the identification of further CD4+subsets (also see chapter immunology). On the one hand, a heterogeneous group of T cells with regulatory properties necessary for maintenance of self-tolerance and with limiting immune reactions was identified and termed regulatory T cells (Tregs). On the other hand, a T cell subset, characterized by its key cytokine

IL-17, which was discovered in murine disease models of chronic (auto) inflammatory conditions and later on verified in humans, was designated Th17 subset (Annunziato and Romagnani 2009a). Another subset, described by our group and highly inducible by vitamin D_3, is characterized by the production of IL-6 and was therefore denominated "Th6"(Azizi-Semrad et al. 2010; Willheim et al. 1999; Thien et al. 2005; Pichler et al. 2002). Very recently, the Th9, which is also an assumed new T-helper cell subset acting as a key player in autoimmune conditions, has been characterized. Whether the latter two subsets picture independent T-helper cell subsets or may be interpreted as implication of the plasticity of the Th system still remains to be elucidated (Veldhoen et al. 2009).

Relatively few data exist about functional subsets of CD8 T cells (Tc), and many of them only refer to the murine system. There are, however, reports about equivalents to Th1-, Th2-, and Th17-cytokine production patterns in CD8 T lymphocytes. These consistently have been termed Tc1, Tc2, and Tc17 and essentially develop under the same conditions as their CD4+ counterparts (Seder et al. 1992; Kelso et al. 1991; Salgame et al. 1991; Byun et al. 1994; Croft et al. 1994; Le Gros and Erard 1994; Kemeny et al. 1994; Noble et al. 1995; Hamada et al. 2009; Huber et al. 2009). Effects of vitamin D_3 on in vitro differentiation of human CD8+ T lymphocytes have been described as very similar to the differentiation of Th1, Th2, and Th6 (Thien et al. 2005).

4.2.4 Th1/Th2 Cells

Th1 cells are characterized by production of their signature cytokine IFNγ and are therefore crucial in the defense against intracellular pathogens by means of macrophage activation, while Th2 cells, characterized by production of IL-4, IL-5, and IL-13, cooperate with B cells in immunoglobulin production (also see chapter Immunology) (Annunziato et Romagnani 2009b). Excessive and pathologic reactions of these distinctive patterns present themselves as the Th1-mediated subgroup of autoimmune diseases at one end of the spectrum and as Th2-mediated allergic conditions at the other end. The decisive role of the local cytokine milieu for Th differentiation has been recognized for the first time in connection with the hypothesized dichotomy of Th1 and Th2 lineage: Whereas IL-12 is a strong inductor of the Th1 response while inhibiting the Th2 pathway, IL-4 is an essential cytokine for the initiation of the Th2 pathway (Demeure et al. 1994; O'Garra and Murphy 1994; Seder and Paul 1994; Thien et al. 2005).

Naive T cells are able to express the VDR upon activation (Baeke et al. 2010b) and vitamin D_3 can affect T cells in two ways: T cells can act as a direct target (Boonstra et al. 2001; Thien et al. 2005; Mahon et al. 2003; Baeke et al. 2010b) or can be influenced by modulation of APC differentiation and their cytokine production (see above Innate Immunity), possibly hampering antigen presentation to naive T cells, leading to inhibition of T cell priming and differentiation. Direct effects of vitamin D_3 on T cells have already been verified on gene level, and over 102 target genes for vitamin D_3 in CD4+ cells have been described (Mahon et al. 2003).

Vitamin D_3 effectively inhibits T cell proliferation (Bikle 2009; Rigby et al. 1984), partly mediated by inhibition of IL-2 production. IL-2, a cytokine secreted by activated T cells, operates in an autocrine fashion and is an indispensable growth and survival factor for T cells (Smith 1998; Smith and Popmihajlov 2008; van Etten and Mathieu 2005). Administration of vitamin D_3 to human or murine T cell cultures inhibits IL-2 production directly (Tsoukas et al. 1984; Müller et al. 1993; Dimeloe et al. 2010; van Etten and Mathieu 2005; Takeuchi et al. 1998; Rigby et al. 1984), resulting in reduced proliferation and differentiation capacity of T cells.

It is well established that vitamin D_3 potently inhibits the induction of the Th1 pathway (van Etten et al. 2005; Boonstra et al. 2001; Willheim et al. 1999; Pichler et al. 2002; Thien et al. 2005), leading to the assumption that local adequate vitamin D_3 levels could prevent Th1-mediated autoimmunity (Cutolo et al. 2007). Th1 cells are characterized by the production of their signature cytokine IFNγ, which is a necessary feedback signal for further activation of macrophages in defense against intracellular pathogens such as *Mycobacteria, Listeria,* or *Leishmania* (Cantorna et al. 2008; van Etten and Mathieu 2005, also see chapter Immunology). It has been shown that vitamin D_3 significantly inhibits the differentiation of (IFNγ producing) Th1 cells (Müller et al. 1996; Willheim et al. 1999; Pichler et al. 2002; Thien et al. 2005) and the IFNγ gene, containing a VDRE, was found to be a direct target of vitamin D_3 (Dimeloe et al. 2010; Cippitelli and Santoni 1998). In fact, vitamin D_3 seems to inhibit especially the IL-12-induced IFN γ production, while the constitutive IFNγ production is only slightly affected (Thien et al. 2005).

While these results from murine and human in vitro studies unequivocally substantiate the suppression of Th1 cytokines, data on Th2 differentiation seem to be contradictory. In an APC-free murine culture system, it was shown that in vitro administration of vitamin D_3 to naive T cells inhibited the Th1 pathway while enhancing Th2 differentiation. Frequency of T cells showing Th2 cytokine expression (IL-4, IL-5, and IL-10) was increased, and the expression of the Th2-associated transcription factors GATA-3 and c-maf was upregulated. Abrogation of these effects by neutralization of IL-4 suggested that they were all mediated by IL-4 (Boonstra et al. 2001). Other in vitro studies, in the murine as well as the human system, showed no response or even suppression of the Th2 lineage by calcitriol (Dimeloe et al. 2010; Pichler et al. 2002; Staeva-Viera and Freedman 2002). Interestingly, one study using a murine culture system showed a suppression of Th2 induction when vitamin D_3 was present during in vitro differentiation and polarization of naive T cells (i.e., addition of IL-4 and anti-IL-12), while in already activated T cells, vitamin D_3 had no effect on IL-4 production (Staeva-Viera and Freedman 2002).

Another in vitro study using human adult peripheral blood mononuclear cells showed a small reduction of IL-2 positive human T cells and an increased proportion of T cells positive for IL-4 and especially for IL-13 on a single-cell level (Willheim et al. 1999). A subsequent study showed that vitamin D_3 only slightly enhances the constitutive IL-4 production in Th2 cells. Nevertheless, when vitamin D_3 was administered in combination with IL-4, a strong co-stimulatory effect on the production of the Th 2 cytokines IL-4 and IL-13 could be observed (Thien et al. 2005).

A possible explanation for the conflicting results regarding Th2 differentiation was recently presented by Dimeloe et al. (2010). They reviewed in vitro vitamin D_3 levels in several studies and hypothesized that low in vitro concentrations of vitamin D_3 (10^{-7}M to 1^{-9}M) induced suppression of the Th1 lineage as well as the Th2 pattern, while high concentrations of vitamin D_3 (1×10^{-6}M) might enhance IL-5 and IL-13 production (Jirapongsananuruk et al. 2000). This seems to corroborate data from epidemiological studies: Hyppönen et al. (2009) found a significant but non-linear positive correlation of serum 25-OH vitamin D levels with serum IgE concentration. In this large British cohort study, both subjects showing very low and very high 25-OH vitamin D_3 serum levels had increased total and specific IgE concentrations. Such a U-shaped correlation might predispose both of them for atopy and allergy, conditions at the Th2 end of the T-helper cell spectrum.

Nevertheless, in studies applying exactly the same "low" vitamin D_3 concentrations (10^{-7}M to 1^{-9}M) in vitro, different effects could also be observed: A complete suppression of Th1 and Th2 cytokine induction was found in human cord blood which exclusively contained naive T cells (Pichler et al. 2002), while in adult blood, Th2 cytokine production was enhanced (Willheim et al. 1999; Thien et al. 2005). Different effects of vitamin D_3 on the naive T cell population compared to the memory T cell subset have already been shown, showing the CD45RA+ (naïve) subset to be less susceptible to inhibitory effects of vitamin D_3 (Müller and Bendtzen 1992, 1996).

As a constant result, independent from suppression or induction of IL-4-producing T cells, a CD4+ T cell population characterized by production of IL-6 is stimulated by in vitro vitamin D_3 administration to human PBMC (Willheim et al. 1999). A strong co-stimulatory effect of IL-4 and vitamin D could be observed, and in adult T cells, this is accompanied by co-expression of IL-4, as has been demonstrated on a single-cell level, whereas naive IL-6+ T lymphocytes remain negative for IL-4 (Pichler et al. 2002; Thien et al. 2005). A distinct IL-6-producing subset could also be detected in the peripheral blood at a low proportion, and since typical transcription factors of the Th2 (GATA-3) and Treg (FoxP3) lineage were not found, the subset was termed "Th6" (Azizi-Semrad et al. 2010). The function of this subset still remains to be elucidated, since peripheral mononuclear cells other than T cells have the capacity to produce great amounts of IL-6 and therefore are eminent for systemic effects. It can be hypothesized that the importance of the Th6 might lay in a paracrine (similar to other cytokines) modulation of immune function, since IL-6 is the cytokine decisive for initiation of the pro-inflammatory Th17 or a tolerance inducing Treg pathway (Workman et al. 2009). The consistent induction of IL-6 in T cells may be a crucial effect of vitamin D, at least in humans, and increase of IL-4 production (or IL-4-producing T cells) might be a mutable concomitant circumstance.

4.2.5 Tregs

Regulatory T cells are a heterogeneous group of CD4+ T cells important for maintenance of immunological tolerance, prevention of autoimmune disease, and limitation of inflammation due to their highly suppressive capacity. The group of regulatory

T cells can be divided into 2 subpopulations according to their origin: Firstly, the natural Tregs (nTregs), characterized by expression of their master transcription factor FOXp3 and high constitutive expression of the IL-2 receptor α chain, CD25, derive from the thymus. The second subset can be induced in the periphery and is therefore called i(nducible)Tregs and can be further subdivided into the Tr1 subset, characterized by production of IL-10 as its signature cytokine and the inducible FOXp3+ subset. Regulatory T cells exert their suppressive function by secretion of the anti-inflammatory cytokines IL-10, TGFβ, as well as by consumption of the T cell survival factor IL-2. They also modulate APC function by the upregulation of CTLA-4, a surface protein expressed by activated T cells, leading to the downregulation of co-stimulatory molecules (CD80 and CD86) on APCs and in consequence limiting T cell activation (Dimeloe et al. 2010; Robinson 2009).

Similar to other cells of the adaptive immune system, vitamin D can either directly exert its influence on the regulatory subset or modulate APC function, the latter resulting in the induction of a tolerogenic DC subset (for details, see above) and thereby modulating T cell response. The effect of vitamin D on the generation of tolerogenic myeloid DC is already described in the upper section of this chapter. These DC are characterized by secretion of large amounts of IL-10, which leads to Treg induction (Penna et al. 2005, 2007; Adorini et al. 2004). Several studies report also on direct effects of vitamin D on Tregs: Barrat et al. (2002) conducted an in vitro study on human CD4+ T cells in the absence of APC and reported on a strong synergistic effect of vitamin D_3 and dexamethasone on development of an IL-10-producing subset staining negative for IL-4, IL-5, or IFNγ. In this in vitro-generated murine, antigen-specific IL-10-producing T cells could prevent experimental autoimmune encephalitis (an animal model for multiple sclerosis), when transferred into mice. A subsequent study of the same group demonstrated that vitamin D and IL-10 restored the generation of an IL-10-producing regulatory CD4+ T cell subsets in PBMC obtained from patients with glucocorticoid-resistant asthma, a condition characterized by impaired induction of IL-10-positive T cells (Xystrakis et al. 2006). The group showed in a following in vitro study that vitamin D_3 alone also induces an IL-10-positive Treg subset showing enhanced expression of TLR9. Induction of this subset could also be demonstrated in vivo after oral administration of calcitriol in healthy volunteers. Binding of TLR 9 agonists resulted in a loss of regulatory function characterized by a decrease in IL-10 production in vitro. The authors hypothesized that this mechanism might permit an adequate immune reaction in the course of infections (Urry et al. 2009). Another group demonstrated a direct strong synergistic effect of vitamin D and IL-2 on human iTreg generation in vitro. After isolation of CD4+CD25-T cells from human PBMC and following stimulation with $1,25(OH)_2D_3$ in combination with IL-2, strong upregulation of FOXp3 and CTLA-4 was observed. At the same time, a reduction in the production of the pro-inflammatory cytokines IL-17, IL-23, and IFNγ could be detected (Jeffery et al. 2009). In the peripheral blood of pregnant women, the Treg population was decreased in vitamin D-deficient (≤19 ng/mL) when compared to insufficient (20-29 ng/mL) and sufficient (≥30 ng/mL) (pregnant) women; analogous alterations of the Treg subset were observed in cord blood (Vijayendra Chary et al. 2015). Moreover, a decreased expression of FOXp3 in placentas from women with vitamin D deficiency was seen.

Taken together with the fact that vitamin D has been shown to mediate transplant tolerance and prevent development of autoimmune disease due to Treg induction in animal models (Adorini and Penna 2008; Gregori et al. 2001), it can be concluded that vitamin D is vital for the maintenance of self-tolerance and prevention of immune-mediated disease.

4.2.6 Th17

Th17 are characterized by expression of their signature cytokine IL-17, an important mediator in host defense by attracting neutrophils and macrophages to the site of inflammation. Moreover, Th17 cells are supposed to play a pathogenic role in the induction, enhancement, and maintenance of various autoreactive disorders and chronic inflammatory conditions, formerly believed to be Th1 mediated. This has been already confirmed in animal disease models such as collagen-induced arthritis or experimental autoimmune encephalomyelitis (EAE), a murine model for multiple sclerosis. Th17 seem also to play a role in psoriasis or inflammatory bowel diseases. Th17 were firstly characterized in mice, and following studies in humans unveiled that human Th17 might differ from their murine counterpart. In both species, RORγT or its human homolog RORC, respectively, was recognized as master transcription factor, and in both human and murine Th17, CCR6 was identified as a major chemokine receptor. The transcription factors RORα and STAT-3 are found in murine Th17, while in humans, co-expression of the Th1 master transcription factor T-bet and RORC can be detected. Primarily discovered in humans, a remarkable proportion of IL-17 + CD4+ cells produce IFNγ additionally to IL-17 (Th17/Th1 cells). Moreover, Th17 exhibit a potential plasticity to Th1, which might be associated with their pathogenic role in autoimmune disorders or chronic inflammation. Murine Th17 originate from naive T cells after stimulation with IL-1β, TGFβ, and IL 6 and are expanded by IL-21 and IL-23, while human Th17 seem to derive from CD4 + CD161+ precursor T cells in the thymus and probably differentiate under exposure to IL1β and IL-23 to Th17. Especially the role of TGFβ in the induction of human Th17 is controversially appraised (Annunziato et Romagnani 2009a, b).

Besides vitamin D-mediated modulation of APC function, which consequently affects T cell priming and therefore Th17 differentiation, several in vitro studies reported on direct effects of vitamin D_3 on Th17 induction and expansion. In a murine in vitro study, it was found that vitamin D_3 suppresses the development of CD4+ IL-17 and IL-22+ T cells in an APC-free culture under Th17-polarizing conditions. Although augmentation of an IL-10-producing subset could be observed at the same time, Th17 suppression could not be attributed to IL-10 signaling, unlike the situation in Th9 (see below). The authors of this study ascribed this inhibitory effect of vitamin D on Th17 partly to the suppression of the Th17 transcription factors (Palmer et al. 2011). However, another murine in vitro study suggested a post-transcriptional inhibition, since at least at physiological concentrations vitamin D only affected Th17 cytokine production at protein and not at transcriptional level (Chang et al. 2010a, b). Another in vivo study conducted in mice could show that orally administered vitamin D inhibited the onset of EAE and diminished the frequency of murine Th17. Migration of Th17 to the central nervous system is regarded as crucial for the initiation of EAE and is mediated by the CCR6 and MIP3α/CCL20 axis. Since the authors observed not only a decreased CCR6 expression in Th17 in response to vitamin D but also a reduced migration capacity in vitro, they concluded that vitamin D might negatively affect migration of pathogenic Th17 to the central nervous system (Chang et al. 2010a).

In human studies, similar results were obtained: It could be shown that the development of Th17 and the IL-17 production in CD4+memory T cells are suppressed in a dose-dependent manner and a strong synergistic effect of vitamin D and all-trans retinoid acid could be observed. The authors also reported on suppression of the master Th17 transcription factor RORγT and enhancement of FOXp3 (Ikeda et al. 2010). Another study examined the effects of vitamin D on PBMC obtained from therapy-naive patients with early rheumatoid arthritis (RA). After stimulation, PBMC from RA patients showed a higher production of IL-17 when compared to healthy controls. When vitamin D was administrated in vitro, a suppressive effect on IL-17 production in these PBMC could be detected and additionally a strong synergistic effect of vitamin D and dexamethasone was observed. Since memory T cells are regarded as the main source for IL-17 in PBMC, the authors assessed IL-17 and IL-22 production in isolated CD4+CD45RO+T cells obtained from RA patients and found increased frequencies positive for these cytokines. Consecutively, in vitro administration of vitamin D diminished TNF α, IL-22, and IL-17 levels in the memory T cell subset (Colin et al. 2010).

Regarding these studies, it can be concluded that vitamin D might prevent and ameliorate Th17-mediated autoimmune conditions by suppressing not only development but also expansion and cytokine production of this Th subset.

4.2.7 Th9 Cells

In 2008, a CD4+T-cell subset characterized by IL-9 production has been described. It could be shown that TGFß and IL-4 murine and human Th9 differentiation in vitro in mice (Veldhoen et al. 2008; Dardalhon et al. 2008; Wong et al. 2010), while IL-10 exerts an inhibitory effect on Th9 induction (Palmer et al. 2011). Th9 cells have been shown to promote experimental colitis and experimental autoimmune encephalitis (EAE) (Jäger et al. 2009; Dardalhon et al. 2008) and are therefore proposed to enhance chronic inflammatory conditions. Regarding vitamin D effects on Th9 development, there is only one study available so far: Palmer et al. (2011) could show that vitamin D_3 inhibits murine Th9 differentiation in vitro. Since abolishment of the IL-10 signaling pathway restored Th9 differentiation by blocking the cytokine or the receptor, the authors concluded that vitamin D exerts its inhibitory effect on Th9 initiation by the induction of an IL-10-positive subset in the T cell culture. It was hypothesized that by this mechanism vitamin D might contribute to the prevention of autoimmune diseases.

4.2.8 B Cells

Similar to T cells, B cells are able to express the VDR as well as the 1α-hydroxylase upon activation (Chen et al. 2007; Heine et al. 2008). Already more than 20 years ago, it was observed that $1,25(OH)_2D_3$ potently inhibited IgG and IgM production in human activated PBMC in a dose-dependent manner (Lemire et al. 1984). However, since then, the research focus in vitamin D-mediated effects has been rather laid on CD4+ T cells, highlighting the significance of impairment of T cell or APC function on B cell differentiation and Ig production (Müller et al. 1991). Nevertheless, there has been strong evidence of direct effects of vitamin D on B cells. This was clearly demonstrated by an early in vitro study on B cell Ig production by using EBV-infected B cells, since EBV-induced activation and Ig production are independent from T cells (Provvedini et al. 1986). Another in vitro study showed that vitamin D inhibits proliferation and Ig production in B cells only when administered before their differentiation into Ig-producing cells (Iho et al. 1986).

More than two decades later, Chen et al. (2007) provided evidence that vitamin D can affect B cells directly by inhibiting proliferation of activated B cells, plasma cell generation, Ig secretion, and class-switch memory cell generation. Moreover, they demonstrated that VDR is expressed constitutively in resting B cells and can be upregulated upon stimulation. In the same study, the authors observed that patients suffering from systemic lupus erythematosus (SLE), an autoimmune disease going along with B cell dysregulation and excessive Ig production, showed lower serum levels of 25 OH D_3 and $1,25(OH)_2D_3$ than healthy controls. Regarding this finding, however, it has to be noticed that SLE frequently involves dermal lesions, and the possible role of sun avoidance by the patients has not been evaluated. The authors postulated that vitamin D might be an important factor for maintenance of B cell homeostasis (Chen et al. 2007).

4.2.9 NKT Cells

Natural killer T cells (NKT) were first identified in mice as a T cell subset sharing some features with NK cells: NKT were found to express an α/β T cell receptor (TCR) with a restricted repertoire as well as CD161 (human)/NK1.1 (mouse), a receptor normally expressed by NK cells. Although NKT are regarded as T lymphocytes, they are not restricted to MHC dependent peptide antigen presentation, but recognize glycolipid structures on exogenous and endogenous ligands presented by

CD1d. By gaining more information on NKT over time, the description as T cells with NK markers seemed to be insufficient, not matching real conditions. Various subsets have been identified, and classification appears difficult and inconsistent in literature. NKT might be assigned to a specific subset by their specific TCR chains, CD161, CD4, and CD8 expression, as well as reactivity to α-galactosylceramid (αGalCer), a synthetic glycolipid resulting in strong activation of NKT cells. Two main types of NKT subsets have been described: The type 1 "classic" NKT express an invariant Vα14- Jα18 (mouse) or Vα24-Jα18 (human) rearranged α chain in their TCR and are therefore also called invariant (i) NKT. The type 2 "non-classic" NKT are able to express diverse TCR. Upon stimulation, NKT cells have the capacity to produce large amounts of different cytokines including those that are assigned to Th1 (IFN-γ) and Th2 (IL-4) cells. NKT might bridge innate and adaptive immunity since they respond rapidly, a feature of innate immune responses, to their specific antigen with secretion of polarizing cytokines determining adaptive immunity pathways. NKT seem to cover a broad spectrum of functions in the immune system by contributing to host defense against pathogens, tumor surveillance, and prevention of autoimmune conditions (Balato et al. 2009; Godfrey et al. 2004).

Although T cell function is modulated by vitamin D, normal T cell development does not require vitamin D or the VDR expression (Cantorna 2011). Remarkably, the situation is different for NKT. In VDR KO mice, not only thymic and peripheral numbers of NKT were reduced but also effector functions of remaining NKT were impaired (Yu and Cantorna 2008). NKT cells remained at an earlier maturation level, lacking expression of NK1.1 and the transcription factor T-bet. Moreover, expression of CD1d in thymocytes was reduced, thus impairing selection of NKT in the thymus (Yu and Cantorna 2008). In a subsequent study, it was shown that vitamin D deficiency in utero resulted in reduced numbers of iNKT cells. This was caused by apoptosis of iNKT precursors (Yu and Cantorna 2011) and could not be corrected by later supplement of vitamin D. Since it is suggested that reduced NKT numbers can contribute to autoimmunity, sufficient vitamin D supply seems a critical factor for maintenance of self-tolerance (Cantorna et al. 2011).

4.2.10 Summary of Vitamin D_3 Effects on Immune Cells

The spectrum of effects of vitamin D has increased in parallel with the discovery of (functional) subsets of cells involved in innate and specific immunity. In a complex network of regulatory and effector mechanisms, vitamin D seems to be integrated with multiple points of application, revealing a subtle modulatory mode of action rather than the previously supposed immunosuppression (also see Fig. 4.3). From the available literature, enhancement of innate defense efficiency (e.g., monocyte function) and control or balance of specific immune reactions (e.g., via dendritic cells, NKT, or Treg) can be subsumed as major tasks of vitamin D. Thereby, vitamin D obviously exerts an important role at the level of cell development and differentiation but also in a negative feedback mode in already initiated and ongoing inflammatory processes. In this respect, conclusions about the role of vitamin D drawn

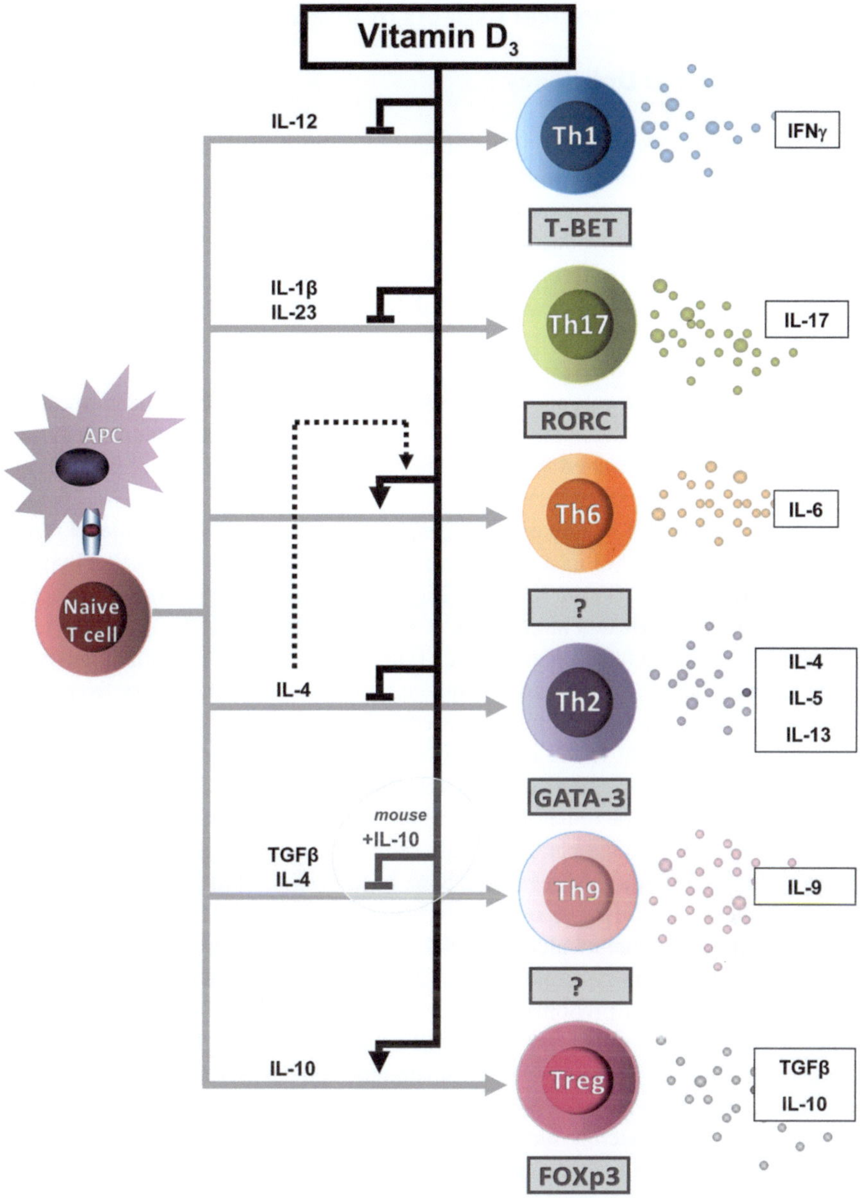

Fig. 4.3 Summary of immunomodulatory effects of vitamin D_3 on T cell response

from the various studies have to be critically evaluated: The point of time within the life span of a specific cell, the phase of an (immunological) process, and, last but not least, the age of the individual seems to play a decisive role for the overall impact of activity or, as a rather frequent phenomenon, deficiency of vitamin D.

4.3 Pathophysiological Aspects of Vitamin D_3

4.3.1 Vitamin D Deficiency

Hypovitaminosis of vitamin D_3 is a widespread phenomenon. The major source of vitamin D is UV-dependent production in the skin (see above), while the contribution of dietary intake appears negligible, at least in countries where dairy products are not fortified. Naturally, due to UV exposition, vitamin D_3 levels are affected by geographic distribution in terms of latitude and altitude. Altered lifestyles including reduced outdoor activity favor vitamin D_3 deficiency as well as the use of sun blockers and skin cancer awareness (Ascherio et al. 2010; Peterlik and Cross 2005; Cantorna and Mahon 2004).

Serum vitamin D_3 levels and, as a consequence, deficiency have been primarily defined with regard to the maintenance of calcium and phosphate homeostasis, and a cutoff could be determined by rise of the parathyroid hormone level (Klaushofer 2010). The most commonly measured parameter in vitamin D_3 metabolism, 25(OH) D_3, reflects the general supply of the organism, while the 1,25$(OH)_2D_3$ level represents the systemic level of its active metabolite, mainly achieved by further hydroxylation in the kidneys. Serum levels of 25(OH)D_3 at 20 nmol/l seem to be efficient in the prevention of rickets in children, but much higher levels are supposed to be necessary for general health maintenance (Dimeloe et al. 2010). Optimal levels are currently a matter of debate, and recommendations for 25(OH)D_3 vary from 30 to 75 nmol/l, although 50–75 nmol/l are considered desirable (Klaushofer 2010). Definition of sufficient levels in regard of bone and calcium homeostasis is less demanding than establishing a serum threshold for maintenance of physiologic immune functions. Bone mineral density and fracture risk assessment are well measurable and useful clinical parameters as control for optimal vitamin D_3 levels in a patient's long-term outcome regarding bone health (Klaushofer 2010). On the contrary, it is the local production of the active metabolite 1,25$(OH)_2D_3$ in an immunological microenvironment which is crucial for regulatory actions exerted by vitamin D_3. Although sufficient systemic levels of 25(OH)D_3 are required, local levels of the active metabolite are predominantly dependent on 1α-hydroxylase activity in immune cells, a process beyond control of the strictly regulated calcium and phosphate feedback loop. Moreover, since the amount of local production quantitatively affects vitamin D_3 levels only as an exception, local vitamin D_3 levels are probably poorly reflected in systemic serum levels and definition of optimal local levels for immunological health maintenance appears difficult.

4.3.2 Vitamin D_3 in Health and Disease

Since it is well demonstrated that vitamin D_3 is important for the maintenance of immunological homeostasis, it can be concluded that vitamin D deficiency consequently leads to a range of disorders. As for research, two different approaches can be chosen: On the one hand, a row of epidemiological studies showed a correlation between vitamin D deficiency or supplementation and several health disorders. On the other hand, a mechanistic approach was based on the hypothesis that a deficiency of vitamin D_3, which in previous studies was found to suppress the Th1 pathway, might favor the occurrence of Th1-mediated autoimmune diseases acting as "environmental factor." On the other hand, since it was believed that vitamin D shifts the immunologic balance toward a Th2 pattern, the influence of vitamin D_3 on allergy development was also investigated. In order to confirm these considerations, several animal disease models were employed.

In several epidemiological studies, measurable parameters such as vitamin D_3 supply, rickets prophylaxis, serum vitamin D_3 levels, and occurrence of immune-mediated disorders were correlated. In the first large epidemiological study on vitamin D_3 in a large Finish birth cohort study, Hyppönen et al. (2001) observed a reduced risk for (autoimmune mediated) type 1 diabetes under vitamin D_3 supplementation for rickets prevention.

A further epidemiological study by Hyppönen et al. raised the question that infant supplementation with vitamin D_3 in order to prevent rickets might be associated with an increased risk for allergic conditions in later life (Hyppönen et al. 2004).

Nevertheless, in addition to modulating autoimmune or allergic diseases, vitamin D appears to play a role in infectious diseases, such as tuberculosis (Rahman et al. 2015; Cao et al. 2015; Esposito and Lelii 2015). Moreover, in children, vitamin D deficiency may also be associated with infections such as otitis media and bronchiolitis (Esposito and Leli 2015).

4.3.3 Vitamin D_3 and Allergic Conditions

Allergy can be considered as an excessive reaction of the immune system toward a normally harmless antigen. The often very casually used term "allergy" mostly refers to a hypersensitivity reaction (type I according to the classification by Gell and Coombs), characterized by a pathologic shift toward a Th2 pattern with excessive IgE production. After "sensitization," the first contact with the respective allergen, specific IgE production is initiated and mast cells are loaded with IgE. Re-exposition to the allergen leads to cross-linking of the membrane-bound IgE molecules by their specific antigen and degranulation of vasoactive mediators such as histamine and leukotrienes takes its course. A late component of the allergic response comprises an inflow of Th2 cells as well as of eosinophils and causes further tissue damage (Dimeloe et al. 2010). Allergy clinically manifests itself as rhinoconjunctivitis, asthma, or anaphylaxis. Also often used term "atopy" refers to a genetic predisposition to allergy and a bias to produce excessive IgE as reaction to specific antigens. Atopy can present itself as atopic dermatitis, allergic rhinoconjunctivitis, or allergic asthma (Fritsch 2009; Herold 2008).

Since it has originally been assumed that in adaptive immune responses vitamin D_3 shifts the balance toward a Th2 pattern by inhibition of the Th1 pathway, the question has arisen whether vitamin D_3 supplementation might favor development of allergies. A large epidemiological Finish birth cohort study showed a possible association between vitamin D supplementation (as prevention for rickets) in infancy and later allergic conditions (Hyppönen et al. 2004). Contrasting this observation, more recent epidemiological studies suggest a preventive effect of vitamin D_3 in pathogenesis of allergic asthma, a chronic obstructive airway disease. Vitamin D_3 seems to be already essential for normal fetal lung development, maturation, and surfactant production (Litonjua 2009). Two epidemiological studies observed a negative correlation between vitamin D_3 intake during pregnancy and risk for asthma in the offspring (Erkkola et al. 2009; Camargo et al. 2007). Additionally, the study by Erkkola et al. (2009) also reports on a decreased risk for allergic rhinitis. The findings by Hyppönen et al. (2009) that both very low and very high 25-OH vitamin D_3 serum levels are associated with increased average and specific IgE concentrations might suggest a predisposition for atopy and allergy, but in this study, clinical manifestation was not assessed (also see above). A recently published large US survey also reported on an association of deficient 25(OH)D_3 levels and higher prevalence of allergic sensitization by measurement of specific IgE levels. Remarkably, the correlation was observed in children and adolescents but could not be defined that clearly in the adult population (Sharief et al. 2011).

Taken together, these results obtained from epidemiological studies raise the question whether a specific time frame in life might be susceptible to an altered vitamin D_3 status. On a cellular level, a bias toward allergy caused by vitamin D supplementation during infancy could not be substantiated: In human PBMC obtained from cord blood vitamin D showed a virtually complete inhibition of Th2 differentiation (Pichler et al. 2002; see above).

Furthermore, several studies showed induction of an IL-10-producing regulatory T cell subset upon vitamin D_3 exposure (see above). Regulatory T cell subsets are capable of limiting potentially detrimental immune reactions as observed in allergic conditions and have been shown to protect against asthma in several studies (Robinson 2009; Dimeloe et al. 2010; Xystrakis et al. 2006). Vitamin D_3 alone, as well as in combination with glucocorticoids, seems to be a strong enhancer of IL-10 production in T cells. Glucocorticoids are considered a first-line therapy in asthma, and their effects are probably mediated by suppression of Th2 cytokines as well as induction of an IL-10-producing T cell subset. In patients suffering from glucocorticoid-resistant asthma, CD4+ T cells fail to produce IL-10. This can be overcome by in vitro administration of vitamin D_3. Moreover, in a pilot study, when glucocorticoid-resistant asthmatic patients were orally supplemented with vitamin D_3, responsiveness to steroids was ameliorated, shown by an increase of IL-10 production in vitro (Xystrakis et al. 2006).

Additionally, research of vitamin D_3 actions on B cells, the cellular source of IgE production, could show that in vitro vitamin D_3 inhibits IgE production stimulated by anti-CD40 mAb plus IL-4 (Heine et al. 2002; Hartmann et al. 2011). In a subsequent study, it was shown that Il-10 production in activated B cells was enhanced by vitamin D_3. IL-10, a suppressive cytokine produced by many immune cells including APC and T cells (see above), enhances an Ig class switch to IgA, IgG, and IgM (Heine et al. 2008).

As a conclusion, recent studies suggest a potentially protective role of vitamin D_3 in allergic conditions. Moreover, it was even suggested that vitamin D analogues might be applied in future allergy therapy and vitamin D_3 might enhance responsiveness to actually applied anti-inflammatory drugs (Hartmann et al. 2011; Xystrakis et al. 2006).

4.3.4 Vitamin D_3 and Autoimmunity

Since various in vitro studies showed that vitamin D_3 suppresses the Th1 pathway, it was also concluded that deficiency might cause the onset or exacerbation of formally believed Th1-mediated autoimmune diseases. The pathophysiological concept of autoimmune diseases mediated by a shift toward a Th1 pattern has been renewed by the discovery of new CD4+T cell subsets (also see above). Th17, a subset capable of entertaining chronic inflammatory conditions and Tregs, vital for the maintenance of self-tolerance, have contributed among them to a far better understanding of autoimmune reactions. In several autoimmune-mediated diseases, suppressive function of Tregs seems to be compromised (Costantino et al. 2008). The role of vitamin D_3 in the prevention of autoimmune conditions has already been extensively defined on a cellular level by its suppressive capacity on the Th1 pathway. Effects of vitamin D_3 on the more recently described subsets also fit well into this concept by inhibition of the Th17 lineage and induction or functional restoration of the regulatory T cell subset.

Regarding autoimmune diseases, mechanisms of origin and onset still remain to be fully elucidated. Gender-related differences to the advantage of females have been observed. Specific HLA constellations might predispose and serve as genetic factor, but additionally mostly unidentified environmental factors seem to be crucial at the onset (Fritsch 2009). The role of vitamin D_3 deficiency is discussed in various immune-mediated diseases with multiple sclerosis (MS), type I diabetes, psoriasis, rheumatoid arthritis, SLE, inflammatory bowel diseases, sclerodermia, pre-eclampsia, Morbus Behçet, and autoimmune iridocyclitis, among them (Hyppönen 2005; Do et al. 2008; Adorini and Penna 2008; Kriegel et al. 2011). Various epidemiological studies indicate that vitamin D_3 deficiency could be related to a higher prevalence of several autoimmune disorders. This notion is in line with a recent meta-analysis that demonstrated lower vitamin D levels and a higher likelihood of vitamin D deficiency in two autoimmune thyroid diseases (Hashimoto thyroiditis and Graves' disease) (Wang et al. 2015).

However, the vulnerable period in life remains to be elucidated when sufficient vitamin D_3 supply is crucial to prevent autoreactive disorders. In several epidemiological studies, decreased serum levels for 25(OH)D_3 or 1,25(OH)D_3 were observed before the onset or in the course of the respective autoimmune disease. This was also a reason to hypothesize that there might be a correlation between vitamin D_3 deficiency and development of autoimmune conditions. Nevertheless, a reverse causation has also to be taken into account, since an altered vitamin D_3 status might be the consequence of the disease and not the cause (Kriegel et al. 2011).

VDR polymorphisms have been attributed to an increased risk for development of various autoimmune diseases in humans, including Hashimoto's thyroiditis, inflammatory bowel disease, Graves' disease, RA, SLE, primary biliary cirrhosis, autoimmune hepatitis, Addison's disease, vitiligo, celiac disease, and type I diabetes, as well as MS. Unfortunately, not all of the detected polymorphisms associated with autoimmune diseases have known functional consequences, and not all of these associations have been reproduced (Kriegel et al. 2011; Cantorna et Mahon 2004).

4.3.4.1 Multiple Sclerosis

MS is a chronic inflammatory autoimmune disease characterized by demyelination in the central nervous system leading to severe neurologic consequences and disability. In addition to genetic susceptibility, exposure to not yet fully identified environmental factors is required for the onset. Apart from Epstein-Barr virus infection and smoking, vitamin D deficiency is suspected as one of these major environmental factors. The prevalence of MS is associated with a remarkable geographical profile: MS onset is less frequent at the equator and increases with either north or south latitude. Furthermore, living at higher altitudes, which is associated with higher UV exposure, negatively influences MS prevalence. Climatic factors with diminished occurrence in very sunny regions also contribute to the risk of MS. Sun exposure at a younger age seems to be especially protective. Migration from a high-latitude/high-prevalence region to a lower-latitude/low-prevalence region during the first two decades seems to be beneficial and vice versa, a fact underlining the crucial role of environmental factors in MS onset. Furthermore, a correlation between low vitamin D_3 serum levels and onset of MS has been described in several studies. The time of birth, with a higher risk of developing MS for the spring born (with a peak in May) and lower risk for autumn (with a peak in November), may reflect the vitamin D_3 supply of the mother in pregnancy. Additionally, dietary factors affecting vitamin D_3 supply also seem to contribute, since a lower prevalence in populations with high consumption of fatty fish has been described (Ascherio et al. 2010; Pierrot-Deseilligny et Souberbielle 2010; Smolders et al. 2008).

MS belongs to the group of Th1-mediated autoimmune diseases (see above). Therefore, it could be concluded that adequate vitamin D_3 supply might prevent MS or ameliorate the disease by suppression of the Th1 lineage. The experimental allergic encephalomyelitis (EAE) mouse model, achieved by vaccination with spinal cord homogenates containing myelin proteins, is commonly used in MS animal studies (Arnson et al. 2007; Smolders et al. 2008). In several studies, the administration of $1{,}25(OH)_2D_3$ before EAE induction avoided its onset, and even administration after immunization prevented the start or at least led to a milder course of the disease. In mice already suffering from EAE vitamin D_3, administration ameliorated the disease and inhibited progression (Smolders et al. 2008). Since it has been discovered that Th17 substantially contribute to Th1-mediated autoimmune diseases as a pro-inflammatory subset, effects of vitamin D_3 on Th17 obtained from EAE mice have also been examined. It could be shown that application of vitamin D_3 not only inhibited Th17 induction but also downregulated CCR6, a chemokine receptor necessary for migration and entry to the central nervous system (Chang et al. 2010a, b).

Data on vitamin D_3-mediated effects on Tregs, other important players in autoimmune diseases, appears conflicting. Since it has been shown in several studies that vitamin D_3 enhances the regulatory T cell compartment (see above), the same effect in MS could be expected. Nevertheless, Chang et al. reported a suppressive effect of vitamin D_3 on Treg in vitro induction in T cells obtained from EAE mice. Sloka et al. recently described that in EAE 1,25-dihydroxyvitamin D decreased clinical severity and reduced the loss of axons (Sloka et al. 2015). In a human study, a correlation of serum 25(OH)D_3 levels in MS patients and suppressive capability of Tregs was found, while an association of vitamin D_3 status and Treg numbers could not be observed. Due to a decline in the IFNγ/IL-4 ratio in patients with high 25(OH) D_3 serum levels, the authors concluded that vitamin D_3 might skew the balance toward a Th2 phenotype. Taken together, they suggest a shift toward a less inflammatory T cell cytokine profile by vitamin D_3 (Smolders et al. 2009, 2010).

4.3.4.2 Type I Diabetes Mellitus (DM I) and Vitamin D_3

Type 1 diabetes mellitus is a chronic autoimmune disease characterized by destruction of pancreatic islet β-cells, the source of endogenous insulin. The progression from insulitis to the clinical onset of DM I occurs usually during childhood and adolescence, at a time point when almost 90 % of the β cells are already destroyed. Typical feature of the insulin deficiency is hyperglycemia, and a severe ketoacidosis may present itself as life-threatening first clinical manifestation. Chronic damage to blood vessels due to hyperglycemia presents as retinopathy, coronary heart disease, stroke, and renal failure as a long-term prospective. Similar to other "Th1"-mediated autoimmune diseases, the pathogenesis of DM I is considered to be triggered by environmental factors acting upon a genetic background (Takiishi et al. 2010; Herold 2008). Several retrospective studies propose vitamin D_3 deficiency as an environmental factor. As to prevalence, a striking geographic distribution similar to MS has been described, emphasizing the role of environmental factors in – autoimmune – disease onset (Szodoray et al. 2008; Takiishi et al. 2010; Hyppönen 2010; Mathieu et Badenhoop 2005). In 2001, Hyppönen et al. published a Finish birth cohort study on a significantly reduced risk for DM I when participants had been supplemented with vitamin D during the first year of their life. A large multicenter study had also observed this protective effect of vitamin D_3 supplementation in infancy (Anonymous 1999). Furthermore, the risk to develop DM I in later life was found to be increased in children suffering from rickets (Hyppönen et al. 2001). Moreover, further epidemiological studies showed that vitamin D_3 obtained from food or cod-liver intake during pregnancy also reduced the risk of DM I in the offspring (Fronczak et al. 2003; Stene and Joner 2003). A case-control study from Italy based on the Piedmont Diabetes Registry found no association between the risk of type 1 diabetes in childhood and 25-hydroxyvitamin D levels at birth; nevertheless, within the subgroup of migrant newborns, babies with 25-hydroxyvitamin D concentrations lower than 2.14 ng/ml had an increased risk of type 1 diabetes (odds ratio: 14.02) (Cadario et al. 2015).

The nonobese diabetic mouse (NOD) model is predominantly used in animal studies examining vitamin D_3 effects in regard to DM I in vivo. Corresponding to other animal autoimmune disease models, treatment with therapeutic doses of vitamin D_3 or analogues resulted in the prevention or delay of development of insulitis or clinically manifested diabetes. Since therapeutic doses of vitamin D_3 were required, hypercalcemia and bone decalcification were observed, leading to a search for effective less-calcemic alternatives (Arnson et al. 2007; Takiishi et al. 2010).

In a randomized, placebo-controlled trial in young subjects with new-onset type 1 diabetes, vitamin D supplementation was found to enhance the suppressive capacity of regulatory T cells (Treiber et al. 2015). When compared to the placebo group, in the vitamin D-supplemented group after 12 months, insulin requirements were lower. The authors conclude that vitamin D has the potential as an immunomodulator of type 1 diabetes, e.g. in combination therapies.

4.3.4.3 Inflammatory Bowel Diseases (IBD)

IBD are chronic recurring inflammatory diseases of the gastrointestinal tract. Crohn's disease and ulcerative colitis are two distinct forms of the IBD. These immune-mediated diseases are characterized by an excessive Th1 reaction to mostly bacterial antigens (Szodoray et al. 2008; Peterlik and Cross 2005; Cantorna et al. 2000). As for other autoimmune-mediated diseases, an encounter of genetic susceptibility and environmental factors is necessary for IBD onset, and vitamin D deficiency might contribute to its development. Similar to MS and DM I, a specific geographic pattern in its prevalence can be observed (Szodoray et al. 2008). IL-10 knockout mice serve as one of the experimental IBD models, since mice spontaneously develop a disorder similar to human IBD, due to an uncontrolled immune reaction toward the intestinal flora. It was shown that vitamin D_3 deficiency aggravated symptoms in the IL-10 KO mice, while supplementation ameliorated the condition caused by the disease (Cantorna et al. 2000). In two other different experimental IBD models, the lack of the VDR led to an aggressive course of the disease (Froicu et al. 2003). Vitamin D_3 deficiency is common in IBD, and a recent study also reported a correlation between $25(OH)D_3$ serum levels and disease activity. Whether vitamin D_3 deficiency is a reason for the onset or a consequence of the local inflammation of the gastrointestinal tract still remains to be elucidated (Ulitsky et al. 2011).

4.3.4.4 Vitamin D_3 and Rheumatoid Arthritis (RA)

RA is an immune-mediated disease characterized by erosive arthritis and joint destruction, potentially leading to disability (Szodoray et al. 2008; Adorini and Penna 2008). As in other immune-mediated diseases, the genetic background, in combination with mostly unidentified environmental factors, is responsible for the onset of RA. Regarding the geographic distribution, the prevalence of RA was found to be higher in the north of Europe than in the southern part. Patients suffering from RA have been reported to show decreased vitamin D_3 serum levels. Moreover, lower vitamin D_3 serum levels have also been correlated with higher disease activity in RA (Cutolo et al. 2007). In one epidemiological survey conducted in elderly women, a negative correlation between dietary or supplementary vitamin D_3 intake and risk for RA could be observed (Merlino et al. 2004). Nevertheless, this study relied solely on a questionnaire on food intake, and serum $25(OH)D_3$ levels, respectively, sun exposition, were not assessed or taken into account. Another study measured vitamin D_3 serum levels without detecting a correlation, thus not corroborating the data from

Merlino et al. (Nielen et al. 2006). A recent meta-analysis based on 24 publications found that patients with RA had lower vitamin D levels than controls; moreover, among the patients, a negative correlation between 25-hydroxyvitamin D concentrations and disease activity index could be established (Lin et al. 2016).

Effects of vitamin D_3 on RA were also examined in animal disease models. After vitamin D_3 supplementation, Cantorna et al. (1998) observed amelioration of symptoms or even prevention of disease progression in murine Lyme arthritis and collagen-induced arthritis. This effect was also achieved in rats with collagen-induced arthritis treated with a vitamin D_3 analogue (Larsson et al. 1998).

4.3.5 Vitamin D_3 in Therapy

As extensively outlined above, vitamin D_3 acts as a modulator on immune functions by shifting the balance within the T cell compartments and influencing adaptive immunity. Therefore, it could be expected that vitamin D_3 might be applied as therapy in various diseases. Unfortunately, therapeutic doses cause severe side effects in form of hypercalcemia, hypocalciuria, nephrocalcinosis, and increased bone resorption (van Etten and Mathieu 2005). Efforts have been laid on the development of low-calcemic analogues with enhanced immunomodulatory effects. So far, vitamin D_3 and VDR agonists have been already successfully applied in vitro and in various animal disease models, but there is still a lack of large clinical studies in humans, probably still due to fear of calcemic side effects.

In humans, vitamin D_3 and analogues are currently applied only topically, as a first-line therapy in psoriasis, a chronic inflammatory immune-mediated disease of the skin. Vitamin D_3 efficacy on psoriasis can be attributed to its pro-differentiating and proliferation-inhibitory effect on keratinocytes as well as its immune modulatory capacity to reduce inflammation (Adorini 2005; Fritsch 2009; Miller and Gallo 2010).

In order to avoid or ameliorate side effects, several strategies have been proposed: Since a number of synthetic analogues, while being less calcemic, still cause severe bone demineralisation, a combination with bone resorption inhibitors was suggested. The most promising strategy would probably be the administration of vitamin D in conjunction with other immunosuppressive or immunomodulatory agents. Synergistic effects with other immunosuppressive drugs have been observed in several studies. Administration of vitamin D_3 analogues in combination with immunosuppressive agents could save the amounts needed on both sides and therefore generally reduce side effects in treatment of autoimmune disorders, graft rejection, or allergy (van Etten and Mathieu 2005).

4.3.6 Conclusion and Perspective

An increasing number of publications suggests a possible role of vitamin D in (auto-)immune diseases, the majority of them focusing on the – rather frequent – deficiency of vitamin D. Experimental data partially substantiate the conclusions drawn from epidemiological studies. However, there remains a hiatus between in vitro results on vitamin D effects, systemic vitamin D levels measured in individuals, and epidemiological data on incidence and course of the disease. As discussed in some of the papers reviewed, there is no definite end-to-end concept of how a specific concentration/deficiency of vitamin D may lead to autoimmunity, and a reverse causation (an altered vitamin D_3 status being the consequence and not the cause of the disease) cannot be unequivocally excluded.

There are two major drawbacks in this context: Firstly, the mechanism of autoimmunity itself is still a matter of debate, so whatever effect of vitamin D on the immune system is discussed, causality for development of autoimmune disease cannot be directly evaluated. Secondly, concentrations of the active metabolite $1\alpha25(OH)D_3$ in different compartments and tissues are reflected only very roughly by systemic serum levels of $25(OH)D_3$ or $1\alpha25(OH)D_3$. Low levels of $25(OH)D_3$, which is also a prerequisite for extrarenal $1\alpha25(OH)D_3$ production, are most likely associated with local vitamin D deficiency, relevant for interference with the immune system. However, the actual local concentration of $1\alpha25(OH)D_3$ cannot be directly measured.

The value of animal models may be limited, since results are not always clear-cut, and their interpretation as accurate image of the human system has to be viewed critically.

However, as mentioned by several authors, improvement of the vitamin D status in a natural way or by supplementation of vitamin D is cheap and easy, and bone metabolism benefits also from ameliorating the widespread deficiency.

The therapeutic application of higher doses of vitamin D_3 is based on promising concepts but is still hampered by side effects and has not become clinical routine so far. Whenever the systemic effects can be successfully separated from bone metabolism, either by the production of less calcemic analogues or by concentrating the active metabolite on the area of interest, a wide field of applications becomes reality.

References

Adams JS, Liu PT, Chun R, Modlin RL, Hewison M (2007b) Vitamin D in defense of the human immune response. Ann N Y Acad Sci 1117:94–105

Adams JS, Ren S, Liu PT, Chun RF, Lagishetty V, Gombart AF, Borregaard N, Modlin RL, Hewison M (2009) Vitamin d-directed rheostatic regulation of monocyte antibacterial responses. J Immunol 182:4289–4295

Adams JS, Sharma OP, Gacad MA, Singer FR (1983) Metabolism of 25-hydroxyvitamin D3 by cultured pulmonary alveolar macrophages in sarcoidosis. J Clin Invest 72:1856–1860

Adorini L (2005) Intervention in autoimmunity: the potential of vitamin D receptor agonists. Cell Immunol 233:115–124

Adorini L, Giarratana N, Penna G (2004) Pharmacological induction of tolerogenic dendritic cells and regulatory T cells. Semin Immunol 16:127–134

Adorini L, Penna G (2008) Control of autoimmune diseases by the vitamin D endocrine system. Nat Clin Pract Rheumatol 4:404–412

Akira S, Uematsu S, Takeuchi O (2006) Pathogen recognition and innate immunity. Cell 124:783–801

Almerighi C, Sinistro A, Cavazza A, Ciaprini C, Rocchi G, Bergamini A (2009) 1Alpha,25-dihydroxyvitamin D3 inhibits CD40L-induced pro-inflammatory and immunomodulatory activity in human monocytes. Cytokine 45:190–197

Annunziato F, Romagnani S (2009a) Do studies in humans better depict Th17 cells? Blood 114:2213–2219

Annunziato F, Romagnani S (2009b) Heterogeneity of human effector CD4+ T cells. Arthritis Res Ther 11:257

Anonymous (1999) Vitamin D supplement in early childhood and risk for type I (insulin-dependent) diabetes mellitus. The EURODIAB Substudy 2 Study Group. 15. Diabetologia 42:51–54

Arnson Y, Amital H, Shoenfeld Y (2007) Vitamin D and autoimmunity: new aetiological and therapeutic considerations. Ann Rheum Dis 66:1137–1142

Ascherio A, Munger KL, Simon KC (2010) Vitamin D and multiple sclerosis. Lancet Neurol 9:599–612

Azizi-Semrad U, Krenbek D, Hofbauer G, Karanikas G, Maldonado-Gonzalez E, Pietschmann P, Willheim M (2010) Cytokine profiling of human peripheral blood CD4+ T lymphocytes reveals a new Th-subpopulation (Th6) characterized by IL-6. Eur Cytokine Netw 21:105–115

Baeke F, Gysemans C, Korf H, Mathieu C (2010a) Vitamin D insufficiency: implications for the immune system. Pediatr Nephrol 25:1597–1606

Baeke F, Korf H, Overbergh L, van Etten E, Verstuyf A, Gysemans C, Mathieu C (2010b) Human T lymphocytes are direct targets of 1,25-dihydroxyvitamin D3 in the immune system. J Steroid Biochem Mol Biol 121:221–227

Balato A, Unutmaz D, Gaspari AA (2009) Natural killer T cells: an unconventional T-cell subset with diverse effector and regulatory functions. J Invest Dermatol 129:1628–1642

Barrat FJ, Cua DJ, Boonstra A, Richards DF, Crain C, Savelkoul HF, de Waal-Malefyt R, Coffman RL, Hawrylowicz CM, O'Garra A (2002) In vitro generation of interleukin 10-producing regulatory CD4(+) T cells is induced by immunosuppressive drugs and inhibited by T helper type 1 (Th1)- and Th2-inducing cytokines. J Exp Med 195(5):603–616

Bikle D (2009) Nonclassic actions of vitamin D. J Clin Endocrinol Metab 94:26–34

Bikle D, Adams JS, Christakos S (2013) Vitamin D: production, metabolism, mechanism of action, and clinical requirements. In: Rosen CJ et al (eds) Primer on the metabolic bone diseases and disorders of mineral metabolism, 8th edn. Wiley-Blackwell, Ames, pp 235–247

Boltz-Nitulescu G, Willheim M, Spittler A, Leutmezer F, Tempfer C, Winkler S (1995) Modulation of IgA, IgE, and IgG Fc receptor expression on human mononuclear phagocytes by 1 α,25-dihydroxyvitamin D3 and cytokines. J Leukoc Biol 58:256–262

Boonstra A, Barrat FJ, Crain C, Heath VL, Savelkoul HF, O'Garra A (2001) 1α,25-Dihydroxyvitamin d3 has a direct effect on naive CD4(+) T cells to enhance the development of Th2 cells. J Immunol 167:4974–4980

Bringhurst FR, Demay MB, Krane SM, Kronenberg HM (2008) Bone and mineral metabolism in health and disease. In: Fauci AS, Braunwald E, Kasper DL, Hauser SL, Longo DL, Jameson JL, Loscalzo J (eds) Harrison's principles of internal medicine, 17th edn. McGraw Hill, New York, pp 2365–2377

Byun DG, Demeure CE, Yang LP, Shu U, Ishihara H, Vezzio N, Gately MK, Delespesse G (1994) In vitro maturation of neonatal human CD8 T lymphocytes into IL-4- and IL-5-producing cells. J Immunol 153:4862–4871

Cadario F, Savastio S, Pagliardini V, Bagnati M, Vidali M, Cerutti F, Rabbone I, Fontana F, Lera R, De Donna V, Valori A, Gruden G, Bona G, Bruno G (2015) Vitamin D levels at birth and risk of type 1 diabetes in childhood: a case–control study. Acta Diabetol 52:1077–1081

Camargo CA Jr, Rifas-Shiman SL, Litonjua AA, Rich-Edwards JW, Weiss ST, Gold DR, Kleinman K, Gillman MW (2007) Maternal intake of vitamin D during pregnancy and risk of recurrent wheeze in children at 3 y of age. Am J Clin Nutr 85:788–795

Cantorna MT (2011) Why do T cells express the vitamin D receptor? Ann NY Acad Sci 1217:77–82

Cantorna MT, Hayes CE, DeLuca HF (1998) 1,25-Dihydroxycholecalciferol inhibits the progression of arthritis in murine models of human arthritis. J Nutr 128:68–72
Cantorna MT, Mahon BD (2004) Mounting evidence for vitamin D as an environmental factor affecting autoimmune disease prevalence. Exp Biol Med 229:1136–1142
Cantorna MT, Munsick C, Bemiss C, Mahon BD (2000) 1,25-Dihydroxycholecalciferol prevents and ameliorates symptoms of experimental murine inflammatory bowel disease. J Nutr 130:2648–2652
Cantorna MT, Yu S, Bruce D (2008) The paradoxical effects of vitamin D on type 1 mediated immunity. Mol Aspects Med 29:369–375
Cao Y, Wang X, Cao Z, Cheng X (2015) Association of vitamin D receptor gene taql polymorphisms with tuberculosis susceptibility: a meta-analysis. Int J Clin Exp Med 8(6):10187–10203
Chang JH, Cha HR, Lee DS, Seo KY, Kweon MN (2010a) 1,25-Dihydroxyvitamin D_3 inhibits the differentiation and migration of T(H)17 cells to protect against experimental autoimmune encephalomyelitis. PLoS One 5:e12925
Chang JM, Kuo MC, Kuo HT, Hwang SJ, Tsai JC, Chen HC, Lai YH (2004) 1-α,25-Dihydroxyvitamin D3 regulates inducible nitric oxide synthase messenger RNA expression and nitric oxide release in macrophage-like RAW 264.7 cells. J Lab Clin Med 143:14–22
Chang SH, Chung Y, Dong C (2010b) Vitamin D suppresses Th17 cytokine production by inducing C/EBP homologous protein (CHOP) expression. J Biol Chem 285:38751–38755
Chen S, Sims GP, Chen XX, Gu YY, Chen S, Lipsky PE (2007) Modulatory effects of 1,25-dihydroxyvitamin D3 on human B cell differentiation. J Immunol 179(3):1634–1647
Chesney RW (2010) Vitamin D and the magic mountain: the anti-infectious role of the vitamin. J Pediatr 156:698–703
Cippitelli M, Santoni A (1998) Vitamin D3: a transcriptional modulator of the interferon-γ gene. Eur J Immunol 28:3017–3030
Colin EM, Asmawidjaja PS, van Hamburg JP, Mus AM, van Driel M, Hazes JM, van Leeuwen JP, Lubberts E (2010) 1,25-dihydroxyvitamin D3 modulates Th17 polarization and interleukin-22 expression by memory T cells from patients with early rheumatoid arthritis. Arthritis Rheum 62:132–142
Costantino CM, Baecher-Allan CM, Hafler DA (2008) Human regulatory T cells and autoimmunity. 43. Eur J Immunol 38:921–924
Croft M, Carter L, Swain SL, Dutton RW (1994) Generation of polarized antigen-specific CD8 effector populations: reciprocal action of interleukin (IL)-4 and IL-12 in promoting type 2 versus type 1 cytokine profiles. J Exp Med 180:1715–1728
Cutolo M, Otsa K, Uprus M, Paolino S, Seriolo B (2007) Vitamin D in rheumatoid arthritis. Autoimmun Rev 7:59–64
dÁmbrosio D (2006) Regulatory T cells: how do they find their space in the immunological arena? 48. Semin Cancer Biol 16:91–97
Dardalhon V, Awasthi A, Kwon H, Galileos G, Gao W, Sobel RA, Mitsdoerffer M, Strom TB, Elyaman W, Ho IC, Khoury S, Oukka M, Kuchroo VK (2008) IL-4 inhibits TGF-beta-induced Foxp3+ T cells, and together with TGF-beta, generates IL-9+ IL-10+ Foxp3(−) effector T cells. Nat Immunol 9:1347–1355
Demeure CE, Wu CY, Shu U, Schneider PV, Heusser C, Yssel H, Delespesse G (1994) In vitro maturation of human neonatal CD4 T lymphocytes. II Cytokines present at priming modulate the development of lymphokine production. J Immunol 152:4775–4782
Dimeloe S, Nanzer A, Ryanna K, Hawrylowicz C (2010) Regulatory T cells, inflammation and the allergic response – the role of glucocorticoids and vitamin D. J Steroid Biochem Mol Biol 120:86–95
Do JE, Kwon SY, Park S, Lee ES (2008) Effects of vitamin D on expression of Toll-like receptors of monocytes from patients with Behcet's disease. Rheumatology 47:840–848
Erkkola M, Kaila M, Nwaru BI, Kronberg-Kippilä C, Ahonen S, Nevalainen J, Veijola R, Pekkanen J, Ilonen J, Simell O, Knip M, Virtanen SM (2009) Maternal vitamin D intake during pregnancy is inversely associated with asthma and allergic rhinitis in 5-year-old children. Clin Exp Allergy 39:875–882

Esposito S, Lelii M (2015) Vitamin D and respiratory tract infections in childhood. BMC Infect Dis 15:487
Fritsch P (2009) Dermatologie und Venerologie für das Studium. Springer, Berlin
Froicu M, Weaver V, Wynn TA, McDowell MA, Welsh JE, Cantorna MT (2003) A crucial role for the vitamin D receptor in experimental inflammatory bowel diseases. Mol Endocrinol 17:2386–2392
Fronczak CM, Barón AE, Chase HP, Ross C, Brady HL, Hoffman M, Eisenbarth GS, Rewers M, Norris JM (2003) In utero dietary exposures and risk of islet autoimmunity in children. Diabetes Care 26:3237–3242
Godfrey DI, MacDonald HR, Kronenberg M, Smyth MJ, Van Kaer L (2004) NKT cells: what's in a name? Nat Rev Immunol 4:231–237
Gregori S, Casorati M, Amuchastegui S, Smiroldo S, Davalli AM, Adorini L (2001) Regulatory T cells induced by 1α,25-dihydroxyvitamin d(3) and mycophenolate mofetil treatment mediate transplantation tolerance. J Immunol 167:1945–1953
Hakim I, Bar-Shavit Z (2003) Modulation of TNF-α expression in bone marrow macrophages: involvement of vitamin D response element. J Cell Biochem 88:986–998
Hamada H, Garcia-Hernandez Mde L, Reome JB, Misra SK, Strutt TM, McKinstry KK, Cooper AM, Swain SL, Dutton RW (2009) Tc17, a unique subset of CD8 T cells that can protect against lethal influenza challenge. J Immunol 182:3469–3481
Hartmann B, Heine G, Babina M, Steinmeyer A, Zügel U, Radbruch A, Worm M (2011) Targeting the vitamin D receptor inhibits the B cell-dependent allergic immune response. Allergy 66:540–548
Heine G, Anton K, Henz BM, Worm M (2002) 1α,25-dihydroxyvitamin D3 inhibits anti-CD40 plus IL-4-mediated IgE production in vitro. Eur J Immunol 32:3395–3404
Heine G, Niesner U, Chang HD, Steinmeyer A, Zügel U, Zuberbier T, Radbruch A, Worm M (2008) 1,25-dihydroxyvitamin D(3) promotes IL-10 production in human B cells. Eur J Immunol 38:2210–2218
Herold G editor (2008) Innere Medizin. Gerd Herold, Köln
Hewison M (2010) Vitamin D and the immune system: new perspectives on an old theme. Endocrinol Metab Clin North Am 39:365–379
Hewison M, Freeman L, Hughes SV, Evans KN, Bland R, Eliopoulos AG, Kilby MD, Moss PA, Chakraverty R (2003) Differential regulation of vitamin D receptor and its ligand in human monocyte-derived dendritic cells. J Immunol 170:5382–5390
Holick MF (2008) Photobiology and noncalcemic actions of Vitamin D. In: Bilezikian JP, Raisz LG, Martin TJ (eds) Principles of bone biology, 3rd edn. Academic, Amsterdam, pp 800–802
Huber M, Heink S, Grothe H, Guralnik A, Reinhard K, Elflein K, Hünig T, Mittrücker HW, Brüstle A, Kamradt T, Lohoff M (2009) A Th17-like developmental process leads to CD8(+) Tc17 cells with reduced cytotoxic activity. Eur J Immunol 39:1716–1725
Hyppönen E (2005) Vitamin D for the prevention of preeclampsia? A hypothesis. Nutr Rev 63:225–232
Hyppönen E (2010) Vitamin D and increasing incidence of type 1 diabetes-evidence for an association? Diabetes Obes Metab 12:737–743
Hyppönen E, Berry DJ, Wjst M, Power C (2009) Serum 25-hydroxyvitamin D and IgE – a significant but nonlinear relationship. Allergy 64:613–620
Hyppönen E, Läärä E, Reunanen A, Järvelin MR, Virtanen SM (2001) Intake of vitamin D and risk of type 1 diabetes: a birth-cohort study. Lancet 358:1500–1503
Hyppönen E, Sovio U, Wjst M, Patel S, Pekkanen J, Hartikainen AL, Järvelinb MR (2004) Infant vitamin D supplementation and allergic conditions in adulthood: northern Finland birth cohort 1966. Ann N Y Acad Sci 1037:84–95
Iho S, Takahashi T, Kura F, Sugiyama H, Hoshino T (1986) The effect of 1,25-dihydroxyvitamin D_3 on in vitro immunoglobulin production in human B cells. J Immunol 136:4427–4431
Ikeda U, Wakita D, Ohkuri T, Chamoto K, Kitamura H, Iwakura Y, Nishimura T (2010) 1α,25-Dihydroxyvitamin D3 and all-trans retinoic acid synergistically inhibit the differentiation and expansion of Th17 cells. Immunol Lett 134:7–16

Jäger A, Dardalhon V, Sobel RA, Bettelli E, Kuchroo VK (2009) Th1, Th17, and Th9 effector cells induce experimental autoimmune encephalomyelitis with different pathological phenotypes. J Immunol 183:7169–7177

Jeffery LE, Burke F, Mura M, Zheng Y, Qureshi OS, Hewison M, Walker LS, Lammas DA, Raza K, Sansom DM (2009) 1,25-Dihydroxyvitamin D3 and IL-2 combine to inhibit T cell production of inflammatory cytokines and promote development of regulatory T cells expressing CTLA-4 and FoxP3. J Immunol 183:5458–5467

Jirapongsananuruk O, Melamed I, Leung DY (2000) Additive immunosuppressive effects of 1,25-dihydroxyvitamin D3 and corticosteroids on TH1, but not TH2, responses. J Allergy Clin Immunol 106:981–985

Kelso A, Troutt AB, Maraskovsky E, Gough NM, Morris L, Pech MH, Thomson JA (1991) Heterogeneity in lymphokine profiles of CD4+ and CD8+ T cells and clones activated in vivo and in vitro. Immunol Rev 123:85–114

Kemeny DM, Noble A, Holmes BJ, Diaz-Sanchez D (1994) Immune regulation: a new role for the CD8+ T cell. Immunol Today 15:107–110

Klaushofer K (2010) Arznei und Vernunft: Osteoporose Knochenbruch-Krankheit, 3. Auflage. Arznei & Vernunft, Vienna

Kreutz M, Andreesen R (1990) Induction of human monocyte to macrophage maturation in vitro by 1,25-dihydroxyvitamin D3. Blood 76:2457–2461

Kriegel MA, Manson JE, Costenbader KH (2011) Does vitamin d affect risk of developing autoimmune disease?: a systematic review. Semin Arthritis Rheum 40:512–531

Krutzik SR, Hewison M, Liu PT, Robles JA, Stenger S, Adams JS, Modlin RL (2008) IL-15 links TLR2/1-induced macrophage differentiation to the vitamin D-dependent antimicrobial pathway. J Immunol 181:7115–7120

Larsson P, Mattsson L, Klareskog L, Johnsson C (1998) A vitamin D analogue (MC 1288) has immunomodulatory properties and suppresses collagen-induced arthritis (CIA) without causing hypercalcaemia. Clin Exp Immunol 114:277–283

Le Gros G, Erard F (1994) Non-cytotoxic, IL-4, IL-S, IL-1O producing CD8+ T cells: their activation and effector functions. Curr Opin Immunol 6:453–457

Lemire JM, Adams JS, Sakai R, Jordan SC (1984) 1 α,25-dihydroxyvitamin D3 suppresses proliferation and immunoglobulin production by normal human peripheral blood mononuclear cells. J Clin Invest 74:657–661

Lin J, Liu J, Davies ML, Weiqian C (2016) Serum vitamin D levels and rheumatoid arthritis disease activity: review and meta-analysis. PLoS One 11(1):e0146351

Litonjua AA (2009) Childhood asthma may be a consequence of vitamin D deficiency. Curr Opin Allergy Clin Immunol 9:202-207

Liu PT, Stenger S, Li H, Wenzel L, Tan BH, Krutzik SR, Ochoa MT, Schauber J, Wu K, Meinken C, Kamen DL, Wagner M, Bals R, Steinmeyer A, Zügel U, Gallo RL, Eisenberg D, Hewison M, Hollis BW, Adams JS, Bloom BR, Modlin RL (2006) Toll-like receptor triggering of a vitamin D-mediated human antimicrobial response. Science 311:1770–1773

Liu PT, Schenk M, Walker VP, Dempsey PW, Kanchanapoomi M, Wheelwright M, Vazirnia A, Zhang X, Steinmeyer A, Zügel U, Hollis BW, Cheng G, Modlin RL (2009) Convergence of IL-1ß and VDR activation pathways in human TLR2/1-induced antimicrobial responses. PLoS One 4:e5810

Mahon BD, Wittke A, Weaver V, Cantorna MT (2003) The targets of vitamin D depend on the differentiation and activation status of CD4 positive T cells. J Cell Biochem 89:922–932

Mathieu C, Badenhoop K (2005) Vitamin D and type 1 diabetes mellitus: state of the art. Trends Endocrinol Metab 16:261–266

Merlino LA, Curtis J, Mikuls TR, Cerhan JR, Criswell LA, Saag KG (2004) Vitamin D intake is inversely associated with rheumatoid arthritis: results from the Iowa Women's Health Study. Arthritis Rheum 50:72–77

Miller J, Gallo RL (2010) Vitamin D and innate immunity. Dermatol Ther 23:13–22

Mora JR, Iwata M, von Andrian UH (2008) Vitamin effects on the immune system: vitamins A and D take centre stage. Nat Rev Immunol 8:685–698

Mosmann TR, Coffman RL (1989) TH1 and TH2 cells: different patterns of lymphokine secretion lead to different functional properties. Annu Rev Immunol 7:145–173

Müller K, Bendtzen K (1992) Inhibition of human T lymphocyte proliferation and cytokine production by 1,25-dihydroxyvitamin D3. Differential effects on CD45RA+ and CD45R0+ cells. Autoimmunity 14:37–43

Müller K, Bendtzen K (1996) 1,25-Dihydroxyvitamin D3 as a natural regulator of human immune functions. J Investig Dermatol Symp Proc 1(1):68–71

Müller K, Heilmann C, Poulsen LK, Barington T, Bendtzen K (1991) The role of monocytes and T cells in 1,25-dihydroxyvitamin D3 mediated inhibition of B cell function in vitro. Immunopharmacology 21:121–128

Müller K, Odum N, Bendtzen K (1993) 1,25-dihydroxyvitamin D3 selectively reduces interleukin-2 levels and proliferation of human T cell lines in vitro. Immunol Lett 35:177–182

Nielen MM, van Schaardenburg D, Lems WF, van de Stadt RJ, de Koning MH, Reesink HW, Habibuw MR, van der Horst-Bruinsma IE, Twisk JW, Dijkmans BA (2006) Vitamin D deficiency does not increase the risk of rheumatoid arthritis: comment on the article by Merlino et al. 121. Arthritis Rheum 54:3719–3720

Noble A, Macary PA, Kemeny DM (1995) IFN-γ and IL-4 regulate the growth and differentiation of CD81 T cells into subpopulations with distinct cytokine profiles. J Immunol 155:2928–2937

Norman AW (2008) 1α,25(OH)$_2$ vitamin D$_3$. In: Bilezikian JP, Raisz LG, Martin TJ (eds) Principles of bone biology, 3rd edn. Academic, Amsterdam, pp 749–771

O'Garra A, Murphy K (1994) Role of cytokines in determining T-lymphocyte function. Curr Opin Immunol 6:458–466

Palmer MT, Lee YK, Maynard CL, Oliver JR, Bikle DD, Jetten AM, Weaver CT (2011) Lineage-specific effects of 1,25-dihydroxyvitamin D(3) on the development of effector CD4 T cells. J Biol Chem 286:997–1004

Pedersen LB, Nashold FE, Spach KM, Hayes CE (2007) 1,25-dihydroxyvitamin D3 reverses experimental autoimmune encephalomyelitis by inhibiting chemokine synthesis and monocyte trafficking. J Neurosci Res 85:2480–2490

Penna G, Amuchastegui S, Giarratana N, Daniel KC, Vulcano M, Sozzani S, Adorini L (2007) 1,25-Dihydroxyvitamin D3 selectively modulates tolerogenic properties in myeloid but not plasmacytoid dendritic cells. J Immunol 178:145–153

Penna G, Roncari A, Amuchastegui S, Daniel KC, Berti E, Colonna M, Adorini L (2005) Expression of the inhibitory receptor ILT3 on dendritic cells is dispensable for induction of CD4+Foxp3+ regulatory T cells by 1,25-dihydroxyvitamin D3. Blood 106:3490–3497

Peterlik M, Cross HS (2006) Dysfunction of the vitamin D endocrine system as common cause for multiple malignant and other chronic diseases. Anticancer Res 26:2581–2588

Peterlik M, Cross HS (2005) Vitamin D and calcium deficits predispose for multiple chronic diseases. Eur J Clin Invest 35(5):290–304

Peterlik M, Boonen S, Cross HS, Lamberg-Allardt C (2009) Vitamin D and calcium insufficiency-related chronic diseases: an emerging world-wide public health problem. Int J Environ Res Public Health 6:2585–2607

Pichler J, Gerstmayr M, Szépfalusi Z, Urbanek R, Peterlik M, Willheim M (2002) 1α,25(OH)2D3 inhibits not only Th1 but also Th2 differentiation in human cord blood T cells. Pediatr Res 52:12–18

Pierrot-Deseilligny C, Souberbielle JC (2010) Is hypovitaminosis D one of the environmental risk factors for multiple sclerosis? Brain 133:1869–1888

Provvedini DM, Manolagas SC (1989) 1α,25-dihydroxyvitamin D3 receptor distribution and effects in subpopulations of normal human T lymphocytes. J Clin Endocrinol Metab 68:774–779

Provvedini DM, Tsoukas CD, Deftos LJ, Manolagas SC (1983) 1,25-dihydroxyvitamin D3 receptors in human leukocytes. Science 221:1181–1183

Provvedini DM, Tsoukas CD, Deftos LJ, Manolagas SC (1986) 1 α,25-Dihydroxyvitamin D3-binding macromolecules in human B lymphocytes: effects on immunoglobulin production. J Immunol 136:2734–2740

Rahman S, Rehn A, Rahman J, Andersson J, Svensson M, Brighenti S (2015) Pulmonary tuberculosis patients with a vitamin D deficiency demonstrate low local expression of the antimicrobial peptide LL-37 but enhanced FoxP3^{+} regulatory T cells and IgG-secreting cells. Clin Immunol 156:85–97

Rigby WF, Stacy T, Fanger MW (1984) Inhibition of T lymphocyte mitogenesis by 1,25-dihydroxyvitamin D3 (calcitriol). J Clin Invest 74:1451–1455

Robinson DS (2009) Regulatory T cells and asthma. Clin Exp Allergy 39:1314–1323

Rockett KA, Brookes R, Udalova I, Vidal V, Hill AV, Kwiatkowski D (1998) 1,25-Dihydroxyvitamin D3 induces nitric oxide synthase and suppresses growth of *Mycobacterium tuberculosis* in a human macrophage-like cell line. Infect Immun 66:5314–5321

Sadeghi K, Wessner B, Laggner U, Ploder M, Tamandl D, Friedl J, Zügel U, Steinmeyer A, Pollak A, Roth E, Boltz-Nitulescu G, Spittler A (2006) Vitamin D3 down-regulates monocyte TLR expression and triggers hyporesponsiveness to pathogen-associated molecular patterns. Eur J Immunol 36:361–370

Salgame P, Abrams JS, Clayberger C, Goldstein H, Convit J, Modlin RL, Bloom BR (1991) Differing lymphokine profiles of functional subsets of human CD4 and CD8 T cell clones. Science 254:279–282

Seder RA, Boulay JL, Finkelman F, Barbier S, Ben-Sasson SZ, Le Gros G, Paul WE (1992) CD8+T Cells can be primed in vitro to produce IL-4. J Immunol 148:1652–1656

Seder RA, Paul WE (1994) Acquisition of lymphokine-producing phenotype by CD4+ T cells. Annu Rev Immunol 12:635–673

Sharief S, Jariwala S, Kumar J, Muntner P, Melamed ML (2011) Vitamin D levels and food and environmental allergies in the United States: results from the National Health and Nutrition Examination Survey 2005–2006. J Allergy Clin Immunol 127:1195–1202

Sloka S, Zhornitsky S, Silva C, Metz LM, Yong VW (2015) 1,25-Dihydroxyvitamin D3 protects against immune-mediated killing of neurons in culture and in experimental autoimmune encephalomyelitis. PLoS One 10(12):e0144084

Smith DJ, Iqbal J, Purewal A, Hamblin AS, Edington N (1998) In vitro reactivation of latent equid herpesvirus-1 from CD5/CD8 leukocytes indirectly by IL-2 or chorionic gonadotrophin. J Gen Virol 79:2997-3004

Smith KA, Popmihajlov Z (2008) The quantal theory of immunity and the interleukin-2-dependent negative feedback regulation of the immune response. Immunol Rev 224:124–140

Smolders J, Damoiseaux J, Menheere P, Hupperts R (2008) Vitamin D as an immune modulator in multiple sclerosis, a review. J Neuroimmunol 194:7–17

Smolders J, Peelen E, Thewissen M, Cohen Tervaert JW, Menheere P, Hupperts R, Damoiseaux J (2010) Safety and T cell modulating effects of high dose vitamin D_3 supplementation in multiple sclerosis. PLoS One 5:e15235

Smolders J, Thewissen M, Peelen E, Menheere P, Tervaert JW, Damoiseaux J, Hupperts R (2009) Vitamin D status is positively correlated with regulatory T cell function in patients with multiple sclerosis. PLoS One 4:e6635

Spittler A, Willheim M, Leutmezer F, Ohler R, Krugluger W, Reissner C, Lucas T, Brodowicz T, Roth E, Boltz-Nitulescu G (1997) Effects of 1 α,25-dihydroxyvitamin D3 and cytokines on the expression of MHC antigens, complement receptors and other antigens on human blood monocytes and U937 cells: role in cell differentiation, activation and phagocytosis. Immunology 90:286–293

Staeva-Vieira TP, Freedman LP (2002) 1,25-dihydroxyvitamin D3 inhibits IFN-γ and IL-4 levels during in vitro polarization of primary murine CD4+ T cells. J Immunol 168:1181–1189

Stene LC, Joner G (2003) Use of cod liver oil during the first year of life is associated with lower risk of childhood-onset type 1 diabetes: a large, population-based, case–control study. Am J Clin Nutr 78:1128–1134

Szodoray P, Nakken B, Gaal J, Jonsson R, Szegedi A, Zold E, Szegedi G, Brun JG, Gesztelyi R, Zeher M, Bodolay E (2008) The complex role of vitamin D in autoimmune diseases. Scand J Immunol 68:261–269

Takeuchi A, Reddy GS, Kobayashi T, Okano T, Park J, Sharma S (1998) Nuclear factor of activated T cells (NFAT) as a molecular target for 1α,25-dihydroxyvitamin D3-mediated effects. J Immunol 160:209–218

Takiishi T, Gysemans C, Bouillon R, Mathieu C (2010) Vitamin D and diabetes. Endocrinol Metab Clin North Am 39:419–446

Thien R, Baier K, Pietschmann P, Peterlik M, Willheim M (2005) Interactions of 1 α,25-dihydroxyvitamin D3 with IL-12 and IL-4 on cytokine expression of human T lymphocytes. J Allergy Clin Immunol 116:683–699

Treiber G, Prietl B, Fröhlich-Reiterer E, Lechner E, Ribitsch A, Fritsch M, Rami-Merhar B, Steigleder-Schweiger C, Graninger W, Borkenstein M, Pieber TR (2015) Cholecalciferol supplementation improves suppressive capacity of regulatory T-cells in young patients with new-onset type 1 diabetes mellitus a randomized clinical trial. Clin Immunol 161:217–224

Tsoukas CD, Provvedini DM, Manolagas SC (1984) 1,25-dihydroxyvitamin D3: a novel immunoregulatory hormone. Science 224:1438–1440

Ulitsky A, Ananthakrishnan AN, Naik A, Skaros S, Zadvornova Y, Binion DG, Issa M (2011) Vitamin D deficiency in patients with inflammatory bowel disease: association with disease activity and quality of life. JPEN J Parenter Enteral Nutr 35:308–316

Unger WW, Laban S, Kleijwegt FS, van der Slik AR, Roep BO (2009) Induction of Treg by monocyte-derived DC modulated by vitamin D3 or dexamethasone: differential role for PD-L1. Eur J Immunol 39:3147–3159

Urry Z, Xystrakis E, Richards DF, McDonald J, Sattar Z, Cousins DJ, Corrigan CJ, Hickman E, Brown Z, Hawrylowicz CM (2009) Ligation of TLR9 induced on human IL-10-secreting Tregs by 1α,25-dihydroxyvitamin D3 abrogates regulatory function. J Clin Invest 119:387–398

van Etten E, Mathieu CJ (2005) Immunoregulation by 1,25-dihydroxyvitamin D3: basic concepts. J Steroid Biochem 97:93–101

Veldhoen M (2009) The role of T helper subsets in autoimmunity and allergy. Curr Opin Immunol 21:606-611

Veldhoen M, Uyttenhove C, van Snick J, Helmby H, Westendorf A, Buer J, Martin B, Wilhelm C, Stockinger B (2008) Transforming growth factor-beta 'reprograms' the differentiation of T helper 2 cells and promotes an interleukin 9-producing subset. Nat Immunol 12:1341–1346

Vijayendra Chary A, Hemalatha R, Seshacharyulu M, Murali MV, Jayaprakash D, Kumar BD (2015) Vitamin D deficiency in pregnant women impairs regulatory T cell function. J Steroid Biochem 147:48–55

Wang J, Lv S, Chen G, Gao C, He J, Zhong H, Xu Y (2015) Meta-analysis of the association between vitamin D and autoimmune thyroid disease. Nutrients 7:2485–2498

Willheim M, Thien R, Schrattbauer K, Bajna E, Holub M, Gruber R, Baier K, Pietschmann P, Reinisch W, Scheiner O, Peterlik M (1999) Regulatory effects of 1α,25-dihydroxyvitamin D3 on the cytokine production of human peripheral blood lymphocytes. J Clin Endocrinol Metab 84:3739–3744

Wong MT, Ye JJ, Alonso MN, Landrigan A, Cheung RK, Engleman E, Utz PJ (2010) Regulation of human Th9 differentiation by type I interferons and IL-21. Immunol Cell Biol 88:624–631

Workman CJ, Szymczak-Workman AL, Collison LW, Pillai MR, Vignali DA (2009) The development and function of regulatory T cells. Cell Mol Life Sci 66:2603

Xystrakis E, Kusumakar S, Boswell S, Peek E, Urry Z, Richards DF, Adikibi T, Pridgeon C, Dallman M, Loke TK, Robinson DS, Barrat FJ, O'Garra A, Lavender P, Lee TH, Corrigan C, Hawrylowicz CM (2006) Reversing the defective induction of IL-10-secreting regulatory T cells in glucocorticoid-resistant asthma patients. J Clin Invest 116:146–155

Yu S, Cantorna MT (2008) The vitamin D receptor is required for iNKT cell development. Proc Natl Acad Sci U S A 105:5207–5212

Yu S, Cantorna MT (2011) Epigenetic reduction in invariant NKT cells following in utero vitamin D deficiency in mice. J Immunol 186:1384–1390

Zarrabeitia MT, Riancho JA, Amado JA, Olmos JM, Gonzalez-Macias J (1992) Effect of calcitriol on the secretion of prostaglandin E2, interleukin 1, and tumor necrosis factor alpha by human monocytes. Bone 13:185–189

Osteoimmunological Aspects of Biomechanics

5

Katharina Kerschan-Schindl and Gerold Ebenbichler

Different endogenous and exogenous factors which interfere with bone health have been identified. Among these, physical activity that relates to regular intermittent mechanical bone loading seems to be one of the major factors controlling bone mass and the prevention of osteoporotic fractures. Moreover, an interaction between bone homeostasis and the immune system which may be modified by regular physical activity exists. Bone and immune cells share a common site of origin, the bone marrow. They are supposed to influence each other not only during maturation; osteoclasts and immune cells have a number of regulatory molecules in common including cytokines, receptors, signalling molecules, and transcription factors, which influence each other.

The aim of this chapter is to review the relationship between muscle activity, mechanical bone loading, and bone remodelling, thereby elucidating the immunological communication pathways intended to modulate osteoblast and osteoclast activities. This chapter will put a major focus on how mechanical bone loading induces bone cell activity intended to either build or abolish bone tissue (mechanoinduction; Fig. 5.1).

5.1 Mechanical Loading and Bone Formation

Galileo has already documented that the shape of the bone would be related to loading. Later, in the nineteenth century, Julius Wolff's work developed a more educated understanding of bone mechanobiology that considered functional

K. Kerschan-Schindl, MD (✉) • G. Ebenbichler, MD
Department of Physical Medicine and Rehabilitation, Medical University of Vienna, Währinger Gürtel 18-20, Vienna 1090, Austria
e-mail: katharina.kerschan@meduniwien.ac.at; katharina.kerschan-schindl@meduniwien.ac.at; gerold.ebenbichler@meduniwien.ac.at

P. Pietschmann (ed.), *Principles of Osteoimmunology*,
DOI 10.1007/978-3-319-34238-2_5

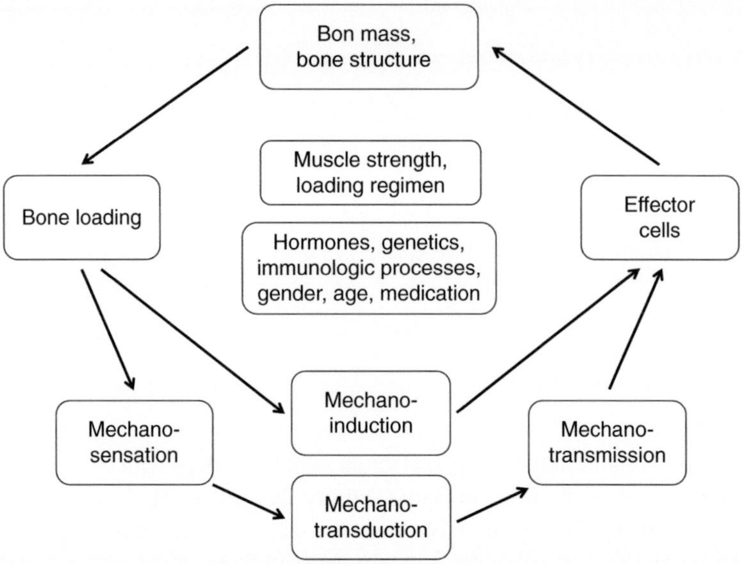

Fig. 5.1 The mechanostat model pictures of bone loading-induced mechanoinduction leading to adaptations in strain-dependent bone mass and bone structure. Note that mechanoinduction is composed of three phases, mechanosensation, mechanotransduction, and mechanotransmission. Regarding mechano-responsiveness, systemic and local factors have a modulatory effect

adaptation of bone structure and mass at the tissue level. Nowadays we know that it is the cells of the musculoskeletal system that are responsible for bone modelling and remodelling.

The peak bone mass in early adulthood is highly dependent on genetic factors. Nevertheless, lifestyle factors like calcium intake and physical activity influence peak bone mass and bone remodelling. While both disuse and inactivity reduce bone mass (Zerwekh et al. 1998), strenuous exercise increases it (Courteix et al. 1998). The earlier a child starts with physical activity, the greater is the benefit on bone mass (Kontulainenen et al. 2002). Such modulations of bone formation/loss are not restricted to young people but also hold true for the elderly population. Both elderly males and females who performed regular physical exercise demonstrated improvements in bone mineral density (Bemben and Bemben 2010; Kemmler et al. 2010). There is increasing evidence that the type of exercise facilitates bone response. Relevant factors would be magnitude of load, number of repetitions/loading cycles, and the rate of strain (for review, see Rubin et al. 2006).

Bone remodelling is a balanced process that is mainly determined by its mechanical demands. It is estimated that approximately one quarter of cancellous bone and 3–4 % of cortical bone are renewed every year. People who lead a sedentary life deprive their bones from sufficient mechanical stimulation. Because of an uncoupling of bone resorption and bone formation, their bone mass and structure will loose towards an osteopenic/osteoporotic fragile bone. In contrast, an active lifestyle that regularly overloads bone increases bone mass and structure and results in strong fracture-resistant bones. Bone strength needs to clearly exceed the mechanical strains to which the bone is exposed in everyday life (Currey 1979). In humans, the limb bones can usually withstand loads that deform them by three to four times the amount they are deformed by peak physiological activity before they fracture (Biewener et al. 1993).

5.1.1 Mechanotransduction

Bone remodelling is sensitive to the magnitude of physical strain, the distribution of the load, and the number of loading cycles (see Rubin et al. 2006). During vigorous activities, for instance, peak strain magnitudes were shown to range from 2000 to 3500 microstrain in animal experiments (Rubin and Lanyon 1984) and to reach up to 2000 microstrain in humans (Burr et al. 1996). In everyday life, very low strains less than 10 microstrain seem to prevail, whereas large strains above 1000 microstrain occur only occasionally (Fritton et al. 2000). Peak strains of about 2000 microstrain may be induced in vigorous locomotion and seem to be strong enough to cause microdamage to bone tissues, thereby stimulating osteoclasts and osteoblasts to remove and replace damaged material (Burr et al. 1985). Microcracking induces pro-apoptotic molecules in osteocytes that are located in close vicinity to the cracks, whereas those osteocytes located slightly more distant from the cracks produce anti-apoptotic molecules. This suggests that bone tissue adjacent to a crack is at risk of apoptotic degeneration, whereas the surrounding tissue induces regeneration (Rochefort et al. 2010). Interestingly, even bone loading as low as 5 microstrain administered at high frequencies of 30 Hz and 20 min daily revealed a significant increase in bone mineral density in sheep hindlimbs (Rubin et al. 2001). Thus, not only high but also low strains seem to be crucial to the maintenance of bone structure and strength.

There is evidence that most cells in the body are able to sense their mechanical environment. Mechanical forces are detected by osteoprogenitor cells which are present not only in the bone marrow but also in the soft mesenchymal tissues exposed to mechanical strain. Dependent on the magnitude of mechanical stress, osteoprogenitors differentiate or transdifferentiate into osteoblast-like cells that express characteristic proteins and form bone matrix (Aubin 1998). However, the most important targets of skeletal loading seem to be the osteocytes (Cowin et al. 1991; Bacabac et al. 2004). They are distributed three-dimensionally throughout the bone and connected to each other via their cytoplasmatic processes within canaliculi in the trabecular and cortical bone. The mechanisms involved in cell activation as a result of mechanical loading are not completely understood yet.

Physical activity loads the musculoskeletal system, thereby causing the processes intended to convert mechanical forces into biochemical signals. Such processes are named *mechanotransduction* (Rubin et al. 2006). Bone loading generates shifts of interstitial fluid out of regions with high compressive strains and shifts back when the load is removed (Burger and Klein-Nulend 1999). Mechanic transduction mediates the release of various pathways that mediate the bone remodelling process via mutually interacting osteoblasts and osteoclasts (*mechanosensing*). Different mechanisms of mechanotransduction have been described: (1) deformation of the hard tissue with strain across the cell substrate, (2) pressure within the intramedullary cavity and within the cortices with transient pressure waves, (3) shear forces through canaliculi which cause drag over cells, and (4) dynamic electric fields as interstitial fluid flows past charged bone crystals (Fig. 5.2).

The mechanosensitivity of osteocytes and, thus, the activation of osteoblasts were supposed to depend on the strain-energy density (SED) rate (Huiskes et al. 2000; Webster et al. 2015). The shear stress induced by interstitial fluid shifts is supposed to be in the order of 0.8–3 Pa and may reach peak pressures up to 3 MPa (Weinbaum et al. 1994; Zhang et al. 1998). Additionally, high-frequency (>30 Hz) low-magnitude loads (<1 MPa) were efficient to elicit cellular response in the bone (Xie et al. 2006). Only intermittent compressive forces but not continuous hydrostatic pressures applied to the bone were shown to increase bone formation (Klein-Nulend et al. 1987). In most of the everyday life activities like walking and performing sports, bone loading is cyclic rather than static.

5.1.2 Sensing Mechanisms and Signalling Pathways Activated by Bone Loading

During bone loading, different pathways are known to be activated by mechanical signals (Fig. 5.3). Cells that convert the mechanical stimulus into a biomechanical signal include osteoblasts, osteoclasts, lining cells on the bone surfaces, osteocytes within the calcified matrix, and mesenchymal precursors within the bone marrow (Judex and Rubin 2010).

In response to mechanical forces, a total of four different osteocyte-mediated sensing and *signalling mechanisms have been identified*. These are (1) the activation of ion channels, (2) the production of integrins, (3) the activation of gap junctions and hemichannels, and (4) an unclear role of primary cilia (Fig. 5.4). For review, also see Klein-Nulend and co-workers (2015).

Ad 1) Three different *ion channels*, the gadolinium-sensitive cation channels (Rawlinson et al. 1996), the L-type voltage-dependent channels (Miyauchi et al. 2000), and the

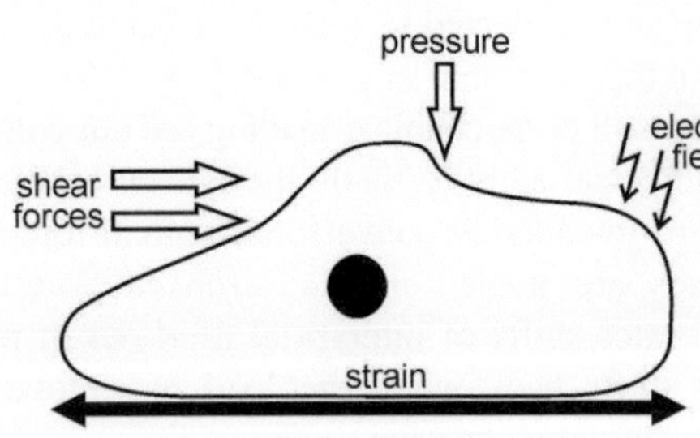

Fig. 5.2 Skeletal loading generates strain across the cell, pressure in the intramedullary cavity and within the cortices, shear forces through canaliculi, and dynamic electric fields (Modified from Rubin et al. 2006)

Fig. 5.3 Three different pathways by which mechanical signals may be sensed by cells within a bone (Modified from Judex and Rubin 2010)

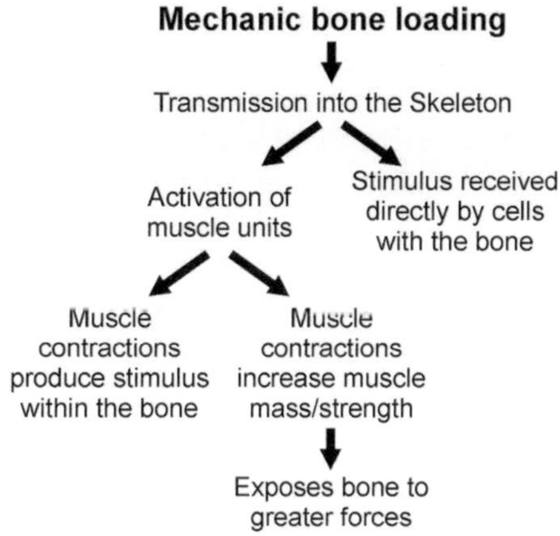

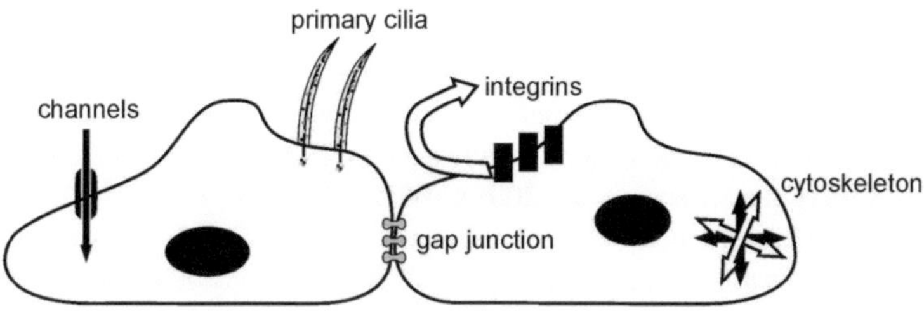

Fig. 5.4 Intracellular signals are activated by different mechanosensors

α, ß, and γ units of epithelium sodium channels (ENaC; Mikuni-Takagaki 1999), have been described to be strain responsive. Additionally, three potassium channels were identified in MLO-Y4 cells, an osteocyte-like cell line (Gu et al. 2001). Their role in mechanotransduction is not yet defined.

Ad 2) A variety of intracellular signalling pathways are activated in osteocytes in an *integrin*-dependent manner. Integrins are membrane-spanning proteins that couple the cell to its extracellular environment. Functional integrins are heterogenous dimers made of α and ß subunits. Ligation of α and ß3 integrin enhances calcium influx in mechanically stretched osteocytes (Miyauchi et al. 2006). Osteocyte apoptosis is prevented by integrin engagement leading to focal adhesion kinase (FAK) activation, Src kinase activity, and phosphorylation of the adaptor protein Shc and, thus, ERK 1/2 activation as well as an interplay with actin filaments and microtubules (Plotkin et al. 2005). Cytoskeleton-mediated signalling in response to fluid flow could occur due to drag forces on the pericellular matrix (You et al. 2008) as tethering elements, presumably composed of integrins connected to the canalicular wall and pericellular matrix to the osteocyte cytoskeleton. Expression of E11, an osteocyte-selective protein, is increased by mechanical load in regions of potential bone remodelling, suggesting dendrite elongation (Zhang et al. 2006).

Ad 3) Osteocytes are connected to one another and to surface osteoblasts via *gap junctions* (Yellowley et al. 2000). Gap junctions are formed when connexins (Cx), especially protein Cx43, join to form an intercellular channel allowing the direct exchange of small molecules; Cx also form *hemichannels* that function independent of gap junctions. Studies are conflicting, but fluid flow-induced release of ATP and PGE_2 may be mediated in part by hemichannels.

Ad 4) Independent of any channels, *primary cilia* may also play a role in the anabolic signalling processes in bone cells (Malone et al. 2007). They are supposed to deflect during fluid flow and possibly interact with extracellular matrix proteins, and thus, integrins on the primary cilium could translate deformations of the extracellular matrix into intracellular signals, thereby amplifying the mechanical stimuli.

5.1.3 Cytokine Response as a Signalling Pathway Activated by Bone Loading

These multiple intracellular signalling pathways are activated after force application. Under physiological mechanical stimuli, osteocytes prevent bone resorption by changing the receptor activator of nuclear factor kappa-B ligand/osteoprotegerin (RANKL/OPG) ratio (You et al. 2008). Wnt ("wingless in Drosophila") gene expression is increased in osteocytes subjected to fluid flow (Santos et al. 2009). This regulatory pathway mediated by load-induced bone adaptation seems to depend on the low-density lipoprotein receptor-related protein (LRP 5/6) and frizzled transmembrane proteins and the gene ß-catenin and the protein sclerostin, which inhibits bone formation by blocking Wnt signalling (Robinson et al. 2006).

Osteocytes exposed to fluid flow also express osteopontin and PGE_2 – both of them are important in the process of bone remodelling (Mc Garry et al. 2005). The activity-induced inhibition of osteoclast formation is at least partially dependent on the activation of the nitric oxide (NO) pathway in the osteocyte (Tan et al. 2007; Vezeridis et al. 2006). It is noteworthy that the intracellular signalling pathways activated after force application effecting on osteoprogenitor differentiation and osteocyte functions still remain limited.

Whereas direct loading of the bone produces cytokines involved in bone remodelling, systemic factors may further contribute to this process. Muscle activity directly stimulates mechanically induced bone remodelling through traction and bending forces. The muscle has been identified as a major source of cytokine production (myokines) that may, via systemic effects, facilitate anabolic bone metabolism.

5.2 Muscular Exercise and Bone Integrity

5.2.1 Activity, Inflammation, and Bone Loss

A relationship between physical activity per se and bone formation has been suggested several centuries ago. Muscle activity seems to be especially beneficial on bone integrity in patients who have a risk of increased bone loss due to sterile, low-grade chronic systemic inflammations. Systemic low-grade inflammation is defined as a two- to fourfold elevation of circulating pro- and anti-inflammatory cytokines, naturally occurring cytokine antagonists, and acute-phase proteins, as well as minor increases in counts of neutrophiles and natural killer cells (Brüünsgaard and Pedersen 2003). Both the lack of physical activity and low-grade inflammation were linked to various chronic diseases like adiposity, type 2 diabetes, and cardiovascular diseases (Booth et al. 2002; Hu et al. 2004; Orsini et al. 2008). For example, the development of type 2 diabetes and insulin resistance is closely correlated with immune cell infiltration and inflammation in white adipose tissue (Hotamisligil 2006). Moreover, regular moderate exercise may reduce systemic inflammation (Gleeson 2007).

The impact of inflammatory diseases on bone metabolism is described elsewhere in detail. In brief, the coupling of osteoclasts and osteoblasts is severely disturbed. Both local and systemic inflammation release cytokines like TNF-α that may directly activate osteoclast differentiation and activity and thereby leading to bone destruction. Interleukin-6 has a special role; it positively influences osteoclast differentiation by inducing the expression of RANKL on osteoblasts and also directly suppresses osteoclast differentiation via an inhibition of the RANK/RANKL pathway (Yoshitake et al. 2008).

Although the systemic mediators of these beneficial exercise effects are unclear, several candidate mechanisms have been identified. Regular muscular activity increases epinephrine, cortisol, growth hormone, and other factors that have immune modulatory effects; all these may influence bone metabolism (Goodman et al. 2015; Brotto and Bonewald 2015). The detailed interactions are not subject of the chapter and will be described elsewhere. However, cytokine secretion intended to reduce inflammation seems also likely to directly improve the coupling of osteoclasts and osteoblasts in a way that bone integrity will be improved.

5.2.2 Anti-inflammatory Effect of Exercise

Recent findings have demonstrated that muscle training induces or increases the systemic level of a number of cytokines with anti-inflammatory properties which differ significantly from that in relation to an acute inflammation. Following an acute inflammation the cytokine cascade consists of the following candidates: TNF-α, IL-1β, IL-6, IL-1 receptor antagonist (IL-1ra), soluble TNF-α receptors (sTNF-R), and IL-10. IL-1ra inhibits IL-1 signal transduction, and sTNF-R represents the naturally occurring inhibitors of TNF-α (Pedersen and Febbraio 2008). As opposed to an acute inflammation, acute bouts of exercise that are not as highly strenuous as marathon running do not increase TNF- α and IL-1 β levels, and IL-6 is usually the first cytokine present in the circulation (Pedersen 2013) (Fig. 5.5). The basal plasma IL-6 concentrations may increase up to 100-fold and depend on the duration and intensity of exercise (Fischer 2006). Exercise-induced IL-6 does seem to be not only released from contracting muscles but also from sources outside the muscle that were stressed during exercise (Fischer et al. 2004). Using the microdialysis method, the Achilles tendons revealed high concentrations of IL-6 following running

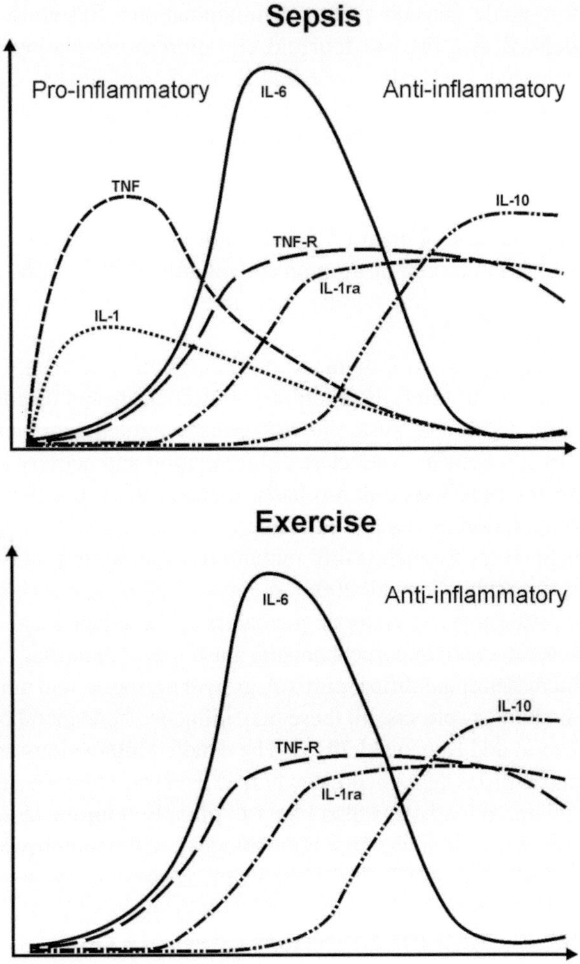

Fig. 5.5 In sepsis, the cytokine cascade within the first few hours consists of TNF-α, IL-1ß, IL-6, IL-1ra, TNF-R, and IL-10. Exercise leads to a different cytokine response: It does not include TNF-α and IL-1ß but shows a marked increase in IL-6, followed by an increase in IL-1ra, TNF-R, and IL-10 (Modified from Petersen and Pedersen 2005)

exercise (Langberg et al. 2002). Furthermore, chronic effects of exercise may differ from those of acute exercise as plasma levels of IL-6 remained unchanged during 1 year of training intervention (Rokling-Andersen et al. 2007). During and following exercise, the high circulating levels of IL-6 are followed by an increase in IL-1ra and IL-10 and sTNF-R. It was further shown that IL-1ra and IL-10 can be induced by IL-6 (Steensberg et al. 2003).

Until now several proteins secreted by skeletal muscle cells have been identified, with the potential to act as hormones, either locally within the muscle tissue or by targeting other distant organs, as reviewed by Pedersen (2013). Such proteins have been termed myokines within the context of skeletal muscle physiology, although they are not exclusively secreted by muscle cells. Muscle contraction may induce acute release of myokines like calprotectin (Mortensen et al. 2008), IL-15, IL-6, and IL-7 which may pertain to homeostatic control under various physiological conditions

(Haugen et al. 2010). Furthermore, several further interleukins have been identified that may be released by exercising muscles and have been reviewed recently (Peake et al. 2015).

Both type 1 and 2 muscle fibres seem to express myokine IL-6, with some studies suggesting that IL-6 expression was more prominent in type 1 fibres (Plomgaard et al. 2005), whereas others demonstrated a higher IL-6 expression in type 2 fibres (Hiscock et al. 2004). IL-6 from the muscle was suggested to be a signal indicating that muscle glycogen stores are reaching critically low levels and that the active muscles' reliance on blood glucose as a source of energy is on the increase (Pedersen 2013). As the increase of IL-6 is followed by an increased level of the anti-inflammatory IL-10 and IL-1ra, it has repeatedly been suggested that IL-6 produced by muscles could be critically involved in mediating the anti-inflammatory environment (Benatti and Pedersen 2015) which subsequently acts both locally and systemically.

During non-strenuous exercise, IL-8 seems to play a role in exercise-induced angiogenesis. One study that measured the arteriovenous concentration difference across nonexhausting concentric bicycle and knee extension exercises observed a transient net release of IL-8 with its peak being between 6 and 9 h after exercise that did not lead to increased plasma IL-8 levels (Akerstrom et al. 2005). These findings seem to contrast those of a more recent study that observed a 32 % increase of IL-8 30 min after treadmill running in both endurance-trained and sedentary men (Landers-Ramos et al. 2014). In another study, healthy male subjects performed 45 min of downhill running. At a total of 3 and 24 h after exercise, up-regulation of IL-6, IL-8, and COX-2 mRNA was significantly increased as compared to baseline, whereas pro-inflammatory cytokines like TNF-α or IL-1 β remained more or less unchanged (Buford et al. 2009). As muscle soreness was significantly correlated with IL-8 at 24 h after exercise, the authors suggested that IL-8 transcription would play a role in inhibiting postexercise muscle soreness through the regulation of angiogenesis (Buford et al. 2009).

IL-15 has recently been discovered as a growth factor that acts independently from IGF-1 and induces an accumulation of myosin heavy-chain proteins in differentiated myotubes (Furmanczyk and Quinn 2003). In contrast to IGF-1, the muscle hypertrophic effect of IL-15 is independent from proliferation or differentiation of skeletal myoblasts (Quinn et al. 2005). The regulatory role of muscle contraction in regard to IL-15 remains unclear. Whereas in human-training studies, IL-15 levels were unchanged after several hours of running (Nieman et al. 2003; Ostrowski et al. 1998), others found clearly increased IL-15 levels after acute resistance exercise (Riechman et al. 2004) and mainly in skeletal muscles dominated by type 2 fibres (Nielsen et al. 2007).

5.3 Therapeutic Exercise: Possible Nonmechanical Effects on Bone

Whereas the mechanical stimulation of the bone is probably the main source for stimulating bone metabolism, several pathways are known to modulate bone metabolism. Among these, several pro-inflammatory cytokines such as IL-1ß, IL-6, and

TNF-α promote osteoclastogenesis and bone resorption in synergy with receptor activator nuclear factor kappa B (RANK) (for review, see Iizuka et al. 2014). TNF-α is the key inflammatory cytokine that directly and indirectly promotes inflammation-associated osteoporosis (Teitelbaum 2007); it may activate the fully differentiated osteoclast independent of RANK signalling (Fuller et al. 2002). IL-1, for instance, may enhance osteoclastogenesis only in the presence of permissive levels of RANKL and mediates a substantial component of TNF-α's osteoclastogenic effect in the bone marrow stromal cells and osteoclast precursors (Wei et al. 2005; Jurado et al. 2010). IL-6 may exhibit positive and negative effects on osteoblast and osteoclast differentiation. IL-6 knock-out mice were significantly protected from joint inflammation and destruction in a mouse model of arthritis (Cuzzocrea et al. 2003). They further were protected from bone loss caused by oestrogen depletion as observed in postmenopausal osteoporosis (Cuzzocrea et al. 2003). Thus, the systemic effects of myokines could be likely candidates to facilitate bone integrity both in healthy and osteoporotic bones.

Pure increases of IL-6 without increases in TNF-α or IL-1 and others in activated muscles would induce a coupling between osteoblasts and osteoclasts that is intended to the apposite bone. Thereby, exercise-induced IL-6 would directly suppress the differentiation of osteoclast progenitors. In the presence of IL-6rx, IL-6 evokes discordant responses for bone homeostasis by both inducing bone formation and increasing the osteoclast-supporting activity in osteoblasts. It was demonstrated that IL-6 and IL-11 exerted a direct inhibition of osteoclast formation (Duplomb et al. 2008; Yoshitake et al. 2008). Thereby the gp-130 signalling pathway upregulates RANK expression on osteoblasts but downregulates this pathway in progenitor osteoclasts. Likewise, IL-6 would be able to directly induce anti-inflammatory cytokines like IL-1ra and IL-10 that would diminish the osteoclast-stimulating effects of TNF-α and IL-1 (Steensberg et al. 2003) (Fig. 5.6). Evidence from a study that examined the effects of exercises with low mechanic impact on bone turnover seems to support the theory of a relationship between immune and skeletal responses in exercise (Mezil et al. 2014). Myokine-induced pathways on osteoblast

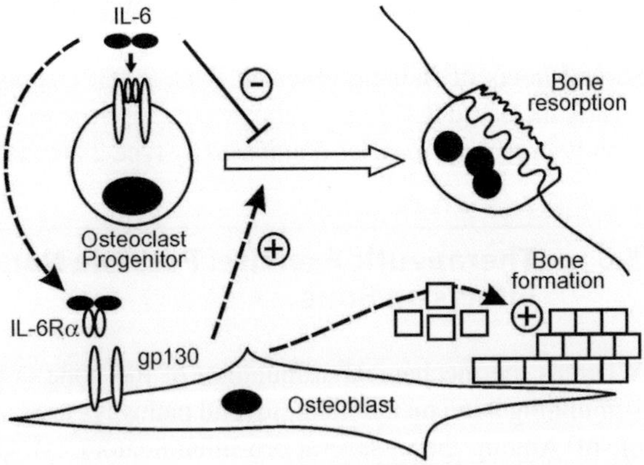

Fig. 5.6 Schematic diagram: effects of IL-6 on osteoclast precursors. IL-6 directly suppresses the differentiation of osteoclast progenitors; in the presence of IL-6Rα, Il-6 both induces bone formation and increases osteoclast-supporting activity of osteoblasts (Modified from Yoshitake et al. 2008)

stimulation by exercise would set off myostatin-induced pathways on osteoclast activation observed in inflammation (Dankbar et al. 2015).

The impact of nonmechanical effects in addition to the mechanical effects of exercises on bone integrity has been examined in several studies. Premenopausal women, for instance, who were engaged in 8 months of running versus weightlifting exercise, revealed similar increases in spine bone mineral density (BMD), whereas only the strength-training group experienced relevant increases in muscle mass (Snow-Harter et al. 1992). Similarly in postmenopausal women, a 9-month impact versus non-impact exercise programme led to similar increases in lumbar spine BMD, but muscle mass increases differed significantly between groups (Kohrt et al. 1997). Such data do not support the notion that only strength training but not endurance training would increase muscle mass (Evans 2004). Furthermore, these studies indirectly support the hypothesis that in addition to the mechanical effects of exercise, the nonmechanical effects seem to contribute to bone integrity in a relevant way.

These nonmechanical effects may be dose dependent. For instance, 1 h of treadmill running at 75 % VO_2 max resulted in large increases in IL-6 and IL-1ra, but not in TNF-α concentrations, compared with those at 55 or 65 % (Scott et al. 2011). Athletes who participate in highly exhausting sports like cycling, long-distance running, and swimming have usually a low BMD (Kohrt et al. 2009), which may be explained by low-grade inflammation and partial deficiency of the immune system associated with over-strenuous exercise performance. Negative effects of excessive running on bone metabolism and respective increases in pro-inflammatory status have been demonstrated in rats (Sipos et al. 2008). In humans, marathon running (Hikida et al. 1983) and especially ultra endurance running (Spartathlon, 246 km) demonstrated increases in systemic inflammation. Spartathlon runners did frequently suffer not only from catabolic muscle processes (Skenderi et al. 2006; Kerschan-Schindl et al. 2015) but also from a catabolic bone metabolism. Thereby, an uncoupling of bone metabolism with increased bone resorption and suppressed bone formation took place (Kerschan-Schindl et al. 2009). However, it is noteworthy to mention that solely exposing the skeleton to muscle loads generated during daily activities and exercises (astronaut exercises on average approximately 2 h per day) is minimally capable or incapable of suppressing bone loss associated with removal of gravitational loading (Judex and Carlson 2009).

5.4 Summary

Osteoporosis is a systemic disease that is associated with increased morbidity, mortality, and healthcare costs. Osteoclasts and osteoblasts are the main regulators of bone homeostasis; however, the key role of the immune system has received little attention. Both mechanical and nonmechanical factors activate bone cells via immunological pathways. The mechanical effects of bone loading are mediated by mechanotransduction. Although the exact mechanosensory mechanisms involved in osteoprogenitor differentiation and osteocyte function remain partly undiscovered, several sensing and signalling pathways have been identified: (1) the activation of ion channels, (2) the production of integrins, (3) the activation of gap junctions and hemichannels, and (4) an unclear role of primary cilia. Under physiological mechanical stimuli, osteocytes prevent bone resorption by changing the RANKL/OPG ratio. Nonmechanical sources of bone remodelling may be, among others, circulating cytokines that are produced by activated muscles (myokines). They may modulate osteoclast differentiation by inducing the expression of RANKL on osteoblasts and/or by directly suppressing osteoclast differentiation via an inhibition of the RANK/RANKL pathway.

References

Akerstrom T, Steensberg A, Keller P, Keller C, Penkowa M, Pedersen BK (2005) Exercise induces interleukin-8 expression in human skeletal muscle. J Physiol 563(Pt 2):507–516

Aubin JE (1998) Bone stem cells. J Cell Biochem Suppl 30–31:73–82

Bacabac RG, Smit TH, Mullender MG, Dijcks SJ, Van Loon JJ, Klein-Nulend J (2004) Nitric oxide production by bone cells is fluid shear stress rate dependent. Biochem Biophys Res Commun 315:823–829

Bemben DA, Bemben MG (2011) Dose–response effect of 40 weeks of resistance training on bone mineral density in older adults. Osteoporos Int 22(1):179–186

Benatti FB, Pedersen BK (2015) Exercise as an anti-inflammatory therapy for rheumatic diseases – myokine regulation. Nat Rev Rheumatol 11:86–97

Biewener AA, Fyhrie D, Parfitt M, Davy T, Schaffler M, Heany R (1993) Safety factors in bone strength. Calcif Tissue Int. 53:S68–S74

Booth FW, Chakravarthy MV, Spangenburg EE (2002) Exercise and gene expression: physiological regulation of the human genome through physical activity. J Physiol 543:399–411

Brotto M, Bonewald L (2015) Bone and muscle: interactions beyond mechanical. Bone 80: 109–114

Brüünsgaard H, Pedersen BK (2003) Age-related inflammatory cytokines and disease. Immunol Allergy Clin North Am 23:15–39

Buford TW, Cooke MB, Shelmadine BD, Hudson GM, Redd L, Willoughby DS (2009) Effects of eccentric treadmill exercise on inflammatory gene expression in human skeletal muscle. Appl Physiol Nutr Metab 34:745–753

Burger EH, Klein-Nulend J (1999) Mechanotransduction in bone-role of the laculocanalicular network. FASEB J 13(Suppl):S101–S112

Burr DB, Martin RB, Schaffler MB, Radin EL (1985) Bone remodeling in response to in vivo fatigue microdamage. J Biomech 18:189–200

Burr DB, Milgrom C, Fyhrie D, Forwood M, Nyska M, Finestone A, Hoshaw S, Saiag E, Simkin A (1996) In vivo measurement of human tibial strains during vigorous activity. Bone 18:405–410

Courteix D, Lespessailles E, Peres SL, Obert P, Germain P, Benhamou CL (1998) Effects of physical training on bone mineral density in prepubertal girls: a comparative study between impact-loading and non-impact-loading sports. Osteoporos Int 8:152–158

Cowin SC, Moss-Salentijn L, Moss ML (1991) Candidate for the mechanosensory system in bone. J Biomech Eng 113:191–197

Currey JD (1979) Mechanical properties of bone tissues with greatly differing functions. J Biomech 12:313–319

Cuzzocrea S, Mazzon E, Dugo L, Genovese T, Di Paola R, Ruggeri Z, Vegeto E, Caputi AP, Van De Loo FA, Puzzolo D, Maggi A (2003) Inducible nitric oxide synthase mediates bone loss in ovariectomized mice. Endocrinology 144:1098–1107

Dankbar B, Fennen M, Brunert D, Hayer S, Frank S, Wehmeyer C, Beckmann D, Paruzel P, Bertrand J, Redlich K, Koers-Wunrau C, Stratis A, Korb-Pap A, Pap T (2015) Myostatin is a

direct regulator of osteoclast differentiation and its inhibition reduces inflammatory joint destruction in mice. Nat Med 21:1085–1090
Duplomb L, Baud'huin M, Charrier C, Berreur M, Trichet V, Blanchard F, Heymann D (2008) Interleukin-6 inhibits receptor activator of nuclear factor κB ligand-induced osteoclastogenesis by diverting cells into the macrophage lineage: key role of Serine727 phosphorylation of signal transducer and activator of transcription 3. Endocrinology 149:3688–3697
Evans WJ (2004) Protein nutrition, exercise and aging. J Am Coll Nutr 23:601S–609S
Fischer CP (2006) Interleukin-6 in acute exercise and training: what is the biological relevance? Exerc Immunol Rev 12:6–33
Fischer CP, Hiscock NJ, Penkowa M, Basu S, Vessby B, Kallner A, Sjöberg LB, Pedersen BK (2004) Supplementation with vitamins C and E inhibits the release of interleukin-6 from contracting human skeletal muscle. J Physiol 558:633–645
Fritton SP, McLeod KJ, Rubin CT (2000) Quantifying the strain history of bone: spatial uniformity and self-similarity of low-magnitude strains. J Biomech 33:317–325
Fuller K, Murphy C, Kirstein B, Fox SW, Chambers TJ (2002) TNFalpha potently activates osteoclasts, through a direct action independent of and strongly synergistic with RANKL. Endocrinology 143:1108–1118
Furmanczyk PS, Quinn LS (2003) Interleukin-15 increases myosin accretion in human skeletal myogenic cultures. Cell Biol Int 27:845–851
Gleeson M (2007) Immune function in sport and exercise. J Appl Physiol 103:693–699
Goodman CA, Hornbrger TA, Robling AG (2015) Bone and skeletal muscle: key players in mechanotransduction and potential overlapping mechanisms. Bone 80:24–36
Gu Y, Preston MR, El Haj AJ, Howl JD, Publicover SJ (2001) Three types of K^+ currents in murine osteocyte-like cells (MLO-Y4). Bone 28:29–37
Haugen F, Norheim F, Lian H, Wensaas AJ, Dueland S, Berg O, Funderud A, Skålhegg BS, Raastad T, Drevon CA (2010) IL-7 is expressed and secreted by human skeletal muscle cells. Am J Physiol Cell Physiol 298:807–816
Hikida RS, Staron RS, Hagerman FC, Sherman WM, Costill DL (1983) Muscle fiber necrosis associated with human marathon runners. J Neurol Sci 59:186–203
Hiscock N, Chan MH, Bisucci T, Darby IA, Febbraio MA (2004) Skeletal myocytes are a source of interleukin-6 mRNA expression and protein release during contraction: evidence of fiber type specificity. FASEB J 18:992–994
Hotamisligil GS (2006) Inflammation and metabolic disorders. Nature 444:860–867
Hu FB, Willett WC, Li T, Stampfer MJ, Colditz GA, Manson JE (2004) Adiposity as compared with physical activity in predicting mortality among women. N Engl J Med 351:2694–2703
Huiskes R, Ruimerman R, van Lenthe GH, Janssen JD (2000) Effects of mechanical forces on maintenance and bone adaptation of form in trabecular bone. Nature 405:704–706
Iizuka K, Machida T, Hirafuji M (2014) Skeletal muscle is an endocrine organ. J Pharmacol Sci 125:125 131
Judex S, Carlson KJ (2009) Is bone's response to mechanical signals dominated by gravitational loading? Med Sci Sports Exerc 41:2037–2043
Judex S, Rubin CT (2010) Is bone formation induced by high-frequency mechanical signals modulated by muscle activity? J Musculoskelet Neuronal Interact 10:3–11
Jurado S, Garcia-Giralt N, Diez-Perez A et al (2010) Effect of IL-1beta, PGE(2), and TGF-beta1 on the expression of OPG and RANKL in normal and osteoporotic primary human osteoblasts. J Cell Biochem 15:304–310
Kemmler W, von Strengel S, Engelke K, Häberle L, Kalender WA (2010) Exercise effects on bone mineral density, falls, coronary risk factors, and health care costs in older women: the randomized controlled senior fitness and prevention (SEFIP) study. Arch Intern Med 170:179–185
Kerschan-Schindl K, Thalmann M, Sodeck GH, Skenderi KP, Matalas AL, Grampp S, Ebner C, Pietschmann P (2009) A 246-km continuous running race causes significant changes in bone metabolism. Bone 45:1079–1083

Kerschan-Schindl K, Thalmann MM, Weiss E, Tsironi M, Föger-Samwald U, Meinhart J, Skenderi KP, Pietschmann P (2015) Changes in serum levels of myokines and Wnt-antagonists after an ultramarathon race. PLoS One 10:e0132478

Klein-Nulend J, Veldhuijzen JP, de Jong M, Burger EH (1987) Increased bone formation and decreased bone resorption in fetal mouse calvaria as a result of intermittent compressive force in vitro. Bone Miner 2:441–448

Klein-Nulend J, van Oers RFM, Bakker AD, Bacabac RG (2015) Bone cell mechanosensitivity, estrogen deficiency, and osteoporosis. J Biomech 48:855–865

Kohrt WM, Eshani AA, Birge SJ Jr (1997) Effects of exercise involving predominantly either joint-reaction or ground-reaction forces on bone mineral density in older women. J Bone Miner Res 12:1253–1261

Kohrt WM, Barry DW, Schwartz RS (2009) Muscle forces or gravity: what predominates mechanical loading on bone? Med Sci Sports Exerc 41:2050–2055

Kontulainenen S, Sievänen H, Kannus P, Pasanen M, Vuori I (2002) Effect of long-term impact-loading on mass, size, and estimated strength of humerus and radius of female racquet-sports players: a peripheral quantitative computed tomography study between young and old starters and controls. J Bone Miner Res 17:2281–2289

Landers-Ramos RQ, Jenkins NT, Spangenburg EE, Hagberg JM, Prior SJ (2014) Circulating angiogenic and inflammatory cytokine responses to acute aerobic exercise in trained and sedentary young men. Eur J Appl Physiol 114:1377–1384

Langberg H, Olesen JL, Gemmer C, Kjaer M (2002) Substantial elevation of interleukin-6 concentration in peritendinous tissue, in contrast to muscle, following prolonged exercise in humans. J Physiol 542:985–990

Malone AM, Anderson CT, Tummala P, Kwon RY, Johnston TR, Stearns T, Jacobs CR (2007) Primary cilia mediate mechanosensing in bone cells by a calcium-independent mechanism. Proc Natl Acad Sci U S A 104:13325–13330

Mc Garry JG, Klein-Nulend J, Prendergast PJ (2005) The effect of cytoskeletal disruption on pulsatile fluid flow-induced nitric oxide and prostaglandin E2 release in osteocytes and osteoblasts. Biochem Biophys Res Commun 330:341–348

Mezil YA, Allison D, Kish K, Ditor D, Ward WE, Tsiani E, Klentrou P (2014) Response of bone turnover markers and cytokines to high-intensity low-impact exercise. Mes Sci Sports Exerc 47:1495–1502

Mikuni-Takagaki Y (1999) Mechanical responses and signal transduction pathways in stretched osteocytes. J Bone Miner Metab 17:57–60

Miyauchi A, Notoya K, Mikuni-Takagaki Y, Takagi Y, Goto M, Miki Y, Fujita T (2000) Parathyroid hormone-activated volume-sensitive calcium influx pathways in mechanically located osteocytes. J Biol Chem 275:3335–3342

Miyauchi A, Gotoh M, Kamioka H, Notoya K, Sekiya H, Takagi Y, Yoshimoto Y, Ishikawa H, Chihara K, Takano-Yamamoto T, Fujita T, Mikuni-Takagaki Y (2006) AlphaVbeta3 integrin ligands enhance volume-sensitive calcium influx in mechanically stretched osteocytes. J Bone Miner Metab 24:498–504

Mortensen OH, Andersen K, Fischer C, Nielsen AR, Nielsen S, Akerström T, Aastrøm MB, Borup R, Pedersen BK (2008) Calprotectin is released from human skeletal muscle tissue during exercise. J Physiol 586:3551–3562

Nielsen AR, Mounier R, Plomgaard P, Mortensen OH, Penkowa M, Speerschneider T, Pilegaard H, Pedersen BK (2007) Expression of interleukin-15 in human skeletal muscle effect of exercise and muscle fibre type composition. J Physiol 584:305–312

Nieman DC, Davis JM, Henson DA, Walberg-Rankin J, Shute M, Dumke CL, Utter AC, Vinci DM, Carson JA, Brown A, Lee WJ, McAnulty SR, McAnulty LS (2003) Carbohydrate ingestion influences skeletal muscle cytokine mRNA and plasma cytokine levels after a 3-h run. J Appl Physiol 94:1917–1925

Orsini N, Bellocco R, Bottai M, Pagano M, Michaelsson K, Wolk A (2008) Combined effects of obesity and physical activity in predicting mortality among men. J Intern Med 264:442–445

Ostrowski K, Hermann C, Bangash A, Schjerling P, Nielsen JN, Pedersen BK (1998) A trauma-like elevation of plasma cytokines in humans in response to treadmill running. J Physiol 513:889–894

Peake JM, Della Gatta P, Suzuki K, Nieman DC (2015) Cytokine expression and secretion by skeletal muscle cells: regulatory mechanisms and exercise effects. Exerc Immunol Rev 21:8–25

Pedersen BK (2013) Muscle as a secretory organ. Compr Physiol 3:1337–1562

Pedersen BK, Febbraio MA (2008) Muscle as an endocrine organ: focus on muscle-derived interleukin-6. Physiol Rev 88:1379–1406

Petersen AMW, Pedersen BK (2005) The anti-inflammatory effect of exercise. J Appl Physiol 98:1154–1162

Plomgaard P, Penkowa M, Pedersen BK (2005) Fiber type specific expression of TNF-alpha, IL-6 and IL-18 in human skeletal muscles. Exerc Immunol Rev 11:53–63

Plotkin LI, Mathov I, Aguirre JI, Parfitt AM, Mangolas SC, Bellido T (2005) Mechanical stimulation prevents osteocyte apoptosis: requirement of integrins, Src kinases, and ERKs. Am J Physiol Cell Physiol 289:C633–C643

Quinn LS, Strait-Bodey L, Anderson BG, Argilés JM, Havel PJ (2005) Interleukin-15 stimulates adiponectin secretion by 3T3-L1 adipocytes: evidence for a skeletal muscle-to-fat signaling pathway. Cell Biol Int 29:449–457

Rawlinson SC, Pitsillides AA, Lanyon LE (1996) Involvement of different ion channels in osteoblasts' and osteocytes' early responses to mechanical strain. Bone 19:609–614

Riechman SE, Balasekaran G, Roth SM, Ferrell RE (2004) Association of interleukin-15 protein and interleukin-15 receptor genetic variation with resistance exercise training responses. J Appl Physiol 97:2214–2219

Robinson JA, Chatterjee-Kishore M, Yaworsky PJ, Cullen DM, Zhao W, Li C, Kharode Y, Sauter L, Babij P, Brown EL, Hill AA, Akhter MP, Johnson ML, Recker RR, Komm BS, Bex FJ (2006) Wnt/ß-catenin signalling is a normal physiological response to mechanical loading in bone. J Biol Chem 281:3170–3178

Rochefort GY, Pallu S, Benhamou CL (2010) Osteocyte: the unrecognized side of bone tissue. Osteoporos Int 21:1457–1469

Rokling-Andersen MH, Reseland JE, Veierød MB, Anderssen SA, Jacobs DR Jr, Urdal P, Jansson JO, Drevon CA (2007) Effects of long-term exercise and diet intervention on plasma adipokine concentrations. Am J Clin Nutr 86:1293–1301

Rubin CT, Lanyon LE (1984) Dynamic strain similarity in vertebrates: an alternative to allometric limb bone scaling. J Theor Biol 107:321–327

Rubin C, Turner AS, Bain S, Mallinckrodt C, McLeod K (2001) Anabolism. Low mechanical signals strengthen long bones. Nature 412:603–604

Rubin J, Rubin C, Jacobs CR (2006) Molecular pathways mediating mechanical signalling in bone. Gene 367:1–16

Santos A, Bakker AD, Zandieh-Doulabi B, Semeins CM, Klein-Nulend J (2009) Pulsating fluid flow modulates gene expression of proteins involved in Wnt signalling pathways in osteocytes. J Orthop Res 27:1280–1287

Scott JP, Sale C, Greeves JP, Casey A, Dutton J, Fraser WD (2011) Effect of exercise intensity on the cytokine response to an acute bout of running. Med Sci Sports Exerc 43:2297–2306

Sipos W, Rauner M, Skalicky M, Viidik A, Hofbauer G, Schett G, Redlich K, Lang S, Pietschmann P (2008) Running has a negative effect on bone metabolism and proinflammatory status in male aged rats. Exp Gerontol 43:578–583

Skenderi KP, Kavouras SA, Anastasiou CA, Yiannakouris N, Matalas AL (2006) Exertional rhabdomyolysis during a 246-km continuous running race. Med Sci Sports Exerc 38: 1054–1057

Snow-Harter C, Bouxsein ML, Lewis BT, Carter DR, Marcus R (1992) Effects of resistance and endurance exercise on bone mineral status of young women: a randomized exercise intervention trial. J Bone Miner Res 7:761–769

Steensberg A, Fischer CP, Keller C, Møller K, Pedersen BK (2003) IL-6 enhances plasma IL-1ra, IL-10, and cortisol in humans. Am J Physiol Endocrinol Metab 285:E433–E437

Tan AD, de Vries TJ, Kuijpers-Jagtman AM, Semeins CM, Everts V, Klein-Nugend J (2007) Osteocytes subjected to fluid flow inhibit osteoclast formation and bone resorption. Bone 41:745–751

Teitelbaum SL (2007) Osteoclasts: what do they do and how do they do it? Am J Pathol 170:427–435

Vezeridis PS, Semeins CM, Chen Q, Klein-Nulend J (2006) Osteocytes subjected to pulsating fluid flow regulate osteoblast proliferation and differentiation. Biochem Biophys Res Commun 348:1082–1088

Webster D, Schulte FA, Lambers FM, Kuhn G, Müller R (2015) Strain energy density gradients in bone marrow predict osteoblast and osteoclast activity: a finite element study. J Biomech 48:866–874

Wei S, Kitaura H, Zhou P, Ross FP, Teitelbaum SL (2005) IL-1 mediates TNF-induced osteoclastogenesis. J Clin Invest 115:282–290

Weinbaum S, Cowin SC, Zeng Y (1994) A model for the excitation of osteocytes by mechanical loading-induced bone fluid shear stresses. J Biomech 27:339–360

Xie L, Jacobson JM, Choi ES, Busa B, Donahue LR, Miller LM, Rubin CT, Judex S (2006) Low-level mechanical vibrations can influence bone resorption and bone formation in the growing skeleton. Bone 39:1059–1066

Yellowley CE, Li Z, Zhou Z, Jacobs CR, Donahue HJ (2000) Functional gap junctions between osteocytic and osteoblastic cells. J Bone Miner Res 15:209–217

Yoshitake F, Itoh S, Narita H, Ishihara K, Ebisu S (2008) Interleukin-6 directly inhibits osteoclast differentiation by suppressing receptor activator of NF-κB signaling pathways. J Biol Chem 283:11535–11540

You L, Temiyasathit S, Lee P, Kim CH, Tummala P, Yao W, Kingery W, Malone AM, Kwon RY, Jacobs CR (2008) Osteocytes as mechanosensors in the inhibition of bone resorption due to mechanical loading. Bone 42:172–179

Zerwekh JE, Ruml LA, Gottschalk F, Pak CYC (1998) The effects of twelve weeks of bed rest on bone histology, biochemical markers of bone turnover, and calcium homeostasis in eleven normal subjects. J Bone Miner Res 13:1594–1601

Zhang D, Weinbaum S, Cowin SC (1998) Electrical signal transmission in a bone cell network: the influence of a discrete gap junction. Ann Biomed Eng 26:644–659

Zhang K, Barragan-Adjemian C, Ye L, Kotha S, Dallas M, Lu Y, Zhao S, Harris M, Harris SE, Feng JQ, Bonewald LF (2006) E11/gp38 selective expression in osteocytes: regulation by mechanical strain and role in dendrite elongation. Mol Cell Biol 26:4539–4552

Utility of the Determination of Biomarkers of Bone Metabolism

6

Barbara Obermayer-Pietsch and Verena Schwetz

6.1 Introduction

Biochemical serum markers of bone metabolism are substances involved in all processes of bone activity. Systemic dispersion is the basis for their biochemical measurements. Both serum/plasma and urine measurements are widely used for clinical parameters of bone turnover, but accurateness and usability have favoured blood analyses over the years. This review will therefore focus on serum/plasma parameters of bone metabolism and their clinical applications and give hints on further applications within biological material.

Targets of biochemical analysis of bone metabolism are enzymes and proteins or their respective metabolites produced during bone formation and degradation. However, direct regulators of osteoblast and/or osteoclast cell function and activity might represent additional parameters of interest, since they may very well reflect dynamic processes in the bone and adjacent tissue.

The usefulness of serum markers of bone metabolism has been widely recognized during the past decades but increased significantly with the technical improvement of biochemical methods and the knowledge of new compounds in bone metabolism. Many of these markers have been tested for practical use, and some of them have entered daily clinical routine applications. The reason for this increasing interest is the option of characterization and diagnosis of metabolic bone diseases but also the improvement of therapy decisions and therapy monitoring by biomarkers.

B. Obermayer-Pietsch, MD (✉) • V. Schwetz, MD, PhD
Division of Endocrinology and Diabetology, Medical University of Graz, Auenbruggerplatz 15, Graz 8036, Austria
e-mail: barbara.obermayer@medunigraz.at; verena.schwetz@meduni-graz.at

P. Pietschmann (ed.), *Principles of Osteoimmunology*,
DOI 10.1007/978-3-319-34238-2_6

Regarding their potential, why did it take a long time to incorporate these biomarkers in international diagnostic and therapeutic recommendations?

First, technical measurements of the bone, e.g. by dual energy x-ray (DXA) absorptiometry, have lead to believe in a precise golden standard measurement method for osteoporosis diagnosis and follow-up. A long-term standardizing process for equipment, interpretation and reference measurements as well as a low number of technical types and a high number of international publications on the technology have guaranteed its place in clinical routine. The World Health Organization (WHO) definition of osteoporosis is based on DXA criteria, but this clearly overestimates the importance of bone density measurements in view of newer more holistic criteria.

Second, a broad spectrum of biomarkers of bone metabolism has been established, but the standardization of laboratory methods has only been done in recent years. Biomarker studies in the past have therefore not been comparable with others and were not suitable to appear in general recommendations. This picture has changed in recent years.

Third, the technical feasibility of laboratory methods for bone biomarkers has increased during the past decade. Not only specialized institutions like academic laboratories, but a number of laboratories have now the experience, equipment and standardized methods, as well as specialists, interpretation tools and a significant economic interest in these markers.

Fourth, preclinical conditions have been crucial for many of these tests. Therefore, long transportation times for biological material or special patient conditions (e.g. time of the day or nutritional requirements) are not feasible or difficult to achieve in general settings. The general implementation of biomarkers of bone metabolism was therefore delayed with regard to its importance. However, the implementation is now accomplished more easily not only in central facilities, and biomarkers of bone metabolism are increasingly used.

Fifth, economical problems in recent years have forced communities to stabilize or even reduce their healthcare costs. Nowadays, bone biomarker measurement tends to be cheaper and more accurate. They are on the way of their establishment together with traditional imaging methods.

This review aims to describe current types of bone biomarkers and the pros and cons of their use in bone diseases. Interpretation, standardization and future aspects will be discussed.

6.2 Overview of Bone Biomarkers

6.2.1 Calciotropic Hormones

Bone biomarkers cannot be interpreted without a view on the individual calciotropic hormones. Given the fact that more than 70 % of the general population tend to have decreased vitamin D levels (Holick 2007), bone metabolism has to be seen in the light of a neglected problem of possible osteomalacia, with wide-ranging consequences.

6.2.1.1 Parathyroid Hormone (PTH)

A long history of analytical attempts for parathyroid hormone exists, starting in 1909, when MacCallum and Voegtlin (MacCallum and Voegtlin 1909) noted that removal of the parathyroid glands quickly resulted in rapidly falling blood calcium levels. Subsequent work in the 1920s to purify the hormone established its central role in calcium homeostasis.

Parathyroid hormone (PTH) or parathyrin is secreted by the parathyroid glands as a polypeptide containing 84 amino acids. It increases the concentration of calcium in the blood by acting upon parathyroid hormone receptors in many parts of the body. PTH half-life is approximately 4 min. It has a molecular mass of 9.4 kDa. PTH is stocked in secretory granules that fuse with the cellular membrane and release PTH in response to a decrease in the extracellular ionized calcium concentration. Most of the PTH is secreted in its intact form, 1-84; however, it can also be secreted as N-terminal truncated fragments or C-terminal fragments after intracellular degradation.

However, identification of different types of PTH fragments and several generations of PTH assays may be responsible for the great variability of PTH values and the difficulties of implementing international recommendations such as those of the National Kidney Foundation/Kidney Disease Outcomes Quality Initiative (Ureña Torres 2006). Calibration attempts are currently on the way, as the generalized use of a unique, true, and accurate intact 1-84 PTH assay is not available yet.

Usability/Utility The measurement of PTH plays an important role in daily clinical routine not only for endocrinologists and nephrologists but also for a variety of other disciplines. Technical development and standardization should be set up with the executive laboratory. Second- and third-generation PTH assays can be aligned by specific factors depending on the assays used (Souberbielle et al. 2010a, b). This will minimize the problem of inter-method variability of PTH measurements, but depending on the type of disease – problems with standardization and reference populations remain to be resolved.

Parathyroid Hormone-Related Protein (PTH-rP)

The parathyroid hormone-related protein (PTH-rP) is structurally related to PTH and seems to play a physiological role in lactation, possibly as a hormone for the mobilization and/or transfer of calcium to the milk. PTH and PTH-rP bind to the same G-protein coupled receptor. It is occasionally secreted by cancer cells (breast cancer, certain types of lung cancer). Cancer-derived PTH-rP appears to play a critical role in bone invasion by tumour cells, mediated by osteoclasts. In the bone, PTH-rP regulates enchondral bone development by maintaining the endochondral growth plate at a constant width (Fritchie et al. 2009).

Usability/Utility PTH-rP testing is more appropriately performed after assessment of PTH. If PTH is not low or low normal, testing for PTH-rP is usually uninformative. The clinical use of PTH-rP is restricted to validated assays and might be a diagnostic challenge for experienced laboratories.

6.2.1.2 Vitamin D

25(OH) Vitamin D

Vitamin D, a secosteroid, is structurally similar to steroids such as testosterone, cholesterol, and cortisol. The molecular weight of vitamin D2 (calciferol) is 396.6 and of vitamin D3 (cholecalciferol) is 384.6. Vitamin D2 originates from irradiation at about 260 nm of ergosterol, while vitamin D3 is formed by irradiation of a provitamin molecule (7-dehydrocholesterol) present in the skin and gut lining cells.

The role of vitamin D in maintaining bone health has been known for decades. Recently, however, the fact that many tissues express vitamin D receptors and are regulated by vitamin D has accentuated the need for valuable vitamin D serum measurements. In addition, 25(OH) vitamin D levels rather than the more locally acting 1,25$(OH)_2$ vitamin D have been associated with an important role in many diseases, including the development of cancer, autoimmune diseases, cardiovascular diseases, diabetes, and infections. Vitamin D deficiency, defined as serum 25(OH) vitamin D levels <30 ng/mL, is very common in our population (Holick 2007).

Measurement of 25(OH) vitamin D is widely used for assessing vitamin D status. For higher analytical throughput, traditional solvent extraction of samples has been replaced by immunological methods. The Vitamin *D E*xternal *Q*uality *A*ssessment *S*cheme (DEQAS) has revealed method-related differences in 25(OH) vitamin D results, raising concerns about the comparability and accuracy of different assays Dachverband Osteologie 2014. However, there is an ongoing debate in view of total/free assays and the potential cross-reactivity of metabolites (Lee et al. 2015). Large standardization programmes are on the way (e.g. the *V*itamin *D S*tandardization *P*rogramme (VDSP), available at ods.od.nih.gov/Research/vdsp.aspx).

Usability/Utility The analytical evaluation of 25(OH) vitamin D assays between different methods (e.g. radioimmunoassays (RIAs), enzyme-immunoassays (ELISAs) or mass spectrometry (MS)) has designated MS as the gold standard. However, most of the existing assays are able to produce sufficiently reliable results for a clinical estimation of 25(OH) vitamin D levels and may therefore monitor not only the initial but also the follow-up values during vitamin D supplementation.

1,25$(OH)_2$ Vitamin D

Calcitriol, also called 1,25-dihydroxycholecalciferol or 1,25$(OH)_2$ vitamin D3, is a hormonally highly active form of vitamin D with three hydroxyl groups. It increases the level of calcium in the blood by increasing the uptake of calcium from the gut into the blood, by decreasing the transfer of calcium from the blood to the urine by the kidneys, and by increasing the release of calcium into the blood from the bone. In turn, 1,25$(OH)_2$ vitamin D declines very late in vitamin D deficiency. Therefore, the measurement of 25(OH) vitamin D better reflects body vitamin D storage. Decreases in 1,25$(OH)_2$ vitamin D are associated with kidney impairment, already with slightly decreased renal function or very rarely with 1α-hydroxylase defects. High levels of 1,25$(OH)_2$ vitamin D have been found in sarcoidosis or idiopathic hypercalciuria. Both of them can lead to hypercalcemia as well as other causes like paraneoplastic syndromes.

Assays often involve partial purification and extraction, e.g. by using coated silica cartridges, to remove other vitamin D metabolites followed by high-performance liquid chromatography, radioimmunoassay, ELISA or MS technologies. Therefore, the routine measurement is often limited to central lab facilities.

Utility The action of 1,25$(OH)_2$ vitamin D on a cellular level depends on local production which has paracrine and autocrine effects in many tissues. Serum measurements of 1,25$(OH)_2$ vitamin D may therefore not reflect this local production and may not be a representative of the cellular levels of the hormone. However, in some patients, e.g. with chronic kidney disease and impairment of 1α-hydroxylase activity, sarcoidosis or hypercalciuria syndromes, 1,25$(OH)_2$ vitamin D determination can help to characterize the actual vitamin D status and needs.

6.2.2 Collagen and Collagen Products

Collagen is a ubiquitous element of human tissue. More than 90% of the organic matrix of the bone is constituted by collagen type I. Its biochemical composition of small, non-steric amino acids (e.g. glycin) facilitates the development of long chains, named helices, which are linked by lysine and hydroxylysine to form the well-known triple structures. Both renal and hepatic functions are crucial in the interpretation of serum collagen product levels, because both interfere with formation, storage and degradation of collagens and their metabolites.

6.2.2.1 CTX and NTX

Crosslaps (carboxy-terminal collagen crosslinks, CTX or ß-(beta)-CTX, 8.5 kDa) are cleavage products of type I C- or N-terminal type 1 collagens (NTX). The assays are based on monoclonal antibodies against the cross-linked site (e.g. Glu-Lys-Ala-His-beta-Asp-Gly-Gly-Arg) connecting collagen I fragments or other parts of the molecules. While NTX have been used in the past in urine and serum, CTX is the currently widely used bone biomarker in a large number of studies and in most clinical labs for the analysis of bone metabolism.

Elevated levels of CTX indicate increased bone resorption and, vice-versa, a decrease is a feature of therapy response in antiresorptive medication. However, changes in bone markers are not entirely disease specific and may reflect alterations in skeletal metabolism independent of the underlying cause, but this has to be interpreted in each individual case as known for all systemic biomarkers. A circadian rhythm is well known, which requires standardized collection as well as age- and sex-defined reference ranges (see below).

The "response" of bone turnover markers to osteotropic therapies has been investigated recently to identify conditions of non-responders and non-compliants for different oral bisphosphonate therapies. The TRIO study (Naylor et al. 2016) showed that reaching the target response for bone markers is associated with respective change in bone mineral density. This concept used both least significant change (LSC) concepts as well as reference interval (RI) responses (Michelsen et al. 2013, see Sect. 6.3.2.5).

Usability/Utility The importance of CTX has even increased since standardization programmes and new reference values as well as least significant change (LSC) intervals have been generated in therapy studies. Serum material is easy to access, and both transport and analysis have been optimized in terms of costs and feasibility. CTX are therefore one of the promising candidates for a routine care parameter in bone diseases and therapy.

6.2.2.2 PICP and PINP

PICP (100 kDa) and PINP (53 kDa) are both propeptides of the C- or N-terminal ends of procollagen type I, respectively. Both peptides are produced in equimolar amounts during collagen synthesis and reflect therefore the process of collagen formation. PINP has found its way in routine labs, mostly due to the excellent reproducibility and international standardization efforts, e.g. in the TRIO study, where reductions in vertebral fractures were paralleled by reductions in PINP and antiresorptive markers (Naylor et al. 2016).

The diagnostic and therapeutic relevance of collagen bone biomarkers is defined only by its origin from mature collagen I. This specificity can be documented by a combination of their primary structure and the modifications during formation, maturation and ageing of the bone matrix.

During bone resorption, collagen markers will normally be elevated but suppressed during and sometime after antiresorptive therapy. This can widely be used as therapy monitoring, e.g. during bisphosphonate therapy. Bone formation or osteoanabolic therapy will produce higher levels of collagen serum products.

Due to the scattering of collagen serum markers, all changes in individual serum levels should refer to a "least significant change" of about 50 % of the initial values (Bieglmayer and Kudlacek 2009).

The involvement of a disintegrin and metalloproteinase with a thrombospondin motifs 2 (ADAMTS2), an extracellular matrix metalloproteinase that cleaves PINP, is of increasing interest in collagen metabolism. Its activity is blocked by tissue inhibitor of matrix metalloproteinase 3 (TIMP3) (Wang et al. 2006). During degradation, collagen products are bound by a scavenger receptor and incorporated into liver cells. Serum levels are therefore dependent on liver function.

Usability/Utility The clinical use of collagen markers ranges from diagnosis of high bone turnover to therapy monitoring of antiresorptive drugs. PINP is now used in a number of routine labs due to standardization efforts and comparative studies in recent years, which have supported the use of collagen markers in a more differentiated interpretation of bone metabolism.

6.2.2.3 Pyridinoline and Deoxypyridinoline

Pyridinoline (PYD) and deoxypyridinoline (DPD) are covalent pyridinium cross-links also produced from the breakdown of collagen during bone resorption. They are released into circulation and excreted into the urine. Both can be detected by RIA and ELISA, methods that afford clinical application and have a better sensitivity compared to the earlier high-performance liquid chromatography (HPLC) methods. However, their clinical relevance has declined during recent years.

The degradation of more mature collagen I material results in a variety of fragments. As specific proteinases cleave collagen I in a specific way, these fragments can be attributed to different degradation processes. Cathepsin K from osteoclast bone cleavage produces crosslaps; matrix metalloproteinases from mononuclear cells tend to produce other fragments, like ICTP (Garnero et al. 2003). Further proteolysis results in smaller fragments with characteristic cross-links such as pyridinolines and deoxypyridinolines. Final fragments are amino acids like hydroxyproline, hydroxylysine and glycosylated molecules.

Usability/Utility PYD and DPD were important during the establishment of bone biomarker analyses in the past. Based on their heterogeneity and lacking standardization, they are not widely used for clinical and scientific purposes.

6.2.3 TRAP 5b

Tartrate resistant acid phosphatase type 5b (TRAP 5b) is a monomeric metalloenzyme protein present in osteoclasts, activating macrophages and dendritic cells. It has a molecular weight of approximately 35 kDa, a basic isoelectric point (7.6–9.5), and optimal activity in acidic conditions. This enzyme is specific for osteoclasts and is activated by RANKL (Liu et al. 2003). Tartrate resistant acid phosphatase (TRAP) is synthesized as a proenzyme and activated by proteolytic cleavage and reduction. The enzyme has two subunits linked by disulphide bonds. It is distinguished from other mammalian acid phosphatases by its resistance to inhibition by tartrate, its molecular weight and its characteristic purple colour. Two closely related isoforms of type 5 TRAP (5a and 5b) can be distinguished. Serum TRAP 5a has an activity tenfold less specific compared to TRAP 5b. The difference in specific activity and pH optima allow immunoassays to be constructed to provide a high degree of selectivity for TRAP 5b.

In osteoclasts, TRAP is localized within the ruffled border area, the lysosomes, the Golgi cisternae and the vesicles. Since the enzyme is not influenced by kidney function, serum levels reflect bone turnover also in renal osteodystrophy (Fahrleitner-Pammer et al. 2008).

Proposed functions of TRAP include osteopontin/bone sialoprotein dephosphorylation, generation of reactive oxygen species, iron transport, cell growth and differentiation. There is some evidence that its activation is induced by cathepsin K by cleaving a small peptide in a side chain (Ljusberg et al. 2005). Its physiological role is not fully understood, but the high phosphatase activity could be related to collagen turnover (Roberts et al. 2007). TRAP 5b is inactivated by the loss of the iron ion in the active centre of the molecule and it is then proteolyzed. This is important, because up to 90 % of the circulating molecule might be inactivated fragments secreted through the liver and kidneys. It is of importance that laboratory-measured TRAP 5b activity does not always correspond to the biological activity of osteoclasts but that it reflects the total number of osteoclasts. Serum also contains TRAP 5a activity, primarily derived from inflammatory macrophages (Chao et al. 2010).

Usability/Utility In the management of patients with severe kidney disease and probably of patients with bone metastases, the determination of serum TRAP 5b will overcome the problems of other biomarkers by providing reliable results. However, substrate stability and the need for frozen transportation may limit its use to central facilities.

6.2.4 ALP and Bone ALP

Alkaline phosphatase (ALP) and bone alkaline phosphatase (bone ALP or bALP) belong to the membrane-associated alkaline phosphate enzymes present in all tissues throughout the entire body with a molecular weight of about 140 kDa. The isoforms are generated by the same gene locus, but differ due to their extent of glycosylation. There are about 15 related isoforms of alkaline phosphatases, mainly occurring in the liver, bile duct, kidney, placenta, and bone, the latter being generated by osteoblasts during bone formation. Normally, ALP activity in the circulation is contributed to by bone and liver isoforms in approximately equal amounts.

A proposed function of this form of the enzyme is matrix mineralization during osteoblast growth; however, mice that lack a functional form of this enzyme show relatively normal skeletal development. This enzyme has been linked directly to hypophosphatasia, a disorder characterized by hypercalcaemia including skeletal defects (Whyte 2010). The character of this disorder varies depending on specific AP mutations determining the age of onset and severity of symptoms.

ALP is linked to the cell membrane by inositol phosphate – the circulating form is therefore a cleaved part of the original molecule. However, ALP can be secreted by exocytosis directly from osteoblasts. Inorganic pyrophosphate is known to inhibit mineralization and might also be able to preserve vessels from calcification. A direct link to vascular calcification has been seen in osteoporotic patients, where ALP levels directly correlated with aortic calcification (Iba et al. 2004).

Usability/Utility In general, total ALP assays are suitable for the assessment of bone turnover in patients with normal liver function, but bone ALP reflects greater specificity for osteoblast function. Assay techniques are widely used and reflect bone formation. Its increase in osteomalacia requires a sufficient knowledge of vitamin D/PTH status of the patient.

6.2.5 Osteocalcin

Osteocalcin, also known as bone gamma-carboxyglutamic acid-containing protein (BGLAP), is the most frequent noncollagenous protein found in the bone and dentin and comprises 49 amino acids. Osteocalcin synthesis has been shown to be regulated by vitamins D and K (Iwamoto et al. 2009). For instance, carboxylation of the three glutamine residues is dependent on vitamin K and crucial for its affinity to hydroxyapatite. Vitamin K deficiency may thus cause higher amounts of uncarboxylated osteocalcin, as it is the case in normal individuals (Rennenberg et al. 2010). Osteocalcin is mainly bound in bone matrix, but 20–30 % may be released systemically. During osteoblast development, osteocalcin is a late marker after ALP and collagen type I.

The function of osteocalcin is not entirely understood, but it is of central importance for bone remodelling as has been shown from knockout experiments and other in vivo tests, both in osteoclasts and osteoblasts.

Osteocalcin has been one of the first important biomarkers of bone metabolism. Circulating osteocalcin is heterogeneous and partly fragmented. Many of these fragments, mainly released during bone resorption, are not detected by current commercial assays.

Based on two decades of research, recent new discoveries have linked osteocalcin and glucose metabolism as well as leptin and testosterone (Schwetz et al. 2012). Crosstalk between the bone and adipose tissue is a new window of importance for osteocalcin and might be intensified in the near future (Kanazawa et al. 2010).

Usability/Utility Even though its assignment to bone formation or bone resorption is either-or, the analysis has been part of many of the diagnostic and therapeutic osteoporosis studies. It has been routinely observed that higher serum osteocalcin levels are relatively well correlated with increases in bone mineral density during treatment with anabolic bone formation drugs for osteoporosis.

6.2.6 RANKL/OPG

The receptor activator for nuclear factor k B ligand/osteoprotegerin (RANKL/OPG) system has been described in several chapters of this book. In brief, osteoprotegerin (OPG) is a member of the tumour necrosis factor (TNF) receptor superfamily. It is a basic glycoprotein comprising 401 amino acid residues arranged into 7 structural domains. It is found as either a 60 kDa monomer or as a 120 kDa dimer linked by disulphide bonds. Osteoprotegerin inhibits the differentiation of osteoclast precursors, related to monocyte/macrophage cells and derived from granulocyte/macrophage-forming colony units and also regulates the resorption of osteoclasts in vitro and in vivo (Terpos 2006). Osteoprotegerin is a receptor activator for nuclear factor k B (RANK) homolog and works by binding to RANKL on osteoblast/stromal cells, thus blocking the RANKL-RANK ligand interaction between these cells and osteoclast precursors. This has the effect of inhibiting the differentiation of the osteoclast precursor into a mature osteoclast.

Receptor Activator for Nuclear Factor k B (RANK) is a type I membrane protein that is expressed on the surface of osteoclasts and is involved in their activation upon ligand binding. RANK is also expressed on dendritic cells and facilitates immune signalling.

Receptor Activator for Nuclear Factor k B Ligand (RANKL) is found on the surface of osteoblasts, but also stromal cells, and T cells.

The potential of the OPG/RANKL determinations have been investigated in several studies. A large prospective population study on fracture risk and OPG/RANKL quotients in 2004 showed an association with higher risk for low trauma fractures and low RANKL levels independent of age, sex, menopausal status and OPG levels (Schett et al. 2004). This was attributed to suppressed bone turnover with accumulation of microdamage and reduced bone quality. However, these results have not been confirmed by further studies. By contrast, low OPG levels have been associated with prevalent vertebral fractures in postmenopausal osteoporosis patients and transplant recipients. However, there are also studies with no differences of OPG levels between patients with or without fractures and recent studies showing an association between high fracture risk and high OPG levels (Li et al. 2009), possibly related to immunological diseases.

Usability/Utility During the past decade, there was no clear answer whether OPG or RANKL determinations might be of further help in the diagnosis and risk determination of osteoporosis. This could be related to the major role of OPG/RANKL not only in bone metabolism but also in immune regulation. However, OPG/RANKL determinations might be useful in special clinical conditions.

6.2.7 New Markers

With the discovery of several new pathways in bone metabolism, a number of interesting substances were identified as promising new candidates for bone serum biomarkers.

In addition to serum proteins, other molecules, such as small single-stranded, non-coding RNAs (ribonucleic acid), named microRNAs are potential new biomarkers with specific profiles in several bone diseases (Hassan et al. 2015; Ulbing and Obermayer-Pietsch 2015).

6.2.7.1 Cathepsin K

Cathepsin K is a lysosomal cysteine protease, a member of the peptidase C1 protein family, which has an optimal enzymatic activity in acidic conditions. It is synthesized as a proenzyme with a molecular weight of 37 kDa and transformed into the mature, active form with a molecular weight of ~27 kDa. Its origin is osteoclasts, but it may also be found in bone giant cell tumours. The enzyme's ability to catabolize elastin, collagen, and gelatin allow it to break down the bone and cartilage. Cathepsin K expression is stimulated by inflammatory cytokines and is associated with the initiation of the osteoclastic brush border, earlier than the appearance of

TRAP during the bone resorption process. Its catabolic activity is regulated by cystatin D, an endogenous inhibitor of cysteine proteases (Holzer et al. 2005).

Usability/Utility Cathepsin K inhibitors, such as odanacatib, show great potential in the treatment of osteoporosis. The use of cathepsin K in clinical routine measurements will depend on the outcome of current studies and on a possible relationship to indication and monitoring in osteoporosis patients as well as in patients with other forms of metabolic bone disease.

6.2.7.2 Matrix Metalloproteinases: MMPs

Matrix metalloproteinases (MMPs) are zinc-dependent endopeptidases belonging to a family of at least 28 secreted or transmembrane proteases. They are collectively capable of processing and degrading various proteins, but also a number of bioactive molecules (Egeblad and Werb 2002). In addition, they are known to be involved in the cleavage of cell surface receptors, the release of apoptotic ligands, and chemokine/cytokine in/activation. MMPs also play a major role in cell proliferation, migration (adhesion/dispersion), differentiation, angiogenesis, apoptosis and host defence.

The major difference to other endopeptidases is their dependence on metal ions as cofactors, their ability to degrade extracellular matrix, and their specific evolutionary highly conserved DNA sequence. The role of MMPs in bone metabolism is far from being entirely understood but will probably play an important role in rheumatic diseases and therapy assignment (Pelletier et al. 2010).

Usability/Utility MMPs have potential effects on the expansion of metastasis but are also considered important in rheumatoid and degenerative arthritis. The usability of these parameters in clinical routine has to be evaluated further in the future.

6.2.7.3 Bone Sialoprotein

Bone sialoprotein (BSP), a glycopeptide with high sialic acid content, is a component of mineralized tissues formed by osteoblasts and osteoclasts. Native BSP has a molecular weight varying between 33 and 80 kDa. It is a significant component of the bone extracellular matrix and has been suggested to constitute approximately 8 % of all noncollagenous proteins in the bone. BSP-1, also called osteopontin, is a cell-binding sialoprotein or integrin-binding sialoprotein and belongs to the "*s*mall *i*ntegrin *b*inding *li*gand *N*-linked *g*lycoprotein" (SIBLINGs) family. It has been demonstrated that BSP is extensively modified post-translationally, which makes the protein highly heterogeneous.

The amount of BSP in the bone and dentin is roughly equal; however, the function of BSP in these mineralized tissues is not known. One possibility is that BSP acts as a nucleus for the formation of the first apatite crystals. As the apatite forms along the collagen fibres within the extracellular matrix, BSP could then help direct, redirect or inhibit the crystal growth. However, BSP also supports osteoclast formation by RANKL and inhibits their apoptosis. Therefore, circulating levels of BSP might correlate to bone metastasis formation. Additional roles of BSP are MMP-2 activation, angiogenesis and protection from complement-mediated cell lysis.

Usability/Utility Bone sialoprotein has been found of value in assessing bone metastases in cancer as well as in some special conditions in metabolic bone disease. These promising results must still be confirmed in large trials. However, combinations of markers have at times helped in assessing cancer stages and lytic bone disease and in monitoring specific treatment modalities.

6.2.7.4 Osteonectin (SPARC)

Osteonectin (also referred to as secreted acidic cysteine-rich protein, SPARC) is an acidic glycoprotein in the bone with a molecular weight of 43 kDa that binds calcium protein at the N-terminal acidic region. It is secreted by osteoblasts during bone formation, initiating mineralization and promoting mineral crystal formation. Osteonectin also shows affinity for collagen in addition to bone mineral calcium. It is a secreted extracellular matrix glycoprotein that plays a vital role in bone mineralization, cell-matrix interactions, and collagen binding.

Osteonectin shows context-specific effects, but generally inhibits adhesion, spreading and proliferation, and promotes collagen matrix formation. It increases the production and activity of matrix metalloproteinases, a function important to cancer cells invading the bone. For endothelial cells, osteonectin disrupts focal adhesions and binds and sequesters platelet-derived growth factor (PDGF) and vascular endothelial growth factor (VEGF). Osteonectin is abundantly expressed in the bone, where it promotes osteoblast differentiation and inhibits adipogenesis. Additional functions of osteonectin have been described as beneficial to tumour cells including angiogenesis, proliferation and migration. A really new topic is osteonectin derived from adipose tissue associated with insulin resistance and diabetic retinopathy, nephropathy as well as adipose tissue fibrosis (Kos and Wilding 2010).

Usability/Utility Overexpression of osteonectin is reported in many human cancers such as breast, prostate and colon cancer and might therefore be of increasing interest for the searching of bone metastases, but also for osteoclast function in metabolic bone disease and for insulin regulation in obesity.

6.2.7.5 Phosphatonins: FGF23

Fibroblast growth factor 23 (FGF23), encoded by the FGF23 gene, is a member of the fibroblast growth factor family which is responsible for phosphate metabolism. The FGF23 gene is located on chromosome 12 and is composed of three exons. Mutations in FGF23 render the protein resistant to proteolytic cleavage. They lead to increased activity of FGF23 and to renal phosphate loss as found in the human disease autosomal dominant hypophosphataemic rickets (Ramon et al. 2010). Some types of tumours also overproduce FGF23, causing tumour-induced osteomalacia. Loss of FGF23 activity has been shown to increase phosphate levels and develop tumour calcinosis.

The study of several renal phosphate wasting disorders resulted in the identification of four factors with the predicted characteristics of phosphatonins, namely, fibroblast growth factor 23 (FGF23), secreted Frizzled related protein-4 (sFRP-4), matrix extracellular phosphoglycoprotein (MEPE), and FGF7. Increased serum phosphate, vitamin D, and probably PTH levels stimulate FGF23 production by the bone (Ramon et al. 2010). During vitamin D replacement therapy, decreased FGF23 concentrations decline further and may again have a favourable impact on bone mineralization by counterregulatory effects on phosphate homeostasis (Uzum et al. 2010).

Usability/Utility The usefulness of FGF23 in clinical settings is not only restricted to the differential diagnosis of tumours and hereditary forms of hypo- and hyperphosphataemia. Several studies on phosphate therapy showed a tight reaction of FGF23 levels on phosphate interventions. The determination of this phosphatonin has therefore been shown to be useful in the follow-up of patients with renal diseases as well as for oncological patients with newly diagnosed disturbances of electrolyte balance.

6.2.7.6 Wnt Associated Proteins and Receptors

The Wnt signalling pathway includes many receptors, agonists and antagonists and has been elucidated during the past years. This signalling system contributes to bone formation and mechanoreception.

Secreted Frizzled-Related Proteins

Secreted frizzled-related proteins (SFRP1 SFRP2, SFRP3, SFRP4, SFRP5) with a molecular weight of about 35 kDa are members of the secreted frizzled-related proteins (SFRP) family that contains a cysteine-rich domain homologous to the putative Wnt-binding site of frizzled proteins. SFRPs act as soluble modulators of Wnt signalling. Each SFRP is ~300 amino acids in length and contains a cysteine-rich domain that shares 30–50 % of sequence homology with frizzled (Fz) receptors. SFRPs are able to bind Wnt proteins and Fz receptors in the extracellular compartment. The interaction between SFRPs and Wnt proteins prevents the latter from binding the Fz receptors. SFRPs are also able to downregulate Wnt signalling by the formation of an inhibitory complex with the frizzled receptors. The Wnt pathway plays a key role in embryonic development, cell differentiation and cell proliferation. It has been shown that the deregulation of this critical developmental pathway occurs in several human tumour entities.

Usability/Utility Due to their frequent involvement in breast cancer, SFRPs have been investigated as bone markers in bone metastasis and metabolic bone disease. However, the clinical use of SFRP determination might increase with some more specific knowledge on its importance in diagnosis of metabolic changes or during therapy regimens.

Dickkopf-1

Dickkopf-1 (DKK1) is another analyte of the Wnt signalling cascade that has been developed to characterize bone metabolism and is encoded by the DKK1 gene. It is a secreted protein with two cysteine-rich regions involved in embryonic development through its inhibition of the Wnt signalling pathway.

Besides a negative correlation to bone mass (Macdonald et al. 2007), elevated levels of DKK1 in bone marrow, plasma and peripheral blood have been shown to be associated with the presence of osteolytic bone lesions in patients with multiple myeloma. The regulation of canonical Wnt signalling by DKK1 may act as a molecular switch mediating the transition from an osteolytic to an osteoblastic response. Therefore, blocking DKK1 activity may prove to be a relevant therapeutic target in the prevention of bone metastasis (Hall and Keller 2006).

Usability/Utility DKK1 determination is based on several available kits and might be of help for bone metastasis detection. Further research and clinical studies will increase the potential of this biomarker.

6.2.7.7 Sclerostin

Sclerostin, the product of the SOST gene, was originally believed to be a non-classical inhibitor of bone morphogenetic protein (BMP) but has finally been identified as binding to LRP5/6 receptors and inhibiting the Wnt signalling pathway. It is produced by osteocytes. Wnt activation under these circumstances is antagonistic to bone formation. Sclerostin may be of increasing importance as a marker of osteocytes, accounting for most of the mature bone cells, however neglected over the past years. Mature osteocytes are sensors for mechanical loading of bone that inhibit osteoclasts and stimulate osteoblasts (Bonewald 2007). Many new results link osteocytes with osteoblastic cell contacts and microfractures (Taylor et al. 2007).

The currently investigated anti-sclerostin antibody for the treatment of osteoporosis codeveloped by Amgen and UCB introduces a new focus of therapy for the future and may require some sclerostin determination. The reaction of sclerostin levels has been recently investigated in several trials. Whereas bisphosphonates had no substantial effects due to a general non-activation of osteocytes, oestrogen replacement therapy (Mirza et al. 2010) and PTH osteoanabolic therapy (Drake et al. 2010) induced significant changes in sclerostin levels.

Usability/Utility Sclerostin inhibitors have been developed as a new generation of medication for osteoporosis. The clinical usefulness of sclerostin measurements will depend on further investigations on diagnostic and monitoring applications for metabolic and oncologic bone diseases.

6.3 Analytics

6.3.1 Preanalytics

The analytical process starts with the selection and preparation of the patients and comprises blood drawing, the performance and validation of the analytical measurement as well as the presentation and interpretation of the gained data.

Many of the recent tests have required extensive work for the specificity and reproducibility of the results under controlled conditions. However, the main problem might lie far before the controlling of laboratory techniques.

A main source of variability is the preanalytical phase, where patient conditions are very difficult to standardize. It is therefore strictly recommended to use biomarkers known to have a robust performance on their way from the patient to analysis and validation.

Some clinical conditions are crucial for the choice of bone biomarkers: renal elimination has been shown for collagen products, crosslaps and osteocalcin. They had been characterized in urine and tests had been routinely used, mainly before serum measurements were made available. Although many publications have used urinary assays, these determinations have been replaced by serum analyses. First, the provision of material was sometimes difficult or not scheduled at the right time. Second, 24 h sampling or even more spot urine measurements were prone to mistakes by patients or staff. Third, interfering admixture by nutrition or medication may have raised problems.

6.3.1.1 Circadian Rhythms

A clear circadian rhythm has been documented both for PTH and biomarkers of bone formation and degradation. Amplitudes of these undulations are different for each marker. The maximum level of most of these biomarkers is reached during sleep. Blood drawing should therefore always be restricted to the same hours in the morning after an at least overnight fast.

6.3.1.2 Nutrition

The dependence of biomarkers of bone metabolism on dietary influences has been known for some years. In an initial study, serum crosslaps levels varied during the day with a maximum at about 05:00 in the morning and a minimum at about 14:00 in the afternoon. The variation had a magnitude of about 40 % around the 24 h mean and was independent of menopause, bone mass, and bed rest (Qvist et al. 2002). In a cross-over study, bone resorption was reduced by intake of glucose, fat, and protein and counteracted by fasting, independent of age and gender. Both exogenous and endogenous insulin stimulation tests induced a decrease in bone resorption by 50 %, but this was only modest when compared with the reduction observed during food intake (Bjarnason et al. 2002). Blood drawing should therefore always be carried out under the same conditions, preferentially in the morning after an overnight fast.

There is an additional role of dietary protein in bone health, which has been controversially discussed. A very recent publication shows an additional modest beneficial long-term effect of protein-rich nutrition on bone markers and bone density (Jesudason and Clifton 2011).

6.3.1.3 Stability

Stability of bone biomarkers has been reported controversially and general statements are difficult. Most of the analytes – except for small well-defined molecules – are very heterogeneous substances and may even differ interindividually. Epitope specificity is therefore often rare and might be subject of change due to sample taking, transportation, and storage.

Cross-reactions between epitopes and antibodies might further influence the stability of the results.

For longer storage, some specific publications describe collagen products such as PINP and serum crosslaps as stable in serum for at least 48 h at room temperature, 7 days at 2–8 °C, and frozen for at least 6 months at −20 to −80 °C (Lomeo and Bolner 2000). In summary, storage of serum and urine samples for bone metabolism markers at 2–8 °C can be delayed for at least 1 week; for long-term storage, (deep) freezing of the sample provides molecular stability for several months for most analytes.

6.3.2 Interpretation

Problems with the analysis per se and problems with the interpretation due to environmental factors are inherent issues not only of bone biomarkers.

6.3.2.1 Disturbances During Analysis

Many of the presented bone biomarkers can be measured by commercially available assays. Most of them use common technologies like RIAs or ELISAs. However, problems with divergent epitope structure such as in collagen metabolites, interference of other molecules or haemolysis, degradation prior to measurement as in enzymes like TRAP 5b and many other influences may contribute to results that may not be reproducible or scatter a lot. Experienced labs have therefore introduced validation and standardization steps during implementation and routine assay performance. Special standard operating procedures and a clear documentation for all analytical steps should therefore be obligatory in all labs performing bone marker analyses.

6.3.2.2 Therapeutic Influences

As stated in the sections on changes during bone therapy, bone biomarkers are very much influenced by all kinds of osteotropic therapies. Whereas bone resorption markers decrease significantly during antiresorptive medication, biomarkers of bone formation and resorption may greatly increase during osteoanabolic therapy, even after prior antiresorptive treatment. Interpretation of lab values has therefore to take into account both therapy regimens and the interval between medication intake and blood drawing.

6.3.2.3 Increasing Standardization

In 2011, the International Osteoporosis Foundation (IOF) and the International Federation of Clinical Chemistry and Laboratory Medicine Working Group (IFCC-WG) (Vasikaran et al. 2011) suggested serum concentrations of PINP and CTX as bone biomarker reference analytes, which should be used in clinical trials and observational studies related to osteoporosis. Many of the biomarker tests currently used have been validated by their customers and should again be optimized in the respective user lab. Standardization of bone biomarkers were done in large sample collection procedures in the USA and Germany, where reference range for one bone formation (PINP) and one bone resorption (CTX) markers and assays used in clinical laboratories have been performed (Bauer et al. 2012; Michelsen et al. 2013). The projects build on the recommendations of the IOF/IFCC Bone Marker Standards Working Group for the most promising markers identified.

Issues of diverging standardization are currently conducted by a number of international consortia. Different marker units used in different countries are well known. 25(OH) vitamin D is an example with the use of ng/ml in Europe and the use of nmol/ml in the USA (conversion: ng/ml × 2,496 = nmol/l). In addition, more complex phenomena have been targeted in terms of bioavailability and technical detection between total vitamin D immunoassays due to various cross-reactivities of their metabolites (Lee et al. 2015). These problems have been recognized, and many institutions are currently working on more general recommendations and standardizations for the most frequently used biomarkers, e.g. again 25(OH) vitamin D for assay and normal value standardization as by the Vitamin D Standardization Programme (VDSP), available at ods.od.nih.gov/Research/vdsp.aspx.

6.3.2.4 Age and Gender

Biomarkers are prone to change, especially during age. As the skeleton evolves and reacts to the age-specific requirements, bone biomarkers may be elevated during phases of growth and more decreased during phases of "rest". The differences in female and male skeletons and disease incidences have been known for decades or a priori and should be characterized separately, except for several regulators such as PTH and 25(OH) vitamin D.

The problem of confining "normal" serum biomarker values includes the availability of higher numbers of disease-free, medication-free, gender-equated and age-distributed persons who have to be phenotyped for as many bone characteristics and concomitant variables as possible. Based on these conditions, recent advances have been made in the interpretation of method-specific reference interval in large, gender- or age-controlled groups (Michelsen et al. 2013). Further efforts using complex biobanking systems as well as international consortia are on the way to optimize these issues.

6.3.2.5 Reference Ranges and Least Significant Change

The most intensively studied group of persons using bone biomarkers are postmenopausal women, being the first group of interest for the diagnosis and treatment of osteoporosis. Premenopausal and male subjects have been identified as diagnostic and therapeutic targets more recently. Recent studies have standardized biomarkers of interest, such as a report of Michelsen et al. on an intensively characterized, large reference population (1107 men, 832 pre- and postmenopausal women) free of bone-related diseases to determine robust reference intervals for serum concentrations of PINP, BAP and CTX in adult men and pre- and postmenopausal women (Michelsen et al. 2013).

Normal values for these age and gender groups have now been established for most of the current parameters, but not for the new generation of more specific bone assays. Children and their normal values are of special interest to researchers and paediatricians, but studies on healthy children are difficult to conduct (Rauchenzauner et al. 2007). However, normal values should be evaluated in the given population and environment wherever possible to be used for routine applications.

Two ways of interpretation of bone marker changes have been established: First, the least significant change (LSC) identifying the minimum change that can be attributed to treatment effect rather than random variation of the marker. The LSC is expressed as % change or absolute units and serves as target for treatment being to reduce the bone marker by at LSC. Alternatively, a second approach is the use of reference intervals (RI). These intervals may also be targets for treatment to decrease bone turnover markers to the lower half of the respective healthy range. However, while this approach would be easily applied in clinical practise, not all patients are above the RI mean before starting treatment and individual changes might better be described by the LSC model (Naylor et al. 2016).

Using these new interpretation tools of bone turnover markers, the International Osteoporosis Foundation (IOF) has proposed that failure to respond to treatment might be defined as:

- Two or more incident fragility fractures
- Lack of a response of bone turnover markers and a significant decrease in BMD

These points indicate a new role of bone turnover markers in the treatment of osteoporosis and other potential diseases of bone metabolism.

6.3.2.6 Representation of Bone Marker Dynamics

Biochemical markers of bone turnover reflect resorptive and reconstructive effects on the skeleton. Although elevated markers are commonly interpreted as a sign of an increased turnover rate, the balance between bone resorption and formation is mostly neglected. Two methods of standardization have been invented: first, the uncoupling index (the difference of z-scores from resorption and formation markers in urine and serum), which estimates balance and not turnover rate (Eastell et al. 1993), and second, more recently, a bone marker plot (Bieglmayer and Kudlacek 2009) based on the combined analysis of serum formation and resorption markers was established, assuming a skewed distribution of the biomarkers. The graphic method is designed to report both on individual- and on group-specific changes in bone metabolism reflected by bone markers.

6.4 Utility of the Determination of Serum Markers of Bone Metabolism

6.4.1 Fracture Prediction

High bone remodelling rates have been associated with an increased risk of fractures (Johansson et al. 2014). Large epidemiologic studies have demonstrated that bone turnover is an independent contributor to fracture risk. Combining a bone biomarker with bone density showed an additive effect on fracture risk, e.g. in the Os des Femmes de Lyon (OFELY) and the Epidemiologie de l'Ostéoporose (EPIDOS) studies (Garnero et al. 2000). In the OFELY study, women in the highest quartile of bone biomarkers had a twofold higher risk of fractures compared with women with low markers of bone turnover.

Recent studies (Vasikaran et al. 2011; Dachverband Osteologie 2014) suggested that bone biomarkers may help to identify individual patients at increased risk of bone fractures. The recently published German guideline for prevention, diagnosis and therapy of osteoporosis (Dachverband Osteologie 2014) recommends the measurement of biomarkers of bone turnover in individual cases.

6.4.2 Therapy Monitoring and Response to Therapy

By using and interpreting bone turnover rates, the biologic action of therapies can be assessed, e.g. the therapeutic efficacy of antiresorptive agents. Bone biomarkers rapidly decrease during bisphosphonate therapy in postmenopausal women, and these changes are associated with increased bone mass and/or fracture rate reduction.

Nonetheless, the notion that fracture risk reduction is associated with a decrease in bone biomarkers during therapy is supported by all studies on oral or intravenous bisphosphonates. Similar results exist for alendronate, risedronate and ibandronate, and more recent data also demonstrate that zoledronic acid, administered once annually for 3 years, resulted in a reduction in vertebral fractures accompanied by a reduction in CTX, bone ALP, and PINP levels (Black et al. 2007). Likewise, similar correlations have been reported for other antiresorptive agents. Biochemical responses to three oral bisphosphonates were assessed in the TRIO study. This open controlled trial randomized 172 patients to alendronate, ibandronate or risedronate, plus calcium and vitamin D supplementation for 2 years (Naylor et al. 2016). Over 70 % of women achieved a target response for serum CTX and PINP, irrespective of the approach used and the same proportion responded by BMD with significant differences to nonresponders, which had smaller increases in BMD and decreases bone turnover markers. Both approaches to response identified similar proportions of women as responders. It was therefore suggested that biochemical assessment of response is a useful tool for the management of women with postmenopausal osteoporosis.

An opposite reaction is seen during osteoanabolic therapy by the increase of bone formation markers, like PINP and bALP, but also by the increase of bone resorption biomarkers, e.g. teriparatide (Obermayer-Pietsch et al. 2008). The changes in bone biomarkers are seen as early as during the first days till the first month after the initiation of therapy (Blumsohn et al. 2011) and are not dependent on prior antiresorptive therapy.

6.4.3 Compliance and Adherence to Therapy

Patients may not always be compliant to therapy regimens, especially over longer periods of treatment as requested in bone therapy. Compliance and adherence to treatment is a potentially useful application of bone biomarkers, particularly in the case of bone resorption inhibitors. Adherence to oral medications is notoriously poor and represents one of the major challenges of reducing the incidence of fractures in the elderly. Bisphosphonates, which must be taken following a strict dosing procedure and the introduction of weekly and monthly oral formulations, have only slightly improved adherence or persistence. Thus, bone biomarkers could be useful in identifying a less than expected suppression of bone turnover, which may suggest either persistence failure (i.e. the patient has discontinued treatment), or that the patient has not been fully compliant with the dosing regimen (Seibel 2006). Such findings could then be discussed with the patient and corrective measures be implemented.

Alternatively, inadequate suppression of bone biomarkers might indicate poor intestinal absorption of oral bisphosphonates. Unfortunately, these highly polar compounds are poorly absorbed by the intestine (Clowes et al. 2004), and low bioavailability may be more frequent than generally thought, especially in elderly patients. This problem might well be a frequent cause of "treatment failure" of oral bisphosphonates. In such cases, a change in treatment modality, e.g., switching from oral to parenteral delivery or changing to a different agent or class of medication, should be considered. A new aspect involves least significant changes (LSC) that can be attributed to treatment effect rather than random variation of the markers. Expressed as % change or absolute units, they may be targets for treatment, e.g. with bisphosphonates, in reducing the bone marker by at LSC (Neylar 2016).

6.4.4 Selection of Pharmacologic Therapy

Based on a profile of bone biomarkers, therapy decisions may be facilitated: patients with accelerated bone turnover tend to lose bone at a faster rate than those with normal turnover; therefore, they should be the best candidates for antiresorptive therapy. For instance, greater reduction in non-vertebral fractures was observed in subjects with high PINP levels after alendronate treatment (Bauer et al. 2006). In turn, low turnover bone disease has to be excluded from antiresorptive treatments and might profit more from osteoanabolic therapy.

6.5 Future Aspects

Bone biomarkers open a big spectrum of disease research in the bone field. New pathways and a challenging number of new biomarkers have been developed recently. Together with new therapeutic applications, a more precise and personalized diagnosis and therapy are therefore possible. Quartiles of bone biomarker values and least significant changes for therapy success will help to use the specific information in everyday practise.

At present, bone biomarkers are more and more used in the clinical setting. Methodological improvements make their use more common beyond the clinical trial environment and combine them with currently used morphological methods. Comparison of least significant change and reference interval approaches for identifying women that reach the target response to oral bisphosphonate therapy and identification of determinants of response are the most promising approaches in recent bone biomarker use.

As discussed in this chapter, bone biomarkers have a variety of potential clinical applications based on their rapid response to treatment, their value in monitoring compliance to medications, and in guiding precise and reliable therapeutic decisions.

The burden of an ageing society will per se increase the need and knowledge for diseases of the skeletal system. Cheap and reliable techniques of individual bone characterization will therefore help to ensure public health.

In summary, bone biomarkers represent a tool to further intensify the implementation of personalized medicine with the focus of medical diagnostics and therapy on individual persons, as we would all want to claim for ourselves.

References

Bauer D, Garnero P, Hochberg MC, for the Fracture Intervention Research Group (2006) Pretreatment levels of bone turnover and the antifracture efficacy of alendronate: the fracture intervention trial. J Bone Miner Res 21:292–299

Bauer D, Krege J, Lane N, Leary E, Libanati C, Miller P, Myers G, Silverman S, Vesper HW, Lee D, Payette M, Randall S (2012) National Bone Health Alliance Bone Turnover Marker Project: current practices and the need for US harmonization, standardization, and common reference ranges. Osteoporos Int 23(10):2425–2433

Bieglmayer C, Kudlacek S (2009) The bone marker plot: an innovative method to assess bone turnover in women. Eur J Clin Invest 39(3):230–238

Bjarnason NH, Henriksen EE, Alexandersen P, Christgau S, Henriksen DB, Christiansen C (2002) Mechanism of circadian variation in bone resorption. Bone 30(1):307–313

Black DM, Delmas PD, Eastell R (2007) Once-yearly zoledronic acid for treatment of postmenopausal osteoporosis. N Engl J Med 356:1809–1822

Blumsohn A, Marin F, Nickelsen T, Brixen K, Sigurdsson G, González de la Vera J, Boonen S, Liu-Léage S, Barker C, Eastell R; EUROFORS Study Group (2011) Early changes in biochemical markers of bone turnover and their relationship with bone mineral density changes after 24 months of treatment with teriparatide. Osteoporos Int 22(6):1935–1946

Bonewald LF (2007) Osteocyte messages from a bony tomb. Cell Metab 5:410–411

Chao T-Y, Yi-Ying W, Janckila AJ (2010) Tartrate-resistant acid phosphatase isoform 5b (TRACP 5b) as a serum maker for cancer with bone metastasis. Clin Chim Acta 411:1553–1564

Clowes JA, Peel NF, Eastell R (2004) The impact of monitoring on adherence and persistence with antiresorptive treatment for postmenopausal osteoporosis: a randomized controlled trial. J Clin Endocrinol Metab 89:1117–1123

Dachverband Osteologie e.V (2014) Prophylaxe, Diagnostik und Therapie der OSTEOPOROSE bei Männern ab dem 60. Lebensjahr und bei postmenopausalen Frauen. S3-Leitlinie des Dachverbands der Deutschsprachigen Wissenschaftlichen Osteologischen Gesellschaften. Available at: http://www.dv-osteologie.org/dvo_leitlinien/osteoporose-leitlinie-2014

Drake MT, Srinivasan B, Moedder U, Peterson J, McCready L, Riggs BL, Dwyer D, Stolina M, Kostenuik PJ, Khosla S (2010) Intermittent PTH treatment decreases circulating sclerostin levels in postmenopausal women. J Clin Endocrinol Metab 95(11):5056–5062

Eastell R, Robins SP, Colwell T, Assiri AM, Riggs BL, Russell RG (1993) Evaluation of bone turnover in type I osteoporosis using biochemical markers specific for both bone formation and bone resorption. Osteoporos Int 3(5):255–260

Egeblad M, Werb Z (2002) New functions for the matrix metalloproteinases in cancer progression. Nat Rev Cancer 2(3):161–174

Fahrleitner-Pammer A, Herberth J, Browning SR, Obermayer-Pietsch B, Wirnsberger G, Holzer H, Dobnig H, Malluche HH (2008) Bone markers predict cardiovascular events in chronic kidney disease. J Bone Miner Res 23(11):1850–1858

Fritchie K, Zedek D, Grenache DG (2009) The clinical utility of parathyroid hormone-related peptide in the assessment of hypercalcemia. Clin Chim Acta 402(1-2):146–149

Garnero P, Sornay-Rendu E, Claustrat B, Delmas PD (2000) Biochemical markers of bone turnover, endogenous hormones and the risk of fractures in postmenopausal women: the OFELY Study. J Bone Miner Res 15:1526–1536

Garnero P, Ferreras M, Karsdal MA, Nicamhlaoibh R, Risteli J, Borel O, Qvist P, Delmas PD, Foged NT, Delaissé JM (2003) The type I collagen fragments ICTP and CTX reveal distinct enzymatic pathways of bone collagen degradation. J Bone Miner Res 18(5):859–867

Hall CL, Keller ET (2006) The role of Wnts in bone metastases. Cancer Metastasis Rev 25:551–558

Hassan MQ, Tye CE, Stein GS, Lian JB (2015) Non-coding RNAs: epigenetic regulators of bone development and homeostasis. Bone 81:746–756

Holick MF (2007) Vitamin D deficiency. N Engl J Med 357(3):266–281

Holzer G, Noske H, Lang T, Holzer L, Willinger U (2005) Soluble cathepsin K: a novel marker for the prediction of nontraumatic fractures? J Lab Clin Med 146(1):13–17

Iba K, Takada J, Yamashita T (2004) The serum level of bone-specific alkaline phosphatase activity is associated with aortic calcification in osteoporosis patients. J Bone Miner Metab 22(6): 594–596

Iwamoto J, Sato Y, Takeda T, Matsumoto H (2009) High-dose vitamin K supplementation reduces fracture incidence in postmenopausal women: a review of the literature. Nutr Res 29(4):221–228

Jesudason D, Clifton P (2011) The interaction between dietary protein and bone health. J Bone Miner Metab 29(1):1–14

Johansson H, Oden A, Kanis JA, McCloskey EV, Morris HA, Cooper C (2014) A meta-analysis of reference markers of bone turnover for prediction of fracture. Calcif Tissue Int 94:560–567

Kanazawa I, Yano S, Yamaguchi T, Notsu Y, Nabika T, Sugimoto T (2010) Relationships between dimethylarginine and the presence of vertebral fractures in type 2 diabetes mellitus. Clin Endocrinol (Oxf) 73(4):463–468

Kos K, Wilding JP (2010) SPARC: a key player in the pathologies associated with obesity and diabetes. Nat Rev Endocrinol 6(4):225–235

Lee JH, Choi JH, Kweon OJ, Park AJ (2015) Discrepancy between vitamin D total immunoassays due to various cross-reactivities. J Bone Metab 22(3):107–112

Li EK, Tam LS, Griffith JF, Zhu TY, Li TK, Li M, Wong KC, Chan M, Lam CW, Chu FS, Wong KK, Leung PC, Kwok A (2009) High prevalence of asymptomatic vertebral fractures in Chinese women with systemic lupus erythematosus. J Rheumatol 36(8):1646–1652

Liu Y, Shi Z, Silveira A, Liu J, Sawadogo M, Yang H, Feng X (2003) Involvement of upstream stimulatory factors 1 and 2 in RANKL-induced transcription of tartrate-resistant acid phosphatase gene during osteoclast differentiation. J Biol Chem 278(23):20603–20611

Ljusberg J, Wang Y, Lång P, Norgård M, Dodds R, Hultenby K, Ek-Rylander B, Andersson G (2005) Proteolytic excision of a repressive loop domain in tartrate-resistant acid phosphatase by cathepsin K in osteoclasts. J Biol Chem 280(31):28370–28381

Lomeo A, Bolner A (2000) Stability of several biochemical markers of bone metabolism. Clin Chem 46:1200–1202

MacCallum SG, Voegtlin C (1909) On the relation of tetany to the parathyroid glands and to calcium metabolism. J Exp Med 11:118–151

Macdonald BT, Joiner DM, Yserman SM, Sharma P, Goldstein SA, He X, Hauschka PV (2007) Bone mass is inversely proportional to Dkk1 levels in mice. Bone 41:331–339

Michelsen J, Wallaschofski H, Friedrich N, Spielhagen C, Rettig R, Ittermann T, Nauck M, Hannemann A (2013) Reference intervals for serum concentrations of three bone turnover markers for men and women. Bone 57(2):399–404

Mirza FS, Padhi ID, Raisz LG, Lorenzo JA (2010) Serum sclerostin levels negatively correlate with parathyroid hormone levels and free estrogen index in postmenopausal women. J Clin Endocrinol Metab 95(4):1991–1997

Naylor KE, Jacques RM, Paggiosi M, Gossiel F, Peel NF, McCloskey EV, Walsh JS, Eastell R (2016) Response of bone turnover markers to three oral bisphosphonate therapies in postmenopausal osteoporosis: the TRIO study. Osteoporos Int 27(1):21–31

Obermayer-Pietsch BM, Marin F, McCloskey EV, Hadji P, Farrerons J, Boonen S, Audran M, Barker C, Anastasilakis AD, Fraser WD, Nickelsen T, EUROFORS Investigators (2008) Effects of two years of daily teriparatide treatment on BMD in postmenopausal women with severe osteoporosis with and without prior antiresorptive treatment. J Bone Miner Res 23(10):1591–1600

Pelletier JP, Raynauld JP, Caron J, Mineau F, Abram F, Dorais M, Haraoui B, Choquette D, Martel-Pelletier J (2010) Decrease in serum level of matrix metalloproteinases is predictive of the disease-modifying effect of osteoarthritis drugs assessed by quantitative MRI in patients with knee osteoarthritis. Ann Rheum Dis 69(12):2095–2101

Qvist P, Christgau S, Pedersen BJ, Schlemmer A, Christiansen C (2002) Circadian variation in the serum concentration of C-terminal telopeptide of type I collagen (serum CTx): effects of gender, age, menopausal status, posture, daylight, serum cortisol, and fasting. Bone 31(1):57–61

Ramon I, Kleynen P, Body JJ, Karmali R (2010) Fibroblast growth factor 23 and its role in phosphate homeostasis. Eur J Endocrinol 162(1):1–10

Rauchenzauner M, Schmid A, Heinz-Erian P, Kapelari K, Falkensammer G, Griesmacher A, Finkenstedt G, Högler W (2007) Sex- and age-specific reference curves for serum markers of bone turnover in healthy children from 2 months to 18 years. J Clin Endocrinol Metab 92(2):443–449

Rennenberg RJ, de Leeuw PW, Kessels AG, Schurgers LJ, Vermeer C, van Engelshoven JM, Kemerink GJ, Kroon AA (2010) Calcium scores and matrix Gla protein levels: association with vitamin K status. Eur J Clin Invest 40(4):344–349

Roberts HC, Knott L, Avery NC, Cox TM, Evans MJ, Hayman AR (2007) Altered collagen in tartrate-resistant acid phosphatase (TRAP)-deficient mice: a role for TRAP in bone collagen metabolism. Calcif Tissue Int 80(6):400–410

Schett G, Kiechl S, Redlich K, Oberhollenzer F, Weger S, Egger G, Mayr A, Jocher J, Xu bQ, Pietschmann P, Teitelbaum S, Smolen J, Willeit J (2004) Soluble RANKL and risk of nontraumatic fracture. JAMA 291(9):1108–1113

Schwetz V, Pieber T, Obermayer-Pietsch B (2012) The endocrine role of the skeleton: background and clinical evidence. Eur J Endocrinol 166(6):959–967

Seibel MJ (2006) Biochemical markers of bone turnover part II: clinical applications in the management of osteoporosis. Clin Biochem Rev 27:123–138

Souberbielle JC, Cavalier E, Jean G (2010a) Interpretation of serum parathyroid hormone concentrations in dialysis patients: what do the KDIGO guidelines change for the clinical laboratory? Clin Chem Lab Med 48(6):769–774

Souberbielle JC, Roth H, Fouque DP (2010b) Parathyroid hormone measurement in CKD. Kidney Int 77(2):93–100

Taylor AF, Saunders MM, Shingle DL, Cimbala JM, Zhou Z, Donahue HJ (2007) Mechanically stimulated osteocytes regulate osteoblastic activity via gap junctions. Am J Physiol 292:C545–C552

Terpos E (2006) Biochemical markers of bone metabolism in multiple myeloma. Cancer Treat Rev 32:15–19

Ulbing M, Obermayer-Pietsch B (2015) MicroRNAs in bone, overview and clinical significance. Osteologie 24:219–224

Ureña Torres P (2006) The need for reliable serum parathyroid hormone measurements. Kidney Int 70(2):240–243

Uzum AK, Salman S, Telci A, Boztepe H, Tanakol R, Alagol F, Ozbey NC (2010) Effects of vitamin D replacement therapy on serum FGF23 concentrations in vitamin D-deficient women in short term. Eur J Endocrinol 163(5):825–831

Vasikaran SR, Eastell O, Bruyere AJ, Foldes P, Garnero A, Griesmacher A, McClung M, Morris HA, Silverman S, Trenti T, Wahl DA, Cooper C, Kanis JA; IOF-IFCC Bone Marker Standards Working Group (2011) Markers of bone turnover for the prediction of fracture risk and monitoring of osteoporosis treatment: a need for international reference standards. Osteoporos Int 22(2):391–420

Wang WM, Ge GX, Lim NH, Nagase H, Greenspan DS (2006) TIMP-3 inhibits the procollagen N-proteinase ADAMTS-2. Biochem J 398(3):515–519

Whyte MP (2010) Physiological role of alkaline phosphatase explored in hypophosphatasia. Ann N Y Acad Sci 1192:190–200

Osteoporosis: Pathophysiology and Clinical Aspects

7

Peter Mikosch

7.1 Osteoporosis, a Worldwide Disease with High Economic Burden

Osteoporosis has become a major health issue over the last years due to the steadily increasing life expectancy. Osteoporosis is estimated to affect 75 million people in the United States, Europe, and Japan and about 200 million women worldwide – approximately one-tenth of women aged 60, one-fifth of women aged 70, two-fifths of women aged 80, and two-thirds of women aged 90. Worldwide, one in three women over age 50 will experience osteoporotic fractures and one in five men aged over 50. Overall, 61 % of osteoporotic fractures occur in women, with a female-to-male ratio of 1.6 (Johnell and Kanis 2006).

For the year 2000, there were an estimated nine million new osteoporotic fractures, of which 1.6 million were at the hip, 1.7 million were at the forearm, and 1.4 million were clinical vertebral fractures. Europe and the Americas accounted for 51 % of all these fractures, while most of the remainder occurred in the Western Pacific region and Southeast Asia (Johnell and Kanis 2006). Based on these fracture data, an osteoporotic fracture occurs every 3 s (Johnell and Kanis 2006). In line with the demographic development, fractures of the humerus, wrist, or hip will occur noticeably more often during the next four decades (Lohmann et al. 2007). The number of patients with hip fractures will increase to 170 % of present-day numbers and in the age group over 80 years to 250 % (Lohmann et al. 2007).

P. Mikosch, MD
Department of Internal Medicine, State Hospital Oberpullendorf, Spitalgasse 23, Oberpullendorf 7350, Austria
e-mail: peter.mikosch@krages.at

P. Pietschmann (ed.), *Principles of Osteoimmunology*,
DOI 10.1007/978-3-319-34238-2_7

In the United States, over ten million have been diagnosed with osteoporosis causing direct medical costs of 17 billion dollars (Tosteson et al. 2005). In Germany, 7.8 million (6.5 million women) were affected by osteoporosis in 2003 (Häussler et al. 2007). At least one clinical fracture was present in 4.3 % of these patients leading to direct costs of 5.4 billion euro, although only 21.7 % of the patients were treated with anti-osteoporotic drugs as shown in a recent study from Germany (Häussler et al. 2007). Considering only osteoporosis-attributable hip fractures, 108,341 occurred in Germany in 2002 resulting in costs of almost three billion euro, which will more than double according to estimations in 2050 (Konnopka et al. 2009). These already tremendous costs of health care linked to osteoporosis are further alarming, as a care gap with underdiagnosis and undertreatment of the entity has been stressed in several studies (Pietschmann et al. 2010; Wright 2007; Haaland et al. 2009; Metge et al. 2008; Papaioannou et al. 2008; Giangregorio et al. 2009; Kamal 2005). Improvements in diagnostic strategies, the diagnostic workup in the context of interdisciplinary settings, are warranted in order to optimize the management and care of patients with osteoporosis. The basis of such an aim has to be set up with a better and broader understanding of the pathophysiology, clinical presentation, interactions with other disorders, and the currently available therapeutic possibilities.

Before discussing the pathophysiological mechanisms on the cellular and intracellular level, a short description of the clinical presentation of patients with osteoporosis and the development of bone mass during lifetime will be presented in order to show the connection between clinical appearance of osteoporosis and its underlying cellular mechanisms.

7.2 Balance of Bone Remodeling During Lifetime and the Development of Bone Fragility

Bones and the skeleton resemble a static organ giving shape to the human body and protection to certain organs such as the brain, heart, and lungs. It is also a niche for mesenchymal and hematopoietic progenitors with the bone marrow being the place of blood production. However, bone is a highly vivid and dynamic organ being constantly molded, shaped, and repaired in order to meet the changing demands throughout life. This process of constant restructuring is termed remodeling, which is a coupled process involving bone resorption by osteoclasts and new bone formation by osteoblasts (Teitelbaum 2007). The action of osteoblasts and osteoclasts in bone remodeling resembles a delicately balanced process that ensures the continual replacement of old bone, weakened by microfractures, with new bone. Failure to reach peak bone mass or the uncoupling of remodeling can result in bone fragility. Bone formation exceeds bone resorption during the years of the growing skeleton in childhood and early adolescence, finally reaching peak bone mass at the age of 30–40 years. In the following years, the balance between bone formation and bone resorption changes toward dominance of bone resorption over bone formation, leading to continuous loss of bone tissue. This loss of bone tissue and also its decline in means of quality and structural integrity with increasing age lead to a state of increased bone fragility.

7.3 Clinical Presentation of Patients with Osteoporosis

In its current definition, osteoporosis is characterized as a disorder of the skeleton with a decrease in bone mass and structural deterioration of bone tissue, leading to bone fragility and increased susceptibility to fractures of the hip, spine, and wrist (NIH Consensus Development Panel on Osteoporosis Prevention, Diagnosis, and Therapy 2001). After achieving peak bone mass at the age of 30–40 years, patients start to steadily loose bone due to the imbalance of bone formation and bone resorption in favor of bone resorption. This process of bone loss is accelerated in women with the onset of menopause. Initially patients will not face any symptoms as long as loss of bone mass, deterioration of microarchitecture, and the decline of quality in bone material will be limited. With ongoing age, however, the stability of bone will deteriorate to such a level that even without severe traumatic events, microcracks within bone will not be fully repaired anymore. The morphological changes associated with osteoporosis start within the central skeleton. The consequences are a slow but steady deformity and loss of height of the vertebrae, which are part of the weight bearing skeleton. The first regions of the vertebral column affected by these changes are the lower parts of the thoracic spine and the upper part of the lumbar spine. The kyphosis of the thoracic vertebral column increases, leading to a forward curving of the body and in severe cases to a clearly visible hump ("Dowager's hump"). In addition, patients are facing a progressive loss of height. The increased kyphosis of the thoracic spine and the loss of height are the first clinical signs of osteoporosis. At this stage, patients may develop back pain which may be due to the changed static situation. Besides a slow and steady loss of vertebral height, the first vertebral fractures may also emerge, followed by severe local back pain. In progressive stages, patients have lost several centimeters of height and usually present severe kyphosis of the thoracic spine. In addition, secondary to these changes, the arms of the patients seem to be elongated in relation to the patient's height and the abdomen, and lungs are compressed, leading to a protruding abdomen and eventually to shortness of breath and pulmonary symptoms of restrictive lung disease. With ongoing age, the progressive loss of bone mass and bone quality will make the bone so fragile that simple falls during daily life will lead to fractures, most commonly hip fractures but also to fractures of the shoulders and the upper- or the forearms. In the elderly, hip fractures are most common and will cause increased morbidity and mortality. Elderly people who have sustained a hip fracture will fully recover only in a small percentage, most face smaller or larger impairments and losses of abilities after a hip fracture. About 20% will die within 1 year after a hip fracture. Very severe cases may even face spontaneous fracture without a fall or any trauma. The deformity of the vertebral column and chronic back pain severely impairs the quality of life of the patients. Patients will often become dependent on assistance in order to perform the daily life activities. Furthermore, anxiety to fall often leads to possible social retraction and isolation, as patients will not be able or willing to leave their homes anymore. Thus, in this respect, osteoporosis then becomes not only a disease affecting the bone but actually affects the life of the patient by severely impairing his mobility, independence, and mental state (Fig. 7.1).

7.4 An Approach to the Mechanisms Leading to Osteoporosis

The quantitative loss of bone mass and the qualitative structural changes of bone leading to osteoporosis are based on complex metabolic changes during life. Estrogen deficiency is the major cause of osteoporosis. However, also other hormonal or inflammatory processes may cause bone loss. Bone tissue and bone cells on the one hand and also cells of the bone marrow on the other hand are closely linked with each other, not only in space but also by physiological and pathophysiological interaction. Hormones, cytokines, and a range of inflammatory mediators all interfere with each other and exert a range of effects in the various organs. Osteoimmunology investigates the complex connections and interactions between immune cells and immune system cytokines and chemokines on the one hand, and bone cells involved in the bone remodeling process on the other hand (Walsh et al. 2006). The common origin of bone and immune stem cells is the key to understand this system and the physiology of bone loss. Knowledge in this rapidly growing field may also facilitate the translation from basic scientific knowledge on bone biology to an improved understanding of different bone disorders, including osteoporosis, periodontal disease, and

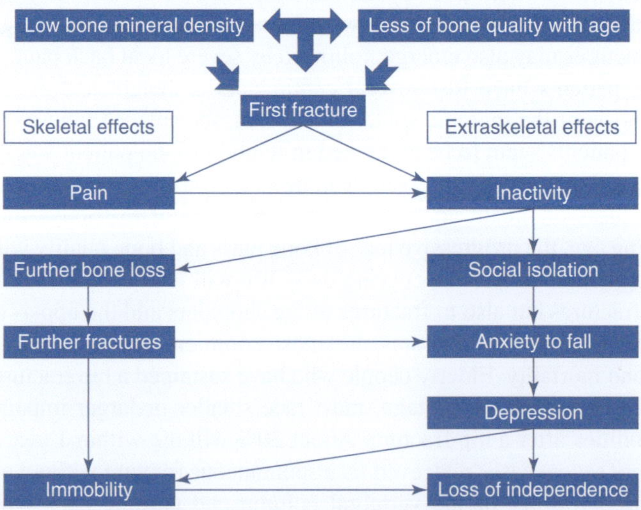

Fig. 7.1 Osteoporosis as a cascade of events affecting not only the skeleton but also activity, social contacts, independence, and the mental status of the patient

rheumatoid arthritis. Furthermore, it also provides the basis for the development and application of targeted therapies of the future (Rothe et al. 2008; Sipos et al. 2008).

7.4.1 Osteoblasts and Osteoclasts: Cellular Promoters of Bone Formation and Degradation

Bone-resorbing cells (osteoclasts) and cells of the immune system both originate in the bone marrow from hematopoietic cells (Fig. 7.2). Bone-forming cells (osteoblasts) are of mesenchymal origin derived from pluripotent mesenchymal stem cells (MSC), which can also give rise to chondrocytes, myoblasts, tenocytes, neurons, and adipocytes (Walsh et al. 2006). During osteoblast differentiation, MSC express increased amounts of phenotypic markers (e.g., alkaline phosphatase, osteocalcin), receptors for bone morphogenetic proteins (BMP), and the Wnt receptor low-density lipoprotein receptor-related proteins (LRP) 5 and 6. Through the activation of these receptors, the progenitor cells are differentiated into osteoblasts with the capacity of bone formation (Yamaguchi et al. 2000).

The counterparts of the bone-forming osteoblasts are the osteoclasts, which origin and develop from precursors of the mononuclear monocyte-macrophage cell line (Fig. 7.2). Osteoclasts resemble highly specialized multinucleated cells being exclusively enabled to resorb bone. Osteoclasts precursors at or close to the bone surface are induced to differentiate into mature osteoclasts after stimulation by macrophage colony-stimulating factor (M-CSF) and receptor for activated nuclear kappa B (RANK) ligand (RANKL). Osteoclasts then attach to bone surface, mediated by integrins. A sealing zone, which is formed by rearrangement of the actin

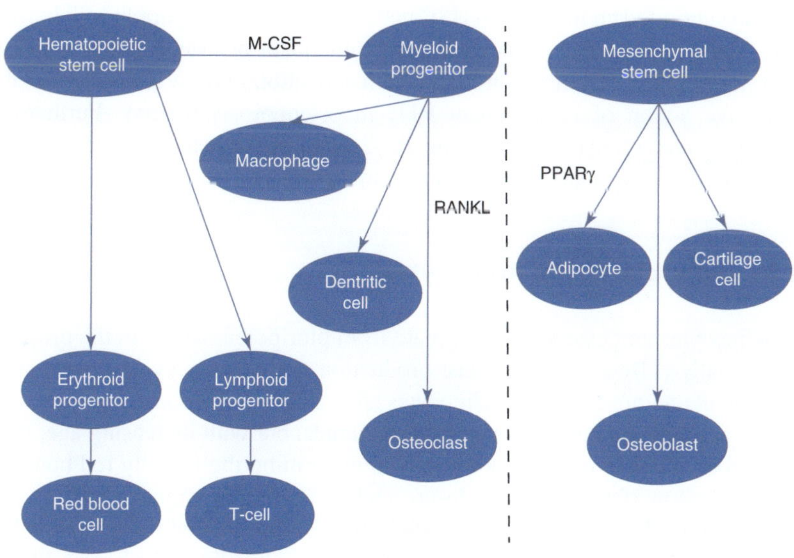

Fig. 7.2 Lineage of osteoclast and osteoblast development

cytoskeleton, occludes the bone surface and the outer rim of the attaching osteoclast, forming a separated resorption pit with the underlying bone surface afterward. During this process, the osteoclasts undergo a structural change into a polarized cell shape. The basal area of the osteoclast enlarges by forming a "ruffled border." The osteoclast can release hydrogen ions through the H+ ATPase into the resorption pit. Proteases, like tartrate-resistant acid phosphatase (TRAP), are then excreted via endocytoplasmatic transport into the resorption pit where they mobilize the hydroxyapatite crystals. In a next step, cathepsin K is secreted by the osteoclast to degrade the organic bone matrix within the acidified fluid. Cathepsin K plays a major role in the proteolytic process of matrix degeneration. The products resorbed from the bone, mainly calcium, phosphate, and fragments of the organic matrix, are taken up by the osteoclast and released into the circulation via its apical surface. Osteoclasts are of essential importance for the homeostasis of calcium and phosphate, and their activity is tightly controlled via different modes including activation by parathyroid hormones (Teitelbaum 2007; Boyle et al. 2003).

7.4.2 Osteoblasts and Stromal Cells: The Connection with the Hematopoietic System

The vast majority of hematopoiesis is located in the bone marrow. Proliferation and differentiation of hematopoietic stem cells (HSC) are controlled by "stromal" cells. The nature of these mesenchymal cells and their mode of action still lack explanation. However, recently it has been pointed out that the stromal cells responsible for HSC regulation in the bone marrow are most likely osteoblasts (Calvi et al. 2003). It could be shown that osteoblastic cells produced high levels of Notch ligand jagged 1 and supported an increase in HSC under the influence of parathyroid hormone (PTH) (Calvi et al. 2001). This regulatory component of osteoblastic cells on the hematopoietic stem cell niche through Notch activation may be the background for the positive action of recombinant PTH in osteoporosis therapy. Furthermore, through this interaction of PTH with the HSC niche, possible therapeutic options for anemia and immunosuppression are now under investigation.

7.4.3 Osteoblasts and Adipocytes

Both adipocytes and osteoblasts originate from pluripotent MSC. In the process of aging, negative effects on osteoblast production and activity evolve, whereas the development and production of adipocytes are enhanced. Adipocytes are rare in the hematopoietic bone marrow of neonatal mammals, but with increasing age, adipocyte number and size increase constantly, transforming the initially red bone marrow into yellow fatty marrow. Intra- and extracellular signals influence the development of the undifferentiated MSC. Runx2 (runt-related transcriptional factor) is essential for the development into the osteoblast linage, whereas PPARγ2 (peroxisome proliferator-activated receptor gamma 2) induces differentiation into adipocytes. PPARγ2 concomitantly induces the differentiation of adipocytes and inhibits the differentiation of osteoblasts (Lecka-Cernik et al. 2002). This makes PPARγ2 a key regulator of the differentiation of osteoblasts. Ectopic expression of recombinant PPARγ2 on osteoblasts suppressed irreversibly runx2 expression and the osteoblast phenotype, transforming the osteoblasts in terminally differentiated adipocytes (Kim et al. 2005).

7.4.4 Osteoclastogenesis and Monocyte Macrophage Interaction

Osteoclasts are derived from the mononuclear monocyte-macrophage cell line, which also generate immune cells, including dendritic cells. Osteoclasts derive from the myeloid-monocyte branch of hematopoietic cells, thus sharing the same precursors as macrophages and myeloid dendritic cells. Depending on the microenvironment to which the precursor cells are exposed, a further development specific to the different cell lines will emerge. The multipotential myeloid progenitor cell population is defined by the surface marker C-Kit and it also expresses the panmyeloid marker CD11b, whereas c-Fms, which is the tyrosine kinase receptor for M-CSF necessary for cell differentiation into osteoclasts, is not expressed. Through interaction of the CD11b+monocytic precursor cells with stem cell factor (SCF), the cells become c-Fms positive (Ziegler-Heitbrock 2007). C-Fms is a key determinant in the development of cells in the monocyte-macrophage lineage (Teitelbaum 2007). The presence of M CSF transforms the early-stage precursor cells to late-stage precursors by increased CD11b expression and upregulation of the expression of receptor activator of NFkB (RANK). RANK ligand (RANKL) will then bind to RANK, inducing a cascade of signaling events which leads to osteoclast formation (Ziegler-Heitbrock 2007). RANK signaling is mediated via TRAF6 (TNF receptor-associated factor 6) in osteoclast precursors. TRAF6 is by itself essential in the regulation of further downstream factors regulating the expression of specific genes necessary for osteoclast differentiation and activation (NFkB, alkaline phosphate-1 mediated by JNK pathway, TGF-β-inducable kinase TAK1, p38 stress kinase) (Wong et al. 1999). NFkB and c-Fos activation induce the transcriptional factor NFATc1, leading to the expression of genes such as TRAP, cathepsin K, and dendritic cell-specific transmembrane protein (DC-STAMP), which are essential for osteoclast formation and function. DC-STAMP is able to induce IL-4 and has been found in osteoclasts, myeloid dendritic cells, and macrophages (Yagi et al. 2006).

7.4.5 The RANK/RANKL/OPG System: Key Regulator of Bone Homeostasis

Bone homeostasis is regulated by a delicate balance between osteoblastic bone formation and osteoclastic bone resorption. Although estrogen is the key sex hormone governing bone homeostasis, the primary regulator of bone remodeling is now being recognized as the RANK/RANKL/OPG system. Osteoclastogenesis is controlled by the ratio of receptor activator of NF-kappaB ligand (RANKL) relative to its decoy receptor, osteoprotegerin (OPG). During normal bone remodeling, marrow stromal cells and osteoblasts produce RANKL, which binds to the transmembrane receptor RANK on osteoclast precursors and induces differentiation and activation as pointed out above (Fig. 7.2). This occurs through the transcription factor, nuclear factor kappa B (NFkB), which is responsible not only for activating osteoclastogenesis but also the body's inflammatory response. Both osteoclast differentiation and the inflammatory process occur via regulation of interleukin-6 (IL-6).

The major role that cytokines play in bone remodeling is demonstrated by the fact that receptors for the proinflammatory cytokines IL-1, IL-6, and tumor necrosis factor-alpha (TNF-α) are present on both osteoclast precursor cells and mature osteoclasts. Estrogen exhibits its nuclear regulatory effects by inhibiting IL-6 activation of NFkB during bone remodeling. Osteoblasts also produce OPG, a soluble decoy receptor that blocks RANKL and maintains control of the remodeling process. OPG is vital to the success of the RANK/RANKL/OPG system of bone homeostasis.

This system determines the success or failure of bone homeostasis. The different inflammatory cytokines, which act in the development of osteoporosis, are in fact essential for adequate bone remodeling. For example, TNF-α-mediated cartilage and bone degradation was IL-1 dependent in a murine arthritis model (Zwerina et al. 2007), and systemic inflammatory bone resorption is fully dependent on IL-1 (Polzer et al. 2010).

At this point, the important interplays between osteoblasts and osteoclasts, and hormones also need to be highlighted with the RANK/RANKL/OPG system being the connecting link between these players. RANKL, which is essential for osteoclast differentiation and activation, is produced by osteoblasts under the influence of vitamin D, parathyroid hormone, and estrogen (Boyle et al. 2003; Rho et al. 2004). Detailed information on the RANK/RANKL/OPG system will be given in a separate chapter (see Chap. 1).

7.4.6 Dendritic Cell Interaction with Bone Cells

Dendritic cells are derived like osteoclasts from the monocyte/macrophage lineage. Dendritic cells develop after exposure to macrophage colony-stimulating factor (M-CSF) and IL-4 (Miyamoto et al. 2001). Dendritic cells are responsible for antigen presentation with selective stimulation of T cells and B cell, thereafter leading to a specific immune response. A direct involvement of dendritic cells in osteoclastogenesis is suspected, as dendritic cells can transdifferentiate into osteoclasts in vitro (Rivollier et al. 2004). Dendritic cells may also be involved in bone loss due to inflammation (Alnaeeli et al. 2007) where they interact with T cells. This is thought to be based on the fact that dendritic cells express RANK and that they mediate an osteoimmunological interaction indirectly by activating T cells to produce RANKL, which then induces osteoclastogenesis (Jones et al. 2002).

7.4.7 T Cells and Osteoporosis

Activated T cells can either stimulate or suppress the formation of bone-resorbing osteoclasts. T-helper cells 1 (Th1) produce IFN-γ and Th2 IL-4, which are cytokines suppressing osteoclastogenesis. Imbalance in the Thl/Th2 adaptive immune response initiated by antigenic stress may play a role in specific cases of osteoporosis. With T cell activation now known to have a major role in RANKL-induced osteoclastogenesis, more research is needed to determine whether early maturational and/or chronic immunological stressing agents contribute to excessive bone loss in later years. In order to resolve this discrepancy, further research was promoted in order to find T-helper cell populations, which may only exert proinflammatory stimuli toward osteoclastogenesis. This means that these T-helper cells should express low levels of IL-4 and IFN-γ but should express TNF-α and RANKL. These criteria were met by a subpopulation of T-helper cells expressing IL-17 (Takayanagi 2009; Sato et al. 2006).

In a study (Won et al. 2011) on TallyHo/JngJ (TH) mice, a polygenic model of type II diabetes, the spontaneous development of osteoporotic features, possibly mediated by IL-17, was observed. Bone mineral density (BMD) was decreased in male TH mice, which displayed hyperglycemia. The bone formation markers osteocalcin (OC) and OPG were decreased, whereas bone resorption markers such as IL-6 and RANKL were significantly elevated in the bone marrow and blood. Furthermore, RANKL expression was increased in CD4+ T cells of TH mice upon T cell receptor stimulation due to enhanced IL-17 production. IL-17 production in CD4+ T cells was directly promoted by treatment with leptin, while IFN-γ production was not. Blockade of IFN-γ further increased RANKL expression and IL-17 production in TH-CD4+ T cells. These results indicate that increased leptin in TH mice may act in conjunction with IL-6 to stimulate IL-17 production in CD4+ T cells and induce RANKL-mediated osteoclastogenesis (Won et al. 2011). Meanwhile, infiltrating Th17 cells have been identified also in rheumatoid arthritis (RA) and periodontitis, highlighting the pathological role of the immune system in these inflammatory disorders (Pene et al. 2008; Gaffen and Hajishengallis 2008). An interesting connection with 1,25-dihydroxyvitamin D (1,25(OH)2D3) on the immunological processes mediated via Th17 cells was shown in patients with RA, thus highlighting the immunomodulating capacity of vitamin D (Colin et al. 2010).

Mononuclear cells and CD4+CD45RO+ (memory) and CD4+CD45RO– (naive) T cells from treatment-naive patients with early RA were stimulated with anti-CD3/anti-CD28 in the absence or presence of various concentrations of 1,25(OH)2D3, dexamethasone, and a combination of both. The presence of 1,25(OH)2D3 reduced IL-17A and IFN-γ levels and increased IL-4 levels. In addition, 1,25(OH)2D3 had favorable effects on TNF-α/IL-4 and IL-17A/IL-4 ratios and prevented the unfavorable effects of dexamethasone on these ratios. Enhanced percentages of IL-17A- and IL-22-expressing CD4+ T cells and IL-17A-expressing memory T cells were observed in mononuclear cells from treatment-naive patients with early RA, as compared with healthy controls. 1,25(OH)2D3, in contrast to dexamethasone, directly modulated human Th17 polarization, accompanied by suppression of IL-17A, IL17F, TNF-α, and IL-22 production by memory T cells (Colin et al. 2010).

Inflammatory bowel diseases alter bone metabolism and are frequently associated with osteopenia, osteoporosis, and increased risk of fractures. Although several mechanisms may contribute to skeletal abnormalities in patients with inflammatory bowel diseases, inflammation and inflammatory mediators such as TNF, IL-1β, and IL-6 may be the most critical (Ghishan and Kiela 2011). In addition to limited nutrient absorption, high antigen load from food allergies or intestinal microbial overgrowth may also contribute to bone loss. Mature osteoclasts gain access to bone surfaces only after mononucleated preosteoclasts have traveled from the circulatory system to the bone, possibly through mechanisms involving transendothelial migration (Saltel et al. 2006). The gut-associated lymphoid tissue normally provides an immunological barrier against disease. When this barrier becomes compromised by endothelial hyperpermeability secondary to chronic inflammation, food allergy, or bacterial overgrowth, nutrient absorption is reduced, and a loss of oral tolerance can initiate a gastrointestinalimmunological stress factor to the bone remodeling process (Saltel et al. 2006).

RANKL not only regulates the function of osteoclasts but also dendritic cells (professional antigen-presenting cells). In chronic inflammation, RANKL promotes dendritic cell survival and the expression of proinflammatory cytokines (Josien et al. 1999). As the gut is overrun by pathogens, professional antigen-presenting cells, through the activation of Toll-like receptors and C-type lectin receptors, are no longer able to silence immune activation (Geijtenbeek et al. 2004) and release proinflammatory cytokines that activate T cells and reduce Tregs. This antigenic stress leads to a Th1-dominant, cell-mediated immune system with increased RANKL, reduced IFN-γ, and a possible uncoupling of bone remodeling (Vojdani and Erde 2006a, b).

7.4.8 Toll-Like Receptors

A healthy gut flora maintains a reduced production of gut-related proinflammatory cytokines. Toll-like receptors are transmembrane receptors found on macrophages, dendritic cells, and some epithelial cells. These receptors exhibit an integral role in the maintenance of oral tolerance. They recognize the molecular patterns of bacteria and cause an inflammatory, destructive response to pathologic microbes and a tolerogenic response to commensal bacteria. An example of how a disease-related genetic polymorphism can be influenced through the reduction of metabolic stress-inducing factors can be seen in the case of Toll-like receptors and IL-1 receptors. In theory, as the cytoplasmic portion of the Toll-like receptor is similar to that of the IL-1 receptor antagonist gene, it could be susceptible to an increased diversion or "switch" of cells from the monocyte-macrophage cell line to form osteoclasts. A reduction of antigen load and oxidative stress independent of the cause (e.g., insulin/glucose imbalance, toxicity, or gut pathogenic microflora) could reduce proinflammatory cytokine-induced chronic inflammation and T cell activation.

7.4.9 Involution of the Thymus Gland and the Start of Bone Loss

Reduced oral tolerance may be a factor in the coincidence between thymus gland involution (and subsequent reduction of naive T cells) and the beginning of bone loss in humans in their mid-30s. Although BMD does not usually decrease significantly until menopause, accelerated bone loss can commence at an earlier age for some individuals. Reduced numbers of naive T cells from chronic systemic inflammation or antigen overload from the gut lead to oligoclonal T cell expansion and increased T cell senescence (Vojdani and Erde 2006c). Senescence reduces a T cell's ability to produce IFN-γ and is a sign of immune aging (Clowes et al. 2005).

The primordial thymus developed as a bud on the immature digestive tract, providing embryological evidence of the uniquely co-dependent and interrelated functions of the thymus gland and the gastrointestinal tract (Cheroutre 2004). As an infant grows, the function of the thymus is to relieve the gut of its primordial function of lymphopoiesis (Cheroutre 2004). With involution of the thymus, the adult gastrointestinal tract remains the source of the least 75% of the body's immune cells (Bengmark 2004). Therefore, it is in the gut that an adult's immune health is maintained or lost. As an individual ages, antigen load often increases and oral tolerance decreases, leading to reduced levels of IL-2 (necessary for the T cell proliferation and differentiation into activated effector cells) and IFN-γ and ultimately to the greater cache of RANKL-expressing (and thus osteoclast-activating) memory cells harbored in the bone marrow.

7.4.10 B Cells and Their Influence on Bone Cells

In comparison to T cells, the link between B cells and osteology has been sparsely investigated so far, but B cells also seem to play a crucial role in the regulation of bone turnover. The source of OPG has in general been attributed to osteoblasts, but it could be demonstrated that B-lymphoid lineage cells are also a major source of endogenous RANKL in bone marrow and support osteoclast differentiation in vitro (Li et al. 2007; Manabe et al. 2001). In addition, B-lymphoid lineage cells in earlier developmental stages may hold a potential to differentiate into osteoclasts when stimulated with M-CSF and soluble RANKL in vitro. Thus, B-lymphoid lineage cells may participate in osteoclastogenesis in two ways: they (1) express RANKL to support osteoclast differentiation and (2) function themselves as osteoclast progenitors (Manabe et al. 2001). Li et al. evaluated in their study the extent of B cell involvement on OPG production. With the use of immunomagnetic isolation of bone marrow B cells and B cell precursor populations, and quantification of their OPG production by enzyme-linked immunosorbent assay (ELISA) and real-time reverse transcriptase-polymerase chain reaction (RT-PCR), cells of the B lineage were found to be responsible for 64% of total bone marrow OPG production, with 45% derived from mature B cells (Li et al. 2007). B cell knockout mice were found to be consistently osteoporotic and deficient in bone marrow OPG, phenomena rescued by B cell reconstitution (Li et al. 2007). In another study by Breuil et al. (Breuil et al. 2010), changes of different B-lymphocyte populations, related to bone mineral density (BMD) and fractures, were evaluated in postmenopausal osteoporosis. Postmenopausal women with osteoporosis had lower numbers of CD19+, CD19+/CD27+, CD19+/CD27+/CD5–/CD38+ and CD19+/CD27+/RANK+, CD4+/CD27+/CD45RA–/RANK+, and CD4+/CD27+/CD45RA–/CD28+ as compared to healthy controls and were positively correlated to BMD. In addition, in postmenopausal women with osteoporosis CD4+ secreted less IFN-γ and B lymphocytes but more GM-CSF under basal conditions. GM-CSF was positively correlated to fracture rate and negatively to BMD. These results suggest a possible role of IFN-γ in the pathophysiology of osteoporosis based on changes in B-lymphocyte populations (Breuil et al. 2010).

At the gene expression level, an in vivo genome-wide expression study on human B cells in relation to osteoporosis seems to confirm the significant role of B cells in the etiology of osteoporosis (Xiao et al. 2008). Real-time RT-PCR showed differential expression of eight genes, including estrogen receptor 1 (ESR1) and mitogen-activated protein kinase 3 (MAPK3). It was assumed that downregulation of ESR1 and MAPK3 in B cells regulates cytokine expression, causing increased osteoclastogenesis or decreased osteoblastogenesis. These results highlight the significance of B cells in the etiology of osteoporosis (Xiao et al. 2008).

7.5 Hormonal Influences on Bone Remodeling

The RANK/RANKL/OPG system is essential for the bone remodeling process and bone homeostasis and is regulated by different factors, which will be presented in the following section. Relevant hormone influences include estrogen, testosterone, parathyroid hormone, and thyroid hormone.

7.5.1 Estrogen

Estrogen has a critical role in the skeletal preservation of both women (Christiansen et al. 1981) and men (Smith et al. 1994). In women, estrogen deficiency appears to be the major determinant in the development of osteoporosis, whereas in men, it has to be regarded as one factor besides others (e.g., level of testosterone, sex hormone-binding globulin). During 3–4 years after the onset of menopause due to estrogen deficiency, a rapid loss of bone can be observed, in particular trabecular bone. During this early postmenopausal period, the estrogen-deprived skeleton exhibits histological features of accelerated bone remodeling with abundant osteoclasts and resorption bays. Markers of bone turnover are frequently elevated during this period reflecting the accelerated remodeling process. The antiresorptive effects of estrogen on osteoclasts are based on the regulatory effects of estrogen on the OPG/RANK/RANKL system. Estrogen shows direct and indirect effects on the skeleton (Fig. 7.3). The direct effects of estrogen on bone are mediated via estrogen receptors on osteoblasts and osteoclasts (Eriksen et al. 1988;

Komm et al. 1988; Oursler et al. 1998). Estrogen by itself induces OPG production in osteoblasts. Due to the antiresorptive effect of OPG in the RANK/RANKL/OPG system, estrogen exerts osteoprotective effects, whereas estrogen deficiency leads to decreased OPG with increased formation and activation of osteoclasts (Saika et al. 2001).

The indirect effects of estrogen (Fig. 7.3) on bone are caused by different cells including marrow stromal cells and cells of the immune system. These also have estrogen receptors but exert influences on bone indirectly by upregulation of OPG due to estrogen exposure. Estrogen induces a downregulation of cytokines which are involved in osteoclast formation, such as IL-1 and TNF-α produced by monocytes and IL-6 and GM-CSF produced by stromal cells and osteoblasts. Through this downregulation of cytokines by estrogen, the expression of RANKL on bone marrow cells is suppressed, whereas in estrogen deficiency, the lack of the suppressive effect on RANKL will lead to accelerated maturation and activation of osteoclasts (Eghbali-Fatourechi et al. 2003). The upregulation of RANKL in estrogen deficiency is also mediated by T cell activation besides the cytokines IL-1, IL-6, and TNF-α. Beyond the downregulation of estrogen on the inflammatory immune response, also the upregulation of immunoglobulin production has to be mentioned. Thus, estrogen exerts a dichotomous impact on the immune system, modulating an immune response.

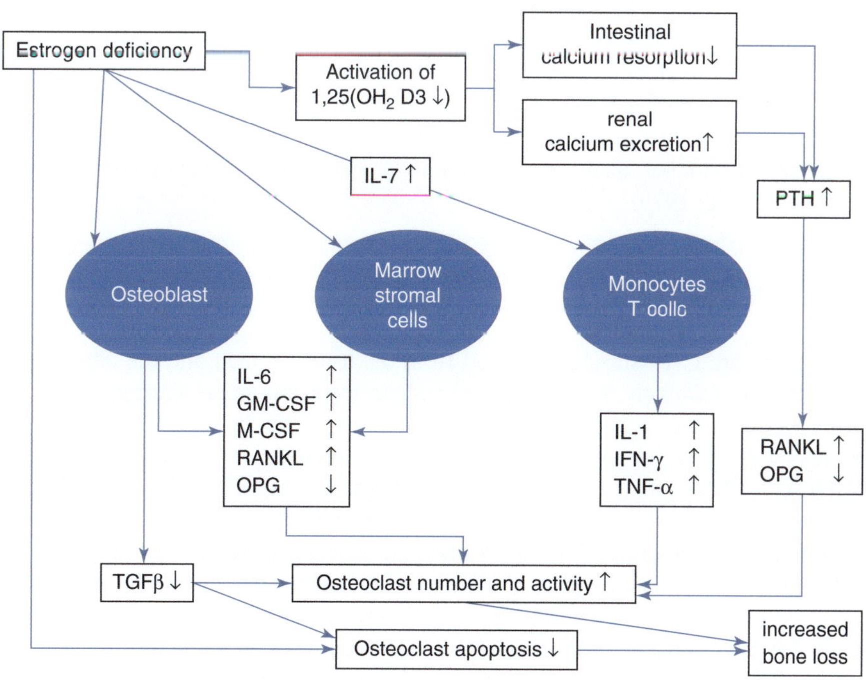

Fig. 7.3 Direct and indirect effects of estrogen deficiency in modulating bone metabolism

In addition, estrogen acts to maintain the appropriate ratio between bone-forming osteoblasts and bone-resorbing osteoclasts, in part through the induction of osteoclast apoptosis, which is mediated via increased levels of TGF-β. In addition, a direct estrogen effect on the extent of osteoclast apoptosis is also demonstrated. Some studies have suggested a role for Fas ligand (FasL) in estrogen-induced osteoclast apoptosis by an autocrine mechanism involving only osteoclasts. Krum et al. described in their study a paracrine mechanism in which estrogen affects osteoclast survival through the upregulation of FasL in osteoblasts (but not osteoclasts), leading to the apoptosis of preosteoclasts (Krum et al. 2008). A cell-type-specific hormone-inducible enhancer located 86 kb downstream of the FasL gene has been characterized as the target of estrogen receptor-alpha induction of FasL expression in osteoblasts. In addition, tamoxifen and raloxifene, two selective estrogen receptor modulators that have protective effects in bone, induce apoptosis in preosteoclasts by the same osteoblast-dependent mechanism. These results demonstrate that estrogen protects bone by inducing a paracrine signal originating in osteoblasts leading to the death of preosteoclasts and offer an important new target for the prevention and treatment of osteoporosis (Krum et al. 2008).

Furthermore, estrogen has extraskeletal effects with influence on calcium homeostasis (Fig. 7.3). In estrogen deficiency, an increased renal calcium excretion and decreased intestinal calcium absorption can be observed (Heaney et al. 1978; McKane et al. 1995; Gennari et al. 1990). This leads to a negative calcium balance which needs to be compensated by mobilization of calcium deposited in bone by upregulation of PTH. In addition, bone sensitivity to PTH exposure is increased in estrogen deficiency (Cosman et al. 1993). Beside the compensatory upregulation of PTH in order to maintain calcium homeostasis, estrogen also seems to have a direct PTH suppressive effect (Arron and Choi 2000). Estrogen also exerts an influence on vitamin D metabolism, as estrogen therapy could increase not only intestinal calcium absorption but also serum vitamin D levels in postmenopausal women (Gallagher et al. 1979).

7.5.2 Testosterone

Androgens are C-19 steroids secreted from the testes in men and the adrenals in both men and women. Testosterone (T) is the most important androgen in men. About 95 % of the amount of testosterone is secreted by the testis. In the adrenal cortex, weakly active androgens are produced such as dehydroepiandrosterone (DHEA), DHEA-sulfate (DHEA-S), and androstenedione. T acts via the androgen receptor (AR) after conversion by the enzyme 5αreductase to the more potent 5α-dihydrotestosterone (DHT) in peripheral tissues. Furthermore, T can also interact with the estrogen receptor α (ERα) and β (ERβ) after being converted into 17β-estradiol (E2) by the P450 aromatase enzyme. In relation to the relative expression of P450 aromatase and 5α-reductase, androgens may preferentially activate the AR or the ERs. AR and ERs are both expressed in bone tissue, thus local metabolism of androgens may have a significant relevance in bone metabolism.

7.5.3 Parathyroid Hormone

Estrogen deficiency induces elevated IL-6 levels which cause increases of PTH. Estrogen deficiency is also associated with decreased intestinal calcium absorption and an increased renal calcium excretion due to reduced levels of 1,25(OH)2 D3, the most potent vitamin D metabolite. The sensitivity of PTH toward bone is also increased with estrogen deficiency. Finally, with increasing age, vitamin D deficiency occurs in most individuals due to impaired skin, gastrointestinal, and renal function in connection with the vitamin D metabolism. Vitamin D deficiency gives further rise to increased PTH levels.

7.5.4 Thyroid Hormone

The hypothalamic-pituitary-thyroid axis plays a key role in skeletal development, acquisition of peak bone mass, and regulation of adult bone turnover. Euthyroid status is essential for maintenance of optimal bone mineralization and strength. In population studies, hypothyroidism and hyperthyroidism have both been associated with an increased risk of fracture. Furthermore, recent studies in healthy euthyroid postmenopausal women indicate that thyroid status in the upper normal range is also associated with low bone mineral density and an increased risk of nonvertebral fracture. Studies in mutant mice have demonstrated that thyroid hormone receptor α is the major mediator of T3 action in bone and that thyroid hormones exert anabolic actions during growth but have catabolic effects on the adult skeleton. Nevertheless, TSH has also been proposed to be a direct negative regulator of bone turnover, although the relative importance of T3 and TSH actions in the skeleton has yet to be clarified.

7.6 Activation of the Immune System in Osteoporosis

However, besides the endocrine effects on osteoblasts and osteoclasts, further influences on bone were discovered, which led to a broader understanding of the development of osteoporosis. In this respect, osteoimmunology could demonstrate that activated lymphocytes also contribute to imbalances in bone remodeling leading to osteoporosis. This perspective changes the picture of osteoporosis as a solely endocrine disorder mainly due to estrogen deficiency but also to a disease caused by inflammatory processes (Pacifici 2008). Furthermore, bone cells, influenced by immune cells themselves, thereafter produce a changed spectrum of cytokines with effects on the immune system. Therefore, the exchange between the immune system and bone cells is of complex nature and bidirectional (Zaidi 2007). The communication between the immune system and bone has to be divided into interactions between immune cells and osteoclasts on the one hand and osteoblasts on the other.

With reduced estrogen levels and/or chronic or recurrent immune activation from either systemic or gastrointestinal origin, there may be a reduction in the body's natural ability to limit the production of RANKL. Associated with estrogen deficiency, there is a progressive proinflammatory status with increased production of IL-1, IL-6, and TNF-α in postmenopausal women. This proinflammatory status is also seen with aging in general and is thus also entitled as inflammaging (Franceschi et al. 2000). Special subsets of T cells seem to be involved in this process (e.g., CD8+ CD57+ subsets), which produce TNF-α and increase in women with osteoporotic spine fractures (Pietschmann et al. 2001). Estrogen deficiency also causes increased IL-7 production, which induces T cell activation. Associated to this T cell activation, there is also an increased production of interferon (IFN)-γ and TNF-α by T cells (Pacifici 2007; Robbie-Ryan et al. 2006). IFN-γ upregulates major histocompatibility complex (MHC) class II molecules on antigen-presenting cells (e.g., bone marrow macrophages and dendritic cells), which causes further T cell activation. Interestingly, although T cell production of IFN-γ is higher in elderly women as compared to young women, in elderly men no increase of IFN-γ production could be seen (Pietschmann et al. 2003). Such gender-specific differences may be part of the explanation why osteoporosis is more frequent in women than men and also exerts rapid bone loss during early postmenopause. These gender differences also underline the pathogenic influence of proinflammatory processes in the pathogenesis of osteoporosis.

Activated T cells produce RANKL, thus being able to target RANK on osteoclast progenitors, and TNF-α, resulting in increased osteoclast activation through a "switch-like" diversion of osteoprogenitor-cell differentiation away from monocyte-macrophage-cell development toward osteoclastogenesis. In this respect, osteoclastic activity, induced by proinflammatory cytokines and activated T cell-induced RANKL, is thought to be modulated by the action of IFN-γ on tumor necrosis factor receptor-associated factor 6 (TRAF-6). TRAF-6 is a RANK adaptor protein that mediates NFkB activation. However, the role of IFN-γ on bone metabolism seems to be variable as the modulating capacity of IFNγ on RANKL is influenced by both vitamin D and estrogen. Under normal conditions without estrogen deficiency, IFN-γ seems to be anti-osteoclastogenic (Rauner et al. 2007), whereas in the context of estrogen deficiency, it has a pro-osteoclastogenic effect (Gao et al. 2007).

7.7 Oxidative Stress in Aging

Aging leads not only to a reduction in sex-hormone production but also to an increase in the general level of proinflammatory cytokines and diminution of the immune system function. This also seems to be mediated, at least in part, by the accumulation of reactive oxygen species (ROS). In ovariectomized mice, estrogen deficiency led to an accumulation of ROS within the bone marrow causing a proinflammatory status with increased levels of TNF-α produced by activated T cells. This upregulation was mediated via the co-stimulatory molecule CD80 on dendritic cells (Messalli et al. 2007). In vivo, free radicals have been shown to increase bone resorption, and oxidative stress reduces BMD in humans. These environmental and/

or age-related catabolic stress-inducing factors contribute to normal bone loss. But when there is a chronic, elevated antigenic load or excessive oxidative stress, which increases proinflammatory cytokine-induced RANKL, the activation of this "switch" in osteoprogenitor-cell differentiation may, independent of age, adversely affect the balance of bone remodeling. It is in this abnormal state that chronic immune activation may alter IFN-γ modulating capacity. When estrogen is deficient, causing RANKL levels to increase, the body's natural ability to limit the transcription factors TRAF 6 and NFkB may be reduced and IFN-γ may exert a pro-osteoclastogenic effect. This uncoupling of the remodeling process results in bone loss. In studies using mice, chronic antigenic load with T cell activation and production of ROS must be present for the low estrogen levels to cause bone loss. It appears that reducing antigenic load and oxidative stress may be equally important as estrogen in maintaining bone health.

7.8 The Gut Microbiota and Oral Tolerance

The gut microbiota (GM) is the commensal bacteria populating our intestine. The GM represents a densely populated system (approximately 10^{13} bacteria) with a diversity of functions still only partly known. However, recent research has unveiled that the GM seems to modulate immunity and also metabolism (Ohlsson and Sjögren 2015). The GM densely interplays with the mucosal immune system, and in healthy individuals, a stable balance is maintained between these components. However, imbalances of the GM may lead to different pathologies including inflammatory bowel disease, colorectal cancer, obesity, type I and type II diabetes, and inflammaging particularly immunosenescence in the elderly. In this respect, the GM has also been shown to have an effect on bone mass and the development of osteoporosis (Ohlsson and Sjögren 2015).

Oral tolerance, the muted immunological response to harmless gut antigens depends on the presence of commensal microorganisms and an intact healthy gut wall. Epithelial cell integrity is maintained by the presence of beneficial organisms such as *Lactobacillus and Bifidobacteria* that do not elicit an inflammatory response. When the normal gastrointestinal flora is maintained, immunological self-tolerance through the activation of T-regulatory cells (Treg) favors a noninflammatory Th2 dominant response to gut microbes. Also Th17 cytokines such as IL-17, IL-22, and IL-23 seem to play an essential role in the regulation of intestinal protection, homeostasis, and intestinally induced proinflammatory effects (Shen and Durum 2010). Pathological bacterial or fungal overgrowth causes inflammation and increased gut permeability that reduces oral tolerance. Focus on the traditional osteo-endocrine explanation for bone homeostasis fails to acknowledge the important role of the immune system in remodeling and the possible role of oral tolerance in maintaining bone health. It is now understood that a high systemic antigen load of bacterial or viral origin and/or a loss of oral tolerance due to pathologic microbial overgrowth (long suspected as major contributing factors in other chronic degenerative diseases) may also contribute to the pathogenesis of bone loss.

In a study by Sjögren et al. (2012), it could be shown that germ-free raised mice exhibited an increased bone mass, a reduced number of osteoclasts per bone surface, and a decreased frequency of CD4+ T cells and CD11b+/GR 1 osteoclast precursor cells in bone marrow leading to a reduced expression of inflammatory cytokines in bone and bone marrow. These alterations could be normalized by colonization with normal gut microbiota. With this perspective, gut microbiota may become even a possible therapeutic option in the future.

Estrogen normally helps to preserve bone by enhancing macrophage production of transforming growth factor ß (TGF–) and limiting CD4+ T cell activation. Reduced levels of estrogen result in an increase in antigen-presenting cells and a reduction in TGF-ß and Treg. This leads to T cell activation and production of pro-inflammatory cytokines and RANKL, which stimulates osteoclastogenesis. By improving gut health and oral tolerance, antigen presentation to T cells is reduced, TGF-ß production is maintained, Tregs are enhanced, and RANKL-induced osteoclastogenesis is limited, even with reduced levels of estrogen.

7.9 Diagnosis of Osteoporosis and Aspects of Osteoimmunology in Clinical Medicine

Diagnosis of osteoporosis is initially based on the evaluation of medical history, including assessment of risk factors for osteoporosis and physical presentation of the patient. X-rays are the primarily used imaging method to diagnose fractures, although in particular within the central skeleton, including the vertebral column, chest, and pelvis, fractures may be frequently overseen by X-rays. For occult fractures, other imaging methods such as computed tomography, magnetic resonance imaging, or bone scintigraphy may add in as diagnostic tools for optimized fracture detection. Imaging methods including X-rays are also of importance in the detection of other bone pathologies such as malignant bone tumors, metastases, hyperostotic bone disorders (e.g., Paget's disease), or degenerative processes. However, osteopenia can be seen by X-ray only after progressive bone loss, thus not allowing early diagnosis of osteoporosis with this diagnostic tool. Osteodensitometry (DXA) allows a quantification of bone loss, and measurements are classified as normal, osteopenia, and osteoporosis based on the amount of bone lost as compared to healthy persons with peak bone mass. Laboratory parameters are also used in the diagnostic evaluation of osteoporosis. In clinical routine, laboratory parameters primarily aims to evaluate the function of central organs such as the liver or the kidneys, which also have a part in metabolic pathways associated with bone. The evaluation of endocrine laboratory parameters allows to estimate the function of endocrine axes with influence on bone. Furthermore, relevant pathological conditions such as malignant myeloma, which can also be the cause of osteoporosis, should be clarified by laboratory assessment. In recent years, a range of bone formation and degradation markers also found their way to clinical practice giving the clinician an insight into the activity of bone turnover. Other parameters, although of high importance for the regulation of bone metabolism (e.g., OPG, RANKL), specific processes (e.g., cathepsin K), or signal transduction (e.g., cytokines), are frequently determined only for study purposes and do not play a role in clinical practice. In this light of clinical medicine, diagnostic evaluation of the osteoimmunological interplay remains superficial. Only major determinants such as hormonal deficits (estrogen or testosterone deficiency, hypothyroidism) or hormonal excess (hyperthyroidism, Cushing syndrome) will be potentially measurable.

7.10 Causes of Osteoporosis from a Clinical Perspective: Postmenopausal Osteoporosis and Secondary Osteoporosis

As outlined above, the mechanisms and interactions of immune cells, cytokines, and hormonal influences on bone cells are complex and interact with each other in a complex network that leads to a balanced formation and resorption of bone tissue under healthy conditions. From a clinical point of view, these interactions on the cellular level will neither be diagnosed nor will they have any impact on the clinical management on a case by case basis. However, the perception that different endocrine disorders or inflammatory processes have an important impact on the pathogenesis of bone loss and its clinical management has to be highlighted. It is the clinicians aim to evaluate, diagnose, and set them into context with pathological processes of the bone. In this respect, the clinician has to see the patient's problems not in an organ-focused perspective but rather in a holistic way in order to percept the full range of possible pathological influences in bone disease. In this respect, a short survey of relevant clinical disorders and its connections to bone pathophysiology and connections to osteoimmunology will be given. Other causes such as drugs, lifestyle factors, diseases with complex backgrounds of osteoporosis development such as anorexia nervosa, or less frequent causes such as organ transplantations are listed in Table 7.1. However, the table does not provide a full range of all known causes of osteoporosis; for further detailed information on different disorders causing osteoporosis, the reader is referred to the literature.

7.10.1 Osteoporosis and Disorders with Chronic Inflammation: Osteoimmunology as a Link

The assessment of different inflammatory diseases such as rheumatoid arthritis, systemic lupus erythematosus, inflammatory bowel disease, celiac disease, cystic fibrosis, and chronic obstructive pulmonary disease is of relevance when diagnosing osteoporosis, as these inflammatory disorders have been associated to pathological bone resorption. As mentioned, the link between osteoclast, M-CSF, and proinflammatory cytokines, especially TNF-α and IL-1, explains the association between inflammation and osteoporosis. Other TNF-related cytokines such as RANK, RANKL, and OPG are also important mediators in inflammatory processes and are critically involved in the pathophysiology of bone loss. These diseases are therefore

Table 7.1 Overview of clinical disorders causing osteoporosis

Endocrine disorders	Complex causes of osteoporosis	Drugs
Hyperthyroidism	Chronic hepatitis	Glucocorticoids
Hyperprolactinemia	Chronic exocrine pancreatitis	L-thyroxine
Estrogen deficiency	Malabsorption	Aromatase inhibitors
Testosterone deficiency	Celiac disease	Herparine
Hyperparathyroidism	Lactose intolerance	Proton pump inhibitors
Growth hormone deficiency	Postgastrectomy	SSRIs
Diabetes mellitus	Anorexia nervosa	Anticonvulsants
	Renal insufficiency	Glitazone
Inflammatory disorders	Mastocytosis	Cytostatics
Rheumatoid arthritis	Multiple myeloma	
Crohn's disease	AIDS	*Lifestyle*
Colitis ulcerosa	Gaucher disease	Inactivity
		Alcoholism
Rare genetic disorders	*Transplantation*	Smoking

AIDS acquired immune deficiency syndrome, *SSRIs* selective serotonin reuptake inhibitors

related to osteoporosis and high fracture risk, independently of other risk factors common to inflammatory diseases such as reduced physical activity, poor nutritional status, hypovitaminosis D, decrease in calcium intake, and glucocorticoid treatment.

7.10.2 Endocrine Disorders

Different endocrine axes including the gonads (estrogen, testosterone), thyroid gland (L-thyroxine), and parathyroid gland (parathyroid hormone) play a crucial role in bone metabolism and were described above in their actions on bone. Hormonal deficiencies, such as estrogen deficiency due to menopause or ovariectomy, hypothyroidism, or hormonal excesses such as hyperthyroidism, Cushing syndrome, and hyperprolactinemia, will influence bone homeostasis. Also other endocrine disorders such as diabetes mellitus were reported to be associated with osteoporosis.

7.10.3 Intestinal Disorders and Liver Diseases

Pathologies of the intestine and the liver currently represent a still underestimated field of causes for osteoporosis (Katz and Weinerman 2010). A broad range of pathologies including Crohn's disease, colitis ulcerosa, celiac disease, pancreatitis, hepatitis, and postgastrectomy may lead to osteoporosis (Katz and Weinerman 2010). The pathophysiological mechanisms are diverse and include reduced calcium absorption, pathologies of the vitamin D metabolism, poor digestion, malabsorption, or reduced production of proteins. In inflammatory bowel disease, inflammatory cytokines such as serum IL-1b, TNF-α, IL-6, and IL-1 are increased, promoting negative effects on bone turnover (Nielsen et al. 2000). In a recent study (Tignor et al. 2010), fracture risks for different frequent gastrointestinal disorders associated with osteoporosis were evaluated: the odds of fracture (odds ratio, 95 % confidence interval) compared with controls, adjusted for age, gender, and race were chronic pancreatitis 2.4 (2.1, 2.9), Crohn's disease 1.7 (1.5, 2.0), gastrectomy 2.5 (1.5, 4.1), cirrhosis 2.6 (2.4, 2.7), and celiac disease 2.7 (2.1, 3.4) (Tignor et al. 2010). These gastrointestinal disorders thus have to be considered as relevant pathological states regarding the development of osteoporosis and should therefore be included into the diagnostic evaluation of osteoporosis.

7.10.4 Renal Disorders

Renal disease can lead to complex pathologies of bone known as renal osteodystrophy. These pathologies include pathological extraosseous calcifications, osteomalacia, and fractures due to osteopenia and osteoporosis. Parathyroid hormone is also frequently increased thus adding aspects of hyperparathyroid-induced bone disease.

7.11 Treatment

The therapy of osteoporosis is first of all based in risk reduction and change of lifestyle as far as necessary. This includes cessation of smoking and alcohol intake, continued physical activity, calcium-enriched food, and the elimination of other disorders with potential deleterious effects on bone (e.g., hyperthyroidism). Furthermore, in particular in the elderly, risks or disorders leading to falls have to be also assessed and eliminated. This means that therapy of osteoporosis can become a very complex venture in order to meet all demands of treatment, and in some circumstances, therapeutic interventions will be aimed not only at bone but also at other organs (e.g., prescription of glasses in case of falls and impaired vision, implantation of a heart pacemaker in case of falls due to rhythmological problems), thus not showing a direct connection to treatment of osteoporosis. In case of fractures, diverse surgical interventions are nowadays available including fracture stabilization by osteosynthesis or vertebroplasty and kyphoplasty, in order to keep immobilization as short as possible and to allow rapid remobilization of the fractured patients. However, specific drug therapy also plays a key role in the management of osteoporosis. The aim of drug therapy in osteoporosis is to reduce fracture risk. This is achieved by influencing bone metabolism in such a way that bone formation exceeds bone resorption, leading in the long term to a net bone gain and greater stability of bone structure. Common to all drug regimens in osteoporosis therapy is the consequent application of calcium and vitamin D. According to current guidelines, 1000 mg of calcium and at least 800 units of vitamin D should be applied. Specific osteoporosis therapies can be divided into therapies influencing bone resorption and bone formation. Therapies such as estrogen, selective estrogen receptor modulators (SERMs), bisphosphonates, calcitonin, and denosumab influence bone metabolism by decreasing osteoclast activity. On the other hand, parathyroid hormones are potent osteoanabolic agents which have shown to lead to significant formation of new bone and bone structure, if intermittently applied via subcutaneous route. Strontium ranelate shows a dual action on bone by both stimulation of osteoblasts and inhibition of osteoclasts, also leading to formation of new bone. All mentioned drugs for osteoporosis have demonstrated fracture reduction for vertebral fractures, some for hip fractures, and other skeletal sites in large multicenter studies.

Considering osteoimmunology, questions arise in which way these drugs may influence bone metabolism on the level of signaling, cytokine expression, and the interplay with the immune system.

7.11.1 Hormone Replacement Therapy

As mentioned above, estrogen limits bone loss through its effects on osteoblast and osteoclast activity. Since estrogen deficiency in menopause is a major cause of osteoporosis in women, substitution of the estrogen deficit may postpone the development of osteoporosis. The interactions between the different bone cells and cells of the bone marrow, including stromal cells and cells of the immune system, their interactions and cytokine expressions effected by estrogen and also its immunomodulatory capacity have been outlined above.

7.11.2 Selective Estrogen Receptor Modulators (SERMs)

The antiresorptive effect of raloxifene might be mediated by changes in several cytokines involved in the bone remodeling process. Serum OPG levels in postmenopausal women significantly increased after treatment with raloxifene (Messalli et al. 2007). On the contrary, in another study, postmenopausal women were treated with 60 mg raloxifene, but serum levels of OPG significantly decreased after 3 and 6 months of therapy and returned to basal levels after 1 year of treatment (Li et al. 2009). There was a significant decrease of RANKL levels and OPG/RANKL ratio after 1 year of raloxifene treatment. According to the authors of the study, this might be due to a reduced number of osteoblasts (Li et al. 2009). On the whole, current results support the hypothesis that raloxifene may inhibit osteoclast activity, at least by partly modulating the OPG-RANKL system (Fernández-García et al. 2008). It could also be demonstrated that raloxifene reduced IL-6 (Ozmen et al. 2007; Gianni et al. 2004), TNF-α (Gianni et al. 2004), and TGF-β1 (Ozmen et al. 2007) and, as pointed out above, SERMs may also induce apoptosis in preosteoclasts (Krum et al. 2008).

LY117018, a raloxifene analogue, can significantly inhibit the generation of osteoclasts in vitro at concentrations between 10 and 12 M and 10–9 M and stimulate osteogenic differentiation at concentrations of 10–14 to 10–7 M. No influence on the proliferation and transcription of RANKL and osteoprotegerin was observed. TNF-α production was suppressed by LY117018, which may add to its anti-osteoclastogenic effect (Wutzl et al. 2010). The effects on estrogen-deficiency osteoporosis of ormeloxifene, another SERM, were studied in retired breeding female rats (Narayana Murthy et al. 2006). Ormeloxifene like estradiol demonstrated its inhibition of estrogen-deficiency osteoporosis effects via inhibition of osteoclastogenesis, apoptosis of osteoclasts, and upregulation of TGF beta-3 expression. Raloxifene – though effective in inhibiting osteoclastogenesis – induced no apoptosis at any concentrations. Thus, raloxifene appears to have a different mechanism of action than ormeloxifene and estradiol (Narayana Murthy et al. 2006).

7.11.3 Bisphosphonates

Bisphosphonates are the drugs most widely used in treating osteoporosis. Bisphosphonates act by binding to the hydroxyapatite in bone tissue. This inhibits the activity of mature osteoclasts and induces both reduced osteoclastogenesis and increased apoptosis of osteoclasts. Aside from these known effects of bisphosphonates, the most drug-sensitive steps have not been determined so far. Risedronate inhibited osteoclast differentiation in coculture of bone marrow cells (BMCs) and osteoblasts and suppressed RANKL-mediated osteoclast differentiation from bone marrow-derived macrophages (BMMs) in a dose-dependent manner without toxicity. Risedronate significantly inhibited expression of c-Fos and nuclear factor of activated T cells (NFAT) c1 induced by RANKL (Kwak et al. 2009). Risedronate significantly reduced the number and degree of differentiation of osteoclast precursors, osteoclast formation, and their vitality and activity after 3 months (D'Amelio et al. 2008). Furthermore, the levels of RANKL and TNF were reduced in cultures and of TNF and OPG in serum (D'Amelio et al. 2008). In a further study on postmenopausal women (Dundar et al. 2009), risedronate significantly decreased serum levels of RANKL and IL-1beta, and the level of OPG significantly increased after 3 and 6 months, but no significant difference was found in TNF-α level (Narayana Murthy et al. 2006). With regard to bone turnover, markers of bone resorption and formation significantly decreased after 6 months (Dundar et al. 2009). A significant reduction of bone resorption markers was observed after 3 months of alendronate (D'Amelio et al. 2010), whereas no significant reduction in the number of osteoclast precursors, osteoclast formation and viability, and cytokine levels was present at that time (D'Amelio et al. 2010). After 1 year of alendronate treatment, reduced osteoclast precursors, osteoclast formation, and serum RANKL were present. This supports the fact that bisphosphonates mainly act on mature bone-resorbing osteoclasts in the short term, whereas its long-term administration diminishes their formation by reducing their precursors and serum RANKL (Kwak et al. 2009; D'Amelio et al. 2008, 2010; Anastasilakis et al. 2008).

Interactions with the immune system were outlined in a recent study (Lim et al. 2010) with a significant increase in mRNA expression and serum levels of IL-6, TNF-alpha, and IFN-gamma shortly after pamidronate infusion. Furthermore, a notable rise in serum C-reactive protein (CRP) was observed over 3 days. In this respect, an intermittent large dose of aminobisphosphonate has an impact on the immunological level, causing acute inflammation. The acute-phase response (APR) is the most frequent side effect after the first dose of intravenous nitrogen-containing bisphosphonates. It has been demonstrated in vitro that nitrogen-containing bisphosphonates stimulate $\gamma\delta$ T cell proliferation and production of cytokines and that vitamin D is able to modulate these. Levels of 25(OH)D were negatively correlated with postdose body temperature and CRP (D'Amelio et al. 2010). An exponential increase in fever and CRP has been found with 25(OH)D levels lower than 30 ng/mL and body temperature less than 37 °C, whereas normal CRP was associated with 25(OH)D levels above 40 ng/mL (Bertoldo et al. 2010). The association between post-N-BPs APR and 25(OH)D suggests an interplay among nitrogen-containing bisphosphonates, 25(OH)D, and the immune system. These results (Lim et al. 2010; Bertoldo et al. 2010) may explain the acute-phase response often seen in patients after receiving nitrogen-containing bisphosphonates for the first time.

7.11.4 Denosumab

Denosumab (RANKL-specific monoclonal antibody) is a specific monoclonal antibody directed against RANKL. By binding to RANKL, denosumab prevents RANKL from binding to its receptor, resulting in a decrease in bone resorption due to reduction in the formation, activity, and survival of osteoclasts. The antiresorptive effect of a single 60-mg injection of denosumab substantially exceeds the effects of alendronate (70 mg weekly) (Tankó 2007; Lewiecki et al. 2007). Denosumab increased BMD at all measured skeletal sites and decreased concentrations of bone turnover markers compared with a placebo at 24 months (Lewiecki et al. 2007). At the lumbar spine, BMD increases with denosumab in the range of 4.13–8.89 %. BMD changes with denosumab 30 mg every 3 months and ≥60 mg every 6 months that were similar to, or in some cases greater than, the use of alendronate (Lewiecki et al. 2007). The role of denosumab in the treatment of osteoporosis is described in more detail in the chapter "Antibodies for the Treatment of Bone Diseases: Clinical Data."

7.11.5 Parathyroid Hormone

Subcutaneous intermittent therapy with teriparatide (rhPTH 1–34) is an established anabolic therapy in osteoporosis. It could be demonstrated that rhPTH 1–34 leads to a rapid and significant increase of RANKL within 1 month, with a persistent increase during the following time of therapy. On the contrary, OPG was suppressed by rhPTH 1–34 after 6 months. OPG normalized to baseline after discontinuation of the rhPTH 1–34 treatment. Furthermore, the cytokines IL-6 and IL-6sR are upregulated by rhPTH 1–34. Thus, rhPTH 1–34 controls bone remodeling primarily through a modulation of the OPG/RANKL/RANK system. Immature cells of the osteoblast lineage are stimulated by rhPTH 1–34, leading to an increase of RANKL, which may bind to its receptor RANK. As OPG is also produced by osteoblasts, this RANKL-RANK interaction is counteracted by OPG by preventing the binding of RANKL to its RANK receptor. It seems that rhPTh initially stimulates osteoblast maturation and function, which then induces an osteoclast activation combined with a shift in the balance of bone formation and bone resorption in favor of bone formation. Interestingly, when in a small uncontrolled study rhPTH 1–34 was combined with the bisphosphonate risedronate, OPG levels remained unchanged, while RANKL decreased gradually after 3 and 6 months of therapy with risedronate and rhPTH 1–34 (Anastasilakis et al. 2008). Thus, when two bone active drugs are given, a further modulation of the OPG/RANKL/RANK system and bone metabolism is induced, which seems to be dissimilar to the effects of the substances when applied singly.

7.11.6 Strontium Ranelate

In a study with ovariectomized goats, strontium (Sr) was co-administered with calcium to investigate the effects of Sr on cytokine expression and cell activities (Zaidi 2007). Serum Sr levels increased six- and tenfold in the Ca + 24 mg/kg/day Sr and Ca + 40 mg/kg/day Sr groups, respectively. In bone Sr increased four- and sixfold in these two groups, and Sr-Ca co-administration considerably increased bone mineral apposition rate. The expression of IGF-1 and runt-related transcription factor 2 (Runx2) was significantly upregulated within the Ca + 40 mg/kg/day Sr treatment group; TNF-α expression was significantly downregulated in the Ca + 40 mg/kg/day Sr group. This indicates that Sr-Ca co-administration increases osteogenic gene expression and stimulates new bone formation.

Strontium (Sr) ranelate reduces fracture risk in postmenopausal women with osteoporosis. Sr ranelate has been proposed as an agonist of the calcium-sensing receptor (CaSR), thus revealing effects on OPG and RANKL expression and cell replication. In postmenopausal women, Sr ranelate increased mRNA and protein levels of OPG and suppressed those of RANKL. Sr ranelate also stimulated osteoblast replication and differentiation and increased cell survival under stress. Knocking down CaSR suppressed the Sr ranelate-induced stimulation of OPG mRNA, the reduction of RANKL mRNA, and an increase in replication, indicating the involvement of CaSR in these responses (Brennan et al. 2009).

7.11.7 Other Therapies Including Upcoming Therapeutic Options

Concerning other therapeutic options, fluoride and calcitonin have to be briefly mentioned. Both substances were used in the past for the treatment of osteoporosis. However, because of questionable treatment effects (calcitonin), a narrow therapeutic window with increased bone mass but not bone strength (fluoride), and the lack of trials showing fracture reduction with these drugs, both were in the meantime outpaced by other more potent drugs as presented above. Concerning osteoimmunological aspects, neither fluoride nor calcitonin therapy revealed any effects on the concentrations of IGF-1, IGF-2, and TGF-β1 in bone matrix extracts from osteoporotic patients (Pepene et al. 2004). However, it could be shown that calcitonin prevented apoptosis of osteocytes and osteoblasts, thus being the possible mechanism of action for its anti-osteoporotic effects (Plotkin et al. 1999).

Today, further new therapies are in the stage of pre- or (already) clinical testing. In comparison to the currently available drugs, which either interact at the hormonal level via hormone receptors (estrogen, SERMs, parathyroid hormone) or by deposition of specific molecules within bone altering bone metabolism (bisphosphonates, Sr ranelate), these new drugs now aim to interact in a specific way with the players of bone metabolism. One example is the cathepsin K-inhibitor odanacatib, which has been shown to interfere specifically with the activation of bone-resorbing processes in active osteoclasts.

Conclusion

Bone cells interact at different points with cells of the immune system. The interactions are modulated by hormonal influences setting up complex mechanisms regulating and influencing bone metabolism. Our understanding of this complex interplay is constantly increasing. The improved understanding of these modes of action already has led to the development of new drugs which specifically interact at certain points of bone metabolism. In this respect, the field of osteoimmunology gains a new dimension, as its complex view on all aspects of bone metabolism gives us not only better insights into bone and bone metabolism at the level of basic research but also becomes a motor in the development of new therapeutic strategies (Pietschmann et al. 2016). With the translational shift of basic research results in osteology and osteoimmunology into clinically applicable treatments, further improvements in the fight of osteoporosis seem to become available (Pietschmann et al. 2016).

References

Alnaeeli M, Park J, Mahamed D, Penninger JM, Teng YT (2007) Dendritic cells at the osteoimmune interface: implications for inflammation-induced bone loss. J Bone Miner Res 22:775–780

Anastasilakis AD, Goulis DG, Polyzos SA, Gerou S, Koukoulis G, Kita M, Avramidis A (2008) Serum osteoprotegerin and RANKL are not specifically altered in women with postmenopausal osteoporosis treated with teriparatide or risedronate: a randomized, controlled trial. Horm Metab Res 40:281–285

Arron JR, Choi Y (2000) Bone versus immune system. Nature 408:535–536

Bengmark S (2004) Acute and "chronic" phase reaction-a mother of disease. Clin Nutr 23:1 256–1266

Bertoldo F, Pancheri S, Zenari S, Boldini S, Giovanazzi B, Zanatta M, Valenti MT, Dalle Carbonare L, Lo Cascio V (2010) Serum 25-hydroxyvitamin D levels modulate the acute-phase response associated with the first nitrogen-containing bisphosphonate infusion. J Bone Miner Res 25:447–454

Boyle WJ, Simonet WS, Lacey DL (2003) Osteoclast differentiation and activation. Nature 423:337–342

Brennan TC, Rybchyn MS, Green W, Atwa S, Conigrave AD, Mason RS (2009) Osteoblasts play key roles in the mechanisms of action of strontium ranelate. Br J Pharmacol 157:1291–1300

Breuil V, Ticchioni M, Testa J, Roux CH, Ferrari P, Breittmayer JP, Albert-Sabonnadière C, Durant J, De Perreti F, Bernard A, Euller-Ziegler L, Carle GF (2010) Immune changes in postmenopausal osteoporosis: the Immunos study. Osteoporos Int 21:805–814

Calvi LM, Sims NA, Hunzelman JL, Knight MC, Giovannetti A, Saxton JM, Kronenberg HM, Baron R, Schipani E (2001) Activated parathyroid hormone/parathyroid hormone-related protein receptor in osteoblastic cells differentially affects cortical and trabecular bone. J Clin Invest 107:277–286

Calvi LM, Adams GB, Weibrecht KW et al (2003) Osteoblastic cells regulate the haematopoietic stem cell niche. Nature 425:841–846

Cheroutre H (2004) Starting at the beginning: new perspectives on the biology of mucosal T cells. Annu Rev Immunol 22:217–246

Christiansen C, Christiansen MS, Transbol I (1981) Bone mass in postmenopausal women after withdrawal of oestrogen/gestagen replacement therapy. Lancet 1:459–461

Clowes JA, Riggs BL, Khosla S (2005) The role of the immune system in the pathophysiology of osteoporosis. Immunol Rev 208:207–227

Colin EM, Asmawidjaja PS, van Hamburg JP, Mus AM, van Driel M, Hazes JM, van Leeuwen JP, Lubberts E (2010) 1,25-dihydroxyvitamin D3 modulates Th17 polarization and interleukin-22 expression by memory T cells from patients with early rheumatoid arthritis. Arthritis Rheum 62:132–142

Cosman F, Shen V, Xie F, Seibel M, Ratcliffe A, Lindsay R (1993) Estrogen protection against bone resorbing effects of parathyroid hormone infusion. Ann Intern Med 118:337–343

D'Amelio P, Gnmaldi A, Di Bella S, Tamone C, Brianza SZ, Ravazzoli MG, Bernabei P, Cristofaro MA, Pescmona GP, Isaia G (2008) Risedronate reduces osteoclast precursors and cytokine production in postmenopausal osteoporotic women. J Bone Miner Res 23:373–379

D'Amelio P, Grimaidi A, Cristofaro MA, Ravazzoli M, Molinatti PA, Pescarmona GP, Isaia GC (2010) Alendronate reduces osteoclast precursors in osteoporosis. Osteoporos Int 21:1741–1750

Dundar U, Kavuncu V, Cifici IH, Evcik D, Solak O, Cakir T (2009) The effect of risedronate treatment on serum cytokines in postmenopausal osteoporosis: a 6-month randomized and controlled study. J Bone Miner Metab 27:464–470

Eghbali-Fatourechi G, Khosla S, Sanyal A, Boyle WJ, Laccy DI, Riggs BL (2003) Role of RANK ligand in mediating increased bone resorption in early postmenopausal women. J Clin Invest 111:1120–1122

Eriksen EF, Colvard DS, Berg NJ, Graham ML, Mann KG, Spelsberg TC, Riggs BL (1988) Evidence of estrogen receptors in normal human osteoblast like cells. Science 241:84–86

Fernández-García D, Muñoz-Torres M, Mezquita-Raya P, de la Higuera M, Alonso G, Reyes-García R, Ochoa AS, Ruiz-Requena ME, Luna JD, Escobar-Jiménez F (2008) Effects of raloxifene therapy on circulating osteoprotegerin and RANK ligand levels in post-menopausal osteoporosis. J Endocrinol Invest 31:416–421

Franceschi C, Bonafè M, Valensin S, Olivieri F, Da Luca M, Ottaviani E, De Benedictitis G (2000) Inflammaging. An evolutionary perspective on immunosenescence. Ann N Y Acad Sci 908:24–254

Gaffen SL, Hajishengallis G (2008) A new inflammatory cytokine on the block: re-thinking periodontal disease and the Th1/Th2 paradigm in the context of Th17 cells and IL-17. J Dent Res 87:817–828

Gallagher JC, Riggs BL, Eisman J, Hamstra A, Arnaud SB, DeLuca HF (1979) Intestinal calcium absorption and serum vitamin D metabolites in normal subjects and osteoporotic patients: effects of age and dietary calcium. J Clin Invest 64:729–736

Gao Y, Grassi F, Ryan MR, Terauchi M, Page K, Yang X, Weitzmann MN, Pacifici R (2007) IFN-γ stimulates osteoclast formation and bone loss in vivo via antigen-driven T cell activation. J Clin Invest 117:12–132

Geijtenbeek TB, van Vliet SJ, Engering A, 't Hart BA, van Kooyk Y (2004) Self- and nonself-recognition by C-type lectins on dendritic cells. Annu Rev Immunol 22:33–54

Gennari C, Agnusdei D, Nardi P, Civitelli R (1990) Estrogen preserves a normal intestinal responsiveness to 1,25-dihydroxyvitamin D3 in oophorectomized women. J Clin Endocrinol Metab 71:1288–1293

Ghishan FK, Kiela PR (2011) Advances in the understanding of mineral and bone metabolism in inflammatory bowel diseases. Am J Physiol Gastrointest Liver Physiol 300:G191–G201

Giangregorio LM, Jantzi M, Papaioannou A, Hirdes J, Maxwell CJ, Poss JW (2009) Osteoporosis management among residents living in long-term care. Osteoporos Int 20:1471–1478

Gianni W, Ricci A, Gazzaniga P, Brama M, Pietropaolo M, Votano S, Patanè F, Aglianò AM, Spera G, Marigliano V, Ammendola S, Agnusdei D, Migliaccio S, Scandurra R (2004) Raloxifene modulates interleukin-6 and tumor necrosis factor-alpha synthesis in vivo: results from a pilot clinical study. J Clin Endocrinol Metab 89:6097–6099

Haaland DA, Cohen DR, Kennedy CC, Khalidi NA, Adachi JD, Papaioannou A (2009) Closing the osteoporosis care gap: increased osteoporosis awareness among geriatrics and rehabilitation teams. BMC Geriatr 9:28. doi:10.1186/1471-2318-9-28

Häussler B, Gothe H, Göl D, Glaeske G, Pientka L, Felsenberg D (2007) Epidemiology, treatment and costs of osteoporosis in Germany – the Bone EVA Study. Osteoporos Int 18: 77–84

Heaney RP, Recker RR, Saville PD (1978) Menopause changes in calcium balance performance. J Lab Clin Med 92:953–963

Johnell O, Kanis JA (2006) An estimate of the worldwide prevalence and disability associated with osteoporotic fractures. Osteoporos Int 17:1726

Jones DH, Kong YY, Penninger JM (2002) Role of RANKL and RANK in bone loss and arthritis. Ann Rheum Dis 61(Suppl):ii32–ii39

Josien R, Wong BR, Li HL, Steinman RM, Choi Y (1999) TRANCE, a TNF family member, is differentially expressed on T cell subsets and induces cytokine production in dendritic cells. J Immunol 162:2562–2568

Kamal HK (2005) Secondary prevention of hip fractures among hospitalized elderly: are we doing enough? J Clin Rheumatol 11:68–71

Katz S, Weinerman S (2010) Osteoporosis and gastrointestinal disease. Gastroenterol Hepatol (N Y) 6:506–517

Kim SW, Her SJ, Kim SY, Shin CS (2005) Ectopic overexpression of adipogenic transcription factors induces transdifferentiation of MC3T3-E1 osteoblasts. Biochem Biophys Res Commun 327:811–819

Komm BS, Terpening CM, Benz DJ, Graeme KA, O'Malley BW, Haussler MR (1988) Estrogen binding receptor mRNA, and biologic response in osteoblast-like osteosarcoma cells. Science 241:81–84

Konnopka A, Jerusel N, König HH (2009) The health and economic consequences of osteopenia- and osteoporosis-attributable hip fractures in Germany: estimation for 2002 and projection until 2050. Osteoporos Int 20:1117–1129

Krum SA, Miranda-Carboni GA, Hauschka PV, Carroll JS, Lane TF, Freedman LP, Brown M (2008) Estrogen protects bone by inducing Fas ligand in osteoblasts to regulate osteoclast survival. EMBO J 27:535–545

Kwak HB, Kirn JY, Kim KJ, Choi MK, Kim JJ, Kirn KM, Shin YI, Lee MS, Kim HS, Kim JW, Chun CH, Cho HJ, Hong GY, Juhng SK, Yoon KH, Park BH, Bae JM, Han JK, Oh J (2009) Risedronate directly inhibits osteoclast differentiation and inflammatory bone loss. Biol Pharm Bull 32:1193–1198

Lecka-Cernik B, Moerman EJ, Grant DF, Lehmann JM, Manolagas SC, Jilka RL (2002) Divergent effects of selective peroxisome proliferator-activated receptor-gamma 2 ligands on adipocyte versus osteoblast differentiation. Endocrinology 143:2376–2384

Lewiecki EM, Miller PD, McClung MR, Cohen SB, Bolognese MA, Liu Y, Wang A, Siddhanti S, Fitzpatrick LA, AMG 162 Bone Loss Study Group (2007) Two-year treatment with denosumab (AMG 162) in a randomized phase 2 study of postmenopausal women with low BMD. J Bone Miner Res 22:1832–1841

Li Y, Toraldo G, Li A, Yang X, Zhang H, Qian WP, Weitzmann MN (2007) B cells and T cells are critical for the preservation of bone homeostasis and attainment of peak bone mass in vivo. Blood 109:3839–3848

Li Z, Lu WW, Chiu PK, Lam RW, Xu B, Cheung KM, Leong JC, Luk KD (2009) Strontium-calcium coadministration stimulates bone matrix osteogenic factor expression and new bone formation in a large animal model. J Orthop Res 27:758–762
Lim MJ, Kwon SR, Park SG, Park W (2010) Acute effects of intravenous administration of pamidronate in patients with osteoporosis. J Korean Med Sci 25:1277–1283
Lohmann R, Haid K, Stöckle U, Raschke M (2007) Epidemiologie und Perspektiven der Alters traumatologie [Epidemiology and perspectives in traumatology of the elderly] Unfallchirurg. 110:553–560
Manabe N, Kawaguchi H, Chikuda H, Miyaura C, Inada M, Nagai R, Nabeshima Y, Nakamura K, Sinclair AM, Scheuermann RH, Kuro-o M (2001) Connection between B lymphocyte and osteoclast differentiation pathways. J Immunol 167:2625–2631
McKane WR, Khosla S, Burritt MF, Kao PC, Wilson DM, Ory SJ, Riggs BL (1995) Mechanism of renal calcium conservation with estrogen replacement therapy in women in early menopause – a clinical research center study. J Clin Endocrinol Metab 80:3458–3464
Messalli EM, Mainini G, Scaffa C, Cafiero A, Salzillo PL, Ragucci A, Cobellis L (2007) Raloxifene therapy interacts with serum osteoprotegerin in postmenopausal women. Maturitas 56:38–44
Metge CJ, Leslie WD, Manness LJ, Yogendran M, Yuen CK, Kvern B, Maximizing Osteoporosis Management in Manitoba Steering Committee (2008) Postfracture care for older women: gaps between optimal care and actual care. Can Fam Physician 54:1270–1276
Miyamoto T, Ohnada O, Arai F, Iwamoto K, Okada S, Tagagi K, Anderson DM, Suda T (2001) Bifurcation of osteoclasts and dendritic cells from common progenitors. Blood 98:2544–2554
Narayana Murthy PS, Sengupta S, Sharma S, Singh MM (2006) Effect of ormeloxifene on ovariectomy-induced bone resorption, osteoclast differentiation and apoptosis and TGF beta-3 expression. J Steroid Biochem Mol Biol 100:117–128
Nielsen OH, Vainer B, Madsen SM, Seidelin JB, Heegaard NH (2000) Established and emerging biological activity markers of inflammatory bowel disease. Am J Gastroenterol 95:359–367
NIH Consensus Development Panel on Osteoporosis Prevention, Diagnosis, and Therapy (2001) Osteoporosis prevention, diagnosis, and therapy. JAMA 285:785–795
Ohlsson C, Sjögren K (2015) Effects of the gut microbiota on bone mass. Trends Endocrinol Metab 26:69–74. doi:10.1016/j.tern2014.11.004
Oursler MI, Osdoby P, Pyfferoen J, Riggs BL, Spelsberg TC (1998) Avian osteoclasts as estrogen target cells. Proc Natl Acad Sci U S A 88:6613–6617
Ozmen B, Kirmaz C, Aydin K, Kafesciler SO, Guclu F, Hekimsoy Z (2007) Influence of the selective oestrogen receptor modulator (raloxifene hydrochloride) on IL-6, TNF-alpha, TGF-beta1 and bone turnover markers in the treatment of postmenopausal osteoporosis. Eur Cytokine Netw 18:148–153
Pacifici R (2007) T cells and post menopausal osteoporosis in murine models. Arthritis Res Ther 9:102
Pacifici R (2008) Estrogen deficiency, T cells and bone loss. Cell Immunol 252:68–80
Papaioannou A, Kennedy CC, Ioannidis G, CaMos Research Group et al (2008) The osteoporosis care gap in men with fragility fractures: the Canadian Multicentre Osteoporosis Study. Osteoporos Int 19:581–587
Pene J, Chevalier S, Preisser L, Venereau E, Guilleux MH, Ghannam S, Moles JP, Danger Y, Ravon E, Lesaux S, Yssel H, Gascan H (2008) Chronically inflamed human tissue are infiltrated by highly differentiated Th17 lymphocytes. J Immunol 180:7423–7430
Pepene CE, Seck T, Diel I, Minne HW, Ziegler R, Pfeilschifter J (2004) Influence of fluor salts, hormone replacement therapy and calcitonin on the concentration of insulin-like growth factor (IGF)-I, IGF-II and transforming growth factor-beta 1 in human iliac crest bone matrix from patients with primary osteoporosis. Eur J Endocrinol 150:81–91
Pietschmann P, Grisar J, Thien R, Willhelm M, Kerschan-Schindl K, Preisinger E, Peter-lik M (2001) Immune phenotype and intracellular cytokine production of peripheral blood mononuclear cells from postmenopausal patients with osteoporotic fractures. Exp Gerontol 36:1749–1759

Pietschmann P, Gollob E, Brosch S, Hahn P, Kudlacek S, Willhelm M, Woloszczuk W, Peterlik M, Tragl KH (2003) The effect of age and gender on cytokine production by human peripheral blood mononuclear cells and markers of bone metabolism. Exp Gerontol 38:1119–1127

Pietschmann P, Azizi-Semrad U, Pils K, Fahrleitner-Pammer A, Resch H, Dobnig H (2010) Pharmacologic undertreatment of osteoporosis in Austrian nursing homes and senior's residences. Wien Klin Wochenschr 160:532–537

Pietschmann P, Mechtcheriakova D, Mescheryakova A, Föger-Samwald U, Ellinger I (2016) Immunology of osteoporosis: a mini-review. Gerontology 62:128–137. doi:10.1159/000431091

Plotkin LI, Weinstein RS, Parfitt AM, Roberson PK, Manolagas SC, Bellido T (1999) Prevention of osteocyte and osteoblast apoptosis by bisphosphonates and calcitonin. J Clin Invest 104:1363–1374

Polzer K, Joosten L, Gasser J, Distler JH, Ruiz G, Baum W, Redlich K, Bobacz K, Smolen JS, van den Berg W, Schett G, Zwerina J (2010) IL-1 is essential for systemic inflammatory bone loss. Ann Rheum Dis 69:284–290

Rauner M, Sipos W, Pietschmann P (2007) Osteoimmunology. Int Arch Allergy Immunol 143:31–48

Rho J, Takami M, Choi Y (2004) Osteoimmunology: interactions of the immune and skeletal systems. Mol Cells 17:1–9

Rivollier A, Mazzorana M, Tebib J, Piperno M, Aitsiselmi T, Rabourdin-Combe C, Jurdic P, Servet-Delprat C (2004) Immature dendritic cell transdifferentiation into osteoclasts: a novel pathway sustained by the rheumatoid arthritis microenvironment. Blood 104:4029–4037

Robbie-Ryan M, Pacifici R, Weitzmann MN (2006) IL-7 drives T cell-mediated bone loss following ovariectomy. Ann N Y Acad Sci 1068:348–351

Rothe A, Power BE, Hudson PJ (2008) Therapeutic advances in rheumatology with the use of recombinant proteins. Nat Clin Pract Rheumatol 4:605–614

Saika M, Inoue D, Kido S, Matsumoto T (2001) 17beta-estradiol stimulates expression of osteoprotegerin by a mouse stromal cell line, ST-2, via estrogen receptor-alpha. Endocrinology 142:2205–2212

Saltel F, Chabadel A, Zhao Y, Lafage-Proust MH, Clézardin P, Jurdic P, Bonnelye E (2006) Transmigration: a new property of mature multinucleated osteoclasts. J Bone Miner Res 21:1913–1923

Sato S, Suematsu A, Okamoto K, Yamaguchi A, Morishita Y, Kadono Y, Tanaka S, Kodama T, Akira S, Iwakura Y, Cua DJ, Takayanagi H (2006) Th17 functions as an osteoclastogenic helper T cell subset that links T cell activation and bone destruction. J Exp Med 203: 2673–2682

Shen W, Durum SK (2010) Synergy of IL-23 and Th17 cytokines: new light on inflammatory bowel disease. Neurochem Res 35:940–946

Sipos W, Pieschmann P, Rauner M (2008) Strategies for novel therapeutic approaches targeting cytokines and signaling pathways of osteoclasto- and osteoblastogenesis in the fight against immune-mediated bone and joint diseases. Curr Med Chem 15:127–136

Sjögren K, Engdahl C, Henning P, Lerner UH, Tremaroli V, Lagerquist MK, Bäckhed F, Ohlsson C (2012) The gut microbiota regulates bone mass in mice. J Bone Miner Res 27:1357–1367

Smith EP, Boyd J, Frank GR, Takahashi H, Cohen RM, Specker B, Williams TC, Lubahn DB, Korach KS (1994) Estrogen resistance caused by a mutation in the estrogen-receptor gene in a man. N Engl J Med 331:1056–1061

Takayanagi H (2009) Osteoimmunology and the effects of the immune system on bone. Nat Rev Rheumatol 5:667–676

Tankó LB (2007) Effect of RANKL-specific denosumab on osteoclast number and function: a potential friend or foe? Curr Opin Investig Drugs 8:830–835

Teitelbaum SL (2007) Osteoclasts: what do they do and how do they do it? Am J Pathol 170:427–435

Tignor AS, Wu BU, Whitlock TL, Lopez R, Repas K, Banks PA, Conwell D (2010) High prevalence of low-trauma fracture in chronic pancreatitis. Am J Gastroenterol 105:2680–2686

Tosteson A, Solomon D, King AB et al (2005) Projections of osteoporosis fractures and costs by skeletal site in the USA. J Bone Miner Res 20:S21
Vojdani A, Erde J (2006a) Regulatory T cells, a potent immunoregulatory target for CAM researchers: the ultimate antagonist (I). Evid Based Complement Alternat Med 3:25–30
Vojdani A, Erde J (2006b) Regulatory T cells, a potent immunoregulatory target for CAM researchers: modulating allergic and infectious disease pathology (II). Evid Based Complement Alternat Med 3:209–215
Vojdani A, Erde J (2006c) Regulatory T cells, a potent immunoregulatory target for CAM researchers: modulating tumor immunity, autoimmunity and alloreactive immunity (III). Evid Based Complement Alternat Med 3:309–316
Walsh MC, Kim N, Kadono Y, Rho J, Lee SY, Lorenzo J, Choi Y (2006) Osteoimmunology: interplay between the immune system and bone metabolism. Annu Rev Immunol 24:33–63
Won HY, Lee JA, Park ZS, Song JS, Kim HY, Jang SM, Yoo SE, Rhee Y, Hwang ES, Bae MA (2011) Prominent bone loss mediated by RANKL and IL-17 produced by CD4+ T Cells in TallyHo/JngJ Mice. PLoS One 6:e18168
Wong BR, Besser D, Kim N, Arron JR, Vologodskaia M, Hanafusa H, Choi Y (1999) TRANCE, a TNF family member, activates Akt/PKB through a signaling complex involving TRAF6 and s-src. Mol Cell 4:1041–1049
Wright RM (2007) Use of osteoporosis medications in older nursing facility residents. J Am Med Dir Assoc 8:453–457
Wutzl A, Gruber R, Brozek W, Hofbauer G, Lernbass I, Brosch S, Pietschmann P (2010) Mechanisms involved in the inhibition of osteoclast generation by the benzothiophene SERM LY117018. Wien Klin Wochenschr 122:626–632
Xiao P, Chen Y, Jiang H, Liu YZ, Pan F, Yang TL, Tang ZH, Larsen JA, Lappe JM, Recker RR, Deng HW (2008) In vivo genome-wide expression study on human circulating B cells suggests a novel ESR1 and MAPK3 network for postmenopausal osteoporosis. J Bone Miner Res 23:644–654
Yagi M, Miyamoto T, Toyama Y, Suda T (2006) Role of DC-STAMP in cellular fusion of osteoclasts and macrophage giant cells. J Bone Miner Metab 24:355–358
Yamaguchi A, Komori T, Suda T (2000) Regulation of osteoblast differentiation mediated by bone morphogenetic proteins, hedgehogs, and Cbfa1. Endocr Rev 21:393–411
Zaidi M (2007) Skeletal remodeling in health and disease. Nat Med 13:791–801
Ziegler-Heitbrock L (2007) The CD14+CD16+ blood monocytes: their role in infection and inflammation. J Leukoc Biol 81:584–592
Zwerina J, Redlich K, Polzer K, Joosten L, Kronke G, Distler J, Hess A, Pundt N, Pap T, Hoffmann O, Gasser J, Scheinecker C, Smolen JS, van den Berg W, Schett G (2007) TNF-induced structural joint damage is mediated by IL-1. Proc Natl Acad Sci U S A 104:11742–11747

Rheumatoid Arthritis and Spondyloarthritis

8

Douglas H.N. White and Roland Kocijan

8.1 Introduction

Recent years have seen extraordinary growth in our knowledge and understanding of the pathogenesis of inflammatory arthritis as reflected in the development of new therapies and changing clinical practice. This chapter will review the most common forms of inflammatory arthritis, including rheumatoid arthritis, the axial spondyloarthritides (radiographic axial spondyloarthritis (r-axSpA) and the non-radiographic axial spondyloarthritis (nr-axSpA)) as well as psoriatic arthritis (PsA) to explore their epidemiology, pathogenesis, clinical features and treatment.

8.2 Rheumatoid Arthritis

8.2.1 Historical Perspective

Rheumatoid arthritis (RA) has been recognised in Europe since the seventeenth century, with Sydenham publishing the first case report in 1676. The artwork of the Dutch painter Peter Paul Rubens (1577–1640) is thought by some to show evidence of hand deformities that can occur in RA (Appelboom et al. 1981). Interestingly, the typical erosive changes of RA have not been found within the European fossil record, yet the characteristic joint damage found with gout, osteoarthritis and ankylosing spondylitis has been well documented

D.H.N. White, M.D., BSc, MBChB, FRACP (✉)
Rheumatology Department, Waikato Hospital, Pembroke Street, Hamilton, New Zealand
e-mail: douglas.white@waikatodhb.health.nz

R. Kocijan, M.D.
Medical Department II, St. Vincent Hospital Vienna, Academic Teaching Hospital of the Medical University Vienna, Vienna, Austria
e-mail: roland.kocijan@bhs.at

P. Pietschmann (ed.), *Principles of Osteoimmunology*,
DOI 10.1007/978-3-319-34238-2_8

(Aceves-Avila et al. 1998). Similar evaluation of skeletal remains from indigenous North Americans has shown these characteristic changes. In these areas, prevalence of RA remains remarkably high at around 5 % (Rothschild et al. 1988). This has led to speculation that perhaps RA was brought from the New World back to the Old World by returning explorers. Currently, there is no direct evidence supporting this.

8.2.2 Epidemiology

Rheumatoid arthritis has a worldwide distribution affecting all ethnic groups, and although all ages can be affected, the peak incidence is between the 4th and 6th decades with females being affected 2–4 times more commonly than males. The gender ratio becomes less pronounced with increasing age. Prevalence varies considerably, but published work suggests that ~0.5–1 % of European and North American adults are affected, with rates being lower in Southern Europe than Northern Europe and highest in native North Americans (Alamanos et al. 2006).

Several authors have suggested that RA appears to be becoming less common and less severe, and there is evidence that the incidence of extra-articular features is declining (Turesson and Matteson 2009). The change in incidence appears to have begun before the advent of aggressive disease management strategies and remains unexplained.

8.2.3 Pathogenesis

Insights into the pathogenesis of RA have been gained through the study of affected tissues, genetic studies and modern molecular approaches. Nevertheless, despite the growth in our understanding of the mechanisms underlying RA, it is not yet possible to unite the different elements into a comprehensive explanation of the heterogeneous phenotype.

As we shall see, RA has features of both T-cell activation with the formation of rheumatoid nodules and also B-cell activation with autoantibody production. Indeed, microscopic examination of synovial tissue from inflamed joints shows evidence of a dense but non-specific infiltration of inflammatory cells including neutrophils, B-cells, T-cells, macrophages and mast cells.

The inflammatory response is coordinated by a complex cytokine network with macrophages being the key secretors of pro-inflammatory cytokines in the inflamed rheumatoid synovium. It is now known that the cellular basis of the inflammatory response changes as the disease progresses with T-cells playing an important role in the early stages of disease (Raza et al. 2005).

Inflammation within the synovium results in the formation of a destructive pannus that may lead to the erosion of bone and consequent deformity and functional impairment. Recent work has suggested that bone oedema on MRI scan can predict future erosion, suggesting that the process of erosion may begin within the bone.

8.2.4 Cellular and Molecular Mechanisms

The first clues to the autoimmune nature of RA came from the discovery of a "rheumatoid factor" in the serum of affected patients by Waaler in 1938. Subsequently rediscovered by Rose in 1948, it was to be another 9 years until this rheumatoid factor was characterised as an antibody that binds to the Fc portion of immunoglobulin (Franklin et al. 1957). Whilst rheumatoid factor is most commonly IgM directed against the Fc portion of IgG, it can also exist in IgA and IgG subtypes. A proposed mechanism of action of rheumatoid factor was put forward in 1973 by Zvaifler in which immune complexes were formed that subsequently fixed complement and released chemoattractant factors to recruit neutrophils and other inflammatory cells to the synovium (Zvaifler 1973). Whilst there is considerable evidence to support this hypothesis, one of the key arguments against it is that rheumatoid factors are found in up to 15 % of the healthy older population and in those with other autoimmune diseases, infection and malignancy without joint involvement.

Subsequently, numerous other autoantibodies have been detected in the sera of patients with rheumatoid arthritis including anti-perinuclear factor and anti-keratin antibodies. Characterisation of these antibodies revealed that they were binding to citrullinated filaggrin (Girbal-Neuhauser et al. 1999), with citrullinated epitopes on fibrinogen and vimentin also acting as targets. Citrullination is a post-translational modification of the amino acid arginine, and the process is thought to have a natural role in apoptosis. The modification is carried out by the enzyme peptidyl arginine deiminase (PAD) in the presence of relatively high calcium concentrations. Commercial assays are now available for anti-citrullinated peptide antibodies (ACPA) which are found in 60–70 % of people with RA and rarely in other diseases. These antibodies can be present for up to two decades before symptoms develop (Jørgensen et al. 2008). It has been shown that ACPAs can directly initiate the differentiation of bone-resorbing osteoclasts, suggesting an independent effect of these antibodies in initiating skeletal damage (Harre et al. 2012). Moreover, IgG sialylation was reported to be a main regulator for the pro-osteoclastogenic potential of immune complexes as only non-sialylated immune complexes stimulated osteoclastogenesis in patients with RA. Current data indicate that RA patients with low IgG and low ACPA sialylation suffer from poorer bone microstructure as compared to those with a high-sialylation status (Harre et al. 2015). Therefore, bone resorption occurs also in the absence of inflammation, as has also been shown in healthy, ACPA-positive individuals with local bone loss (Kleyer et al. 2014). Moreover, systemic bone loss and secondary osteoporosis in RA seem to be triggered by ACPAs (Kocijan et al. 2014b). In addition, it has been shown that in patients with ACPA positivity, tapering and even stopping of antirheumatic treatment are associated with a higher risk for relapse (Haschka et al. 2016).

The abundance of Th1 cytokines such as IFN-γ and the relative lack of the Th2 cytokines IL-4, IL-5 and IL-12 support the hypothesis that RA is primarily a Th1 disease. However, in recent years, our understanding has changed with the discovery of a new subclass of regulatory T-cell that produces IL-17, the Th17-cell. Production of IL-17 by these cells is driven by IL-23 which shares a common subunit with IL-12. Increased concentrations of IL-17 and IL-23 are found in the sera of patients with RA compared to controls with osteoarthritis, thereby supporting their role in the disease. In addition, mice deficient in IL-23 are resistant to developing arthritis in the collagen-induced arthritis model. The Th17-cells produce TNF-α, IL-6, IL-17, IL-22, and GM-CSF, cytokines known to be important in the inflammatory response. IL-17 is an important stimulator of further cytokine production including IL-1β, IL-6, IL-23, IL-8, GM-CSF, G-CSF, VEGF, and COX-2, thereby amplifying the immune response (Furuzawa-Carballeda et al. 2007; Lundy et al. 2007). IL-17 and Th17 cells have therefore become an important area of research and offer new therapeutic targets. The role of the activated macrophage in the synovium of affected joints is crucial to the maintenance of chronic inflammation. In addition to interaction with T-cells and fibroblasts, the macrophage is a potent effector cell that produces pro-inflammatory cytokines, expresses toll-like receptors (TLRs) and is involved in antigen processing and presentation. They also have phagocytic ability and are involved in tissue remodelling. Interestingly, the macrophage also responds to oestrogen concentrations, such that a high concentration (as in pregnancy) inhibits IL-1 secretion (Cutolo and Lahita 2005). This may be responsible in part for the improvement many women experience during their pregnancies.

The presence of autoantibodies, together with the formation of germinal centre-like structures in the synovium of affected joints and the good therapeutic response to B-cell depletion, suggests that B-cell dysfunction is also important in pathogenesis. B-cells have several roles in both humoral and innate arms of the immune system including antigen presentation and antibody and cytokine production. The link between humoral and innate responses is evidenced by the expression of TLRs on B-cells. These receptors can bind to hypomethylated CpG sequences in bacterial or mitochondrial DNA, single-stranded RNA or bacterial cell wall components, and it has been shown that mitochondrial DNA from apoptotic synovial cells can stimulate potentially autoreactive B-cells through this mechanism (Leadbetter et al. 2002).

Central to the development of autoimmunity is the breakdown of tolerance. Given that the majority of patients develop RA at an age when thymic function has severely declined or ceased entirely, the defect is more likely to be with peripheral tolerance rather than central tolerance. The precipitating event in RA is not yet known, but the T-cell repertoire in those with RA is altered in a number of respects from those without RA, including evidence of early senescence as evidenced by reduced telomere lengths (Colmegna et al. 2008). In addition, the T-cell repertoire appears to be reduced by a factor of 10 in those with RA compared to controls without RA (Wagner et al. 1998). Since the proliferation of naive T-cells is dependent on antigenic stimulation, over time, this will lead to the development of a peripheral T-cell repertoire with an increasing affinity for self. Thus, it is hypothesised that defective thymic selection coupled with peripheral selection over time predisposes the susceptible individual to the development of autoimmunity.

8.2.5 Genetics

Concordance rates in monozygotic twins indicate that approximately 50 % of the variation in prevalence of RA is genetic and that 30 % of this is attributable to the HLA-DR locus. Experiments first performed in 1969 identified a region on chromosome 6, now known to code genes within the major histocompatibility complex (MHC). Further work has mapped the linked region precisely to the third hypervariable region of the HLA-DRβ chain (Nepom 1989). The precise amino acid sequence between positions 70 and 74 appears to be particularly important as variations in the sequence both increase and decrease the risk of ACPA-positive RA. The amino acid sequence DERAA appears in ~30 % of healthy controls but in only 15 % of patients with RA and tends to be associated with less erosive disease when present, whereas the sequence QKRAA, QRAAA or RRRAA appears to increase the risk of ACPA-positive RA. Thus, it appears that the amino acids in positions 70 and 71 modulate the T-cell response such that the amino acids arginine (R), glutamine (Q) or leucine (K) increase the risk and alanine (A) or glutamic acid (E) is protective (van der Helm-van Mil et al. 2005). Work continues to establish exactly how these differences confer variable risk. This sequence is common to several HLA-DR alleles including DR*0101, DR*0102, DR*0401, DR*0404, DR*0405, DR*0408, DR*1001 and DR*1402 and has been termed the "shared epitope" by Gregersen et al. (1987). Individuals who are heterozygous for one of these alleles tend to have more severe, erosive disease. The effect is further intensified by homozygosity (Weyand et al. 1992).

Further work on the HLA region has found an association between the HLA-DR3 locus and ACPA-negative RA (van der Helm-van Mil et al. 2007). The precise mechanism by which genes at this locus influence disease is not known, although it is conceivable that DR3 polymorphisms could be predisposed to production of an as yet unidentified antibody.

Genetic factors independent of the HLA region have also been identified. The C>T single nucleotide polymorphism at position 1858, causing a missense mutation in the protein tyrosine phosphatase (PTPN22) gene, has been linked to ACPA-positive RA (Wesoly et al. 2005). The polymorphism has been validated in Canadian, North American and European populations, but does not appear to exist in Asians. Protein tyrosine phosphatase exerts a negative feedback regulation in T-cell receptor signalling; it binds to the regulatory kinase, Csk, and this complex is responsible for the dephosphorylation of the Lck protein at position 394 and its phosphorylation at position 505, thereby terminating the T-cell receptor signal. The C>T 1858 polymorphism appears to directly modify the phosphorylase activity or affect the binding of PTPN22 to Csk (Cloutier and Veillette 1999). Interestingly, the polymorphism has also been found in a number of other autoimmune diseases including type I diabetes, Graves' disease, SLE, JIA and vitiligo and appears to be a gain-of-function mutation that is hypothesised to impair thymic selection of autoreactive T-cells (Bottini et al. 2006). The genetic factors above are neither necessary nor sufficient for the development of RA; however, they do indicate that these pathways are important for disease susceptibility in individual patients.

8.2.6 Environment

Genetic factors alone are unable to account for the susceptibility to RA, and attention has focussed on environmental triggers for the disease. Pathogens including Epstein-Barr virus, Parvovirus B19 and mycobacteria have been investigated. To date, pathogen-derived antigens have not been discovered and evidence for molecular mimicry is lacking.

The most consistent environmental factor is cigarette smoking, wherein there appears to be a relationship between the number of pack-years of cigarettes consumed and the risk of developing RA, which can be as much as 21-fold above non-smokers (Klareskog et al. 2006, 2007). Importantly, smoking increases the risk for ACPA-positive RA, but not ACPA-negative RA, and the risk is further increased in those who possess the shared epitope. There are two hypotheses that might explain the link between smoking and ACPA-positive RA. Firstly, the detection of citrullinated peptide in the alveolar fluid of smokers has led to the hypothesis that smoking induces apoptosis in the lung, generating citrullinated peptides that are then recognised more strongly by those with the shared epitope who then develop RA. Secondly, smoking is known to increase the levels of tetrachlorodibenzo-P-dioxin (TCDD) which has been shown to upregulate IL1β, IL-6 and IL-8 production through binding to the aryl hydrocarbon receptor. Recently, work on the pharmacokinetics of methotrexate, the drug used most commonly to treat RA, has shown that smoking also reduces intracellular methotrexate polyglutamate levels, which may account, in part, for the unfavourable outcome in smokers (Stamp et al. 2009).

Other environmental factors that have received attention include coffee and alcohol consumption, periodontitis, exposure to mineral oils and body mass index; however, these factors are not as consistent as cigarette smoking and should be interpreted with caution. Additionally, pregnancy does appear to be a risk factor in as many as 12 % of females developing RA do so within 1 year after pregnancy.

Besides the risk of development of RA, the presence of autoantibodies, environmental and genetic factors such as smoking, HLA-DRB genotype and low socioeconomic status has been identified as poor prognostic factors in RA (Manfredsdottir et al. 2006; Kaltenhauser et al. 2007; Harrison et al. 2005; Hetland et al. 2009; Sanmartí et al. 2007).

8.2.7 Mechanisms of Bone Erosion

In RA there appears to be a coupling between the processes of inflammation and bone erosion. TNF-α is capable of binding to two receptors, designated TNFR1 (or p55) and TNFR2 (or p75); the former appears to have greater activity as it is directly coupled to a death domain that can induce apoptosis. Both receptors for TNF-α, but mostly TNFR1, are found on osteoclast precursors, osteoclasts as well as on osteoblasts. These cells are also capable of secreting TNF-α in response to external stimulation.

The osteoclast is the key cell type involved in the destruction of bone within the inflamed rheumatoid joint. Osteoclast differentiation required the cytokines macrophage colony-stimulating factor (M-CSF) and the receptor activator of NF-kB (RANKL). M-CSF promotes proliferation and survival of the monocyte lineage through activation of a tyrosine kinase (cFms) and RANKL through binding to its receptor. RANK leads to activation of the transcriptions factors NFATc1, AP-1 and NF-kB required for osteoclast differentiation. Since RANKL is part of the TNF superfamily and RANK shares many of the TNF-α signalling properties, it is conceivable that the elevated levels of TNF-α found in inflammatory arthritis contribute to the increased osteoclast differentiation (Abu-Amer et al. 2000; Kobayashi et al. 2000). This is strengthened by the finding that the in vitro differentiation of osteoclast precursors lacking RANK could be driven by TNF-α and TGF-β (Kim et al. 2005). It would appear, however, that in vivo osteoclast differentiation required IL-1 as TNF-α alone is unable to activate the transcription factor TRAF-6, a necessary condition for the formation of the actin ring structure required for bone resorption (Nakamura et al. 2002). The central role of TNF-α in the regulation of bone metabolism is presented in Fig. 8.1.

In RA, the source of RANKL is mainly from stromal cells including fibroblasts and osteoblasts, stimulated in turn by the action of the inflammatory cytokines, IL-1, IL-6, TNF-α and IL-17. In addition, RANKL is produced by CD4+ regulatory T-cells recruited to the inflamed synovium. The effects of these cytokines on osteoclast progenitors are amplified by upregulation of RANK through the direct action of IL-1β and TNF-α on the progenitor cells.

The process of erosion results as a consequence of both increased bone resorption and reduced bone healing. This suggests alterations in osteoblast function, and indeed, TNF-α is a potent inhibitor of osteoblastogenesis through inhibition of the transcription factors Runx2 and Osterix (Lu et al. 2006). Additionally, TNF-α inhibits Wnt signalling, a major pathway regulating osteoblast differentiation (Baron et al. 2006). Members of the Wnt family of cytokines bind to complex membrane-bound receptors incorporating the Frizzled protein and LRP5 and 6. TNF-α is able to interfere with this process at multiple levels by inducing *secretable* Frizzled-related proteins and by production of Dickkopf-1 and sclerostin that interfere with binding of LRP5 and 6 to the Frizzled receptor (Diarra et al. 2007). Increased serum levels of Dickkopf-1 and an association to bone erosions and osteoporosis were reported in patients with RA (Rossini et al. 2015).

In summary, the effects of systemic inflammation are to enhance bone resorption through osteoclast differentiation and activation and to impair bone healing by inhibition of osteoblast differentiation. Our increased understanding of the molecular mechanisms of bone erosion and repair has opened up novel therapeutic targets to prevent bone erosion in RA. The results of a 12-month study of denosumab, a monoclonal antibody directed against RANKL in RA, have shown reduced evidence of erosive damage on MRI scan as early as 6 months after treatment (Cohen et al. 2008). Consistent with its mechanism of action, this antibody has no effect on measures of disease activity in RA. Similar effects were shown for the sclerostin antibody, which did not affect joint swelling or synovitis but blocked and reversed

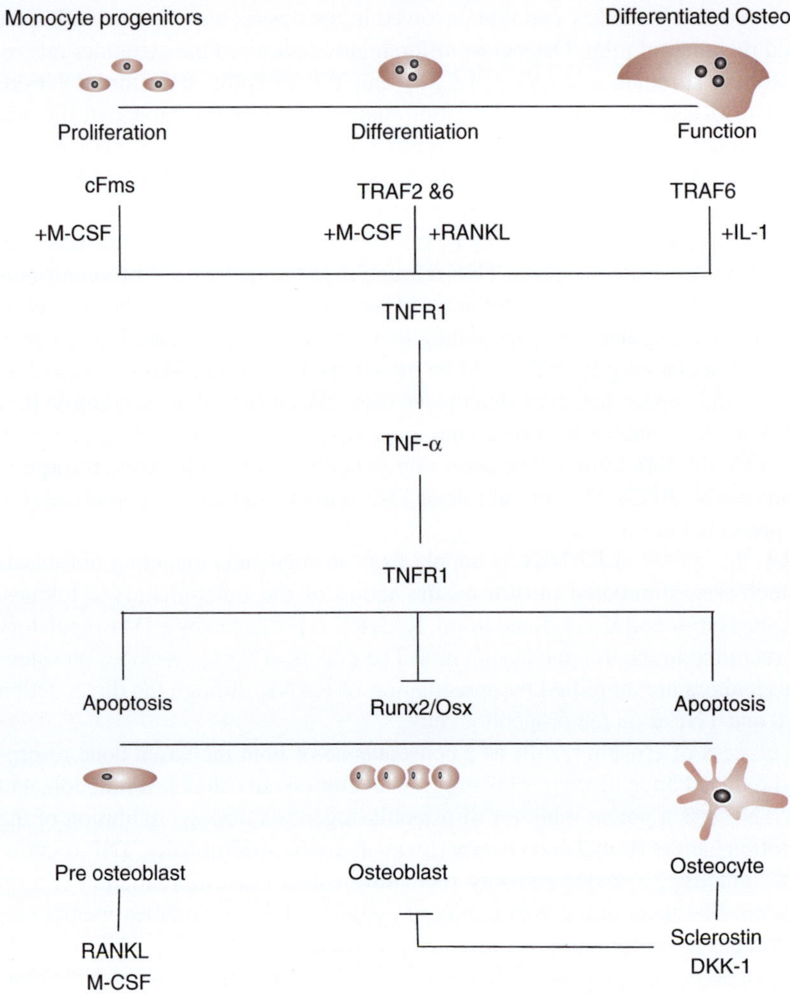

Fig. 8.1 The central role of TNF-α in bone metabolism (Adapted from David and Schett (2010))

periarticular bone loss in patients with RA. The sclerostin antibody arrested the progression of bone erosions and showed positive effects on the articular cartilage (Chen et al. 2013).

8.2.8 Diagnosis and Presentation

The new criteria set for the classification of RA were introduced in 2010 by the American College of Rheumatology (ACR) and the European League Against Rheumatism (EULAR). Therefore, the diagnosis of RA can be made in the (i) presence of synovitis in at least one joint, (ii) the absence of alternative diagnosis and

Table 8.1 The 2010 American College of Rheumatology/European League Against Rheumatism classification criteria for RA

	Criterion	Score
A	Joint involvement	
	1 large joint	0
	2–10 large joints	1
	1–3 small joints (with or without large joint involvement)	2
	4–10 small joints (with or without large joint involvement)	3
	>10 joints (including at least 1 small joint)	5
B	Serology	
	Negative RF and negative ACPA	0
	Low-positive RF or low-positive ACPA	2
	High-positive RF or high-positive ACPA	3
C	Acute phase reactants	
	Normal CRP and ESR	0
	Abnormal CRP or abnormal ESR	1
D	Duration of symptoms	
	<6 weeks	0
	>6 weeks	1

The new criteria should not be applied if the symptoms are explained by another disease; joint involvement includes either tenderness or swelling and can include imaging assessment. RA is defined as a score of > 6/10

(iii) a score of 6 or more points out of the following items: number and site of affected joints, serologic parameters, elevation of acute-phase proteins and duration of symptoms (Aletaha et al. 2010). The new classification criteria especially focus on early diagnosis of disease to prevent longstanding effects related to RA (see Table 8.1). These replace the 1987 criteria which had poor sensitivity to early disease.

Presentation may be preceded by a period of non-specific malaise with wide-spread aches and pains and fatigue. Symptoms can evolve slowly or start suddenly and can affect all joints from the beginning or spread to affect different joints as disease progresses. Classically, the patient will describe symptoms with an "inflammatory rhythm" where their symptoms are worst early in the mornings or overnight and improve during the course of the day.

Disease is not confined solely to the joints, and there are well-described extra-articular features (Table 8.2). The incidence of cardiovascular disease and lymphoma is increased in RA. Indeed, the major cause of premature mortality in RA is cardiovascular disease. The incidence of extra-articular features appears to be falling, possibly as a consequence of earlier diagnosis and more aggressive initial management; however, the decline appears to have begun before these management changes were well established.

Table 8.2 Extra-articular manifestations of RA

Organ system	Extra-articular manifestation
Eye	Scleritis, episcleritis and scleromalacia perforans, keratoconjunctivits sicca
Skin	Rheumatoid nodules, vasculitis, pyoderma gangrenosum, panniculitis
Lung	Pulmonary fibrosis, rheumatoid nodules, pleural effusion
Haematological	Anaemia, Felty syndrome
Nervous system	Vasculitis, peripheral neuropathy, atlanto-axial subluxation
Kidney	Amyloidosis
Cardiovascular	Accelerated atherosclerosis, pericarditis

8.2.9 Treatment of RA

Based on the treat-to-target (T2T) recommendations, the treatment aim is defined as remission with low disease activity as an acceptable alternative goal. The long-term health-related quality of life, the prevention of structural damage and the normalisation of function and social participation were defined as primary goals in patients with rheumatoid arthritis (Smolen et al. 2010). Should the patient's current therapy prove insufficient, therapeutic adaptation to reach treatment goals is recommended after 3–6 months. Follow-up examinations should include assessment of disease activity and joint counts (Smolen et al. 2016).

Pharmacological management of RA can be thought of in terms of agents to relieve symptoms and agents to suppress the underlying disease. Many patients benefit from the use of non-steroidal anti-inflammatory drugs (NSAIDs) which can be helpful for relief of pain and stiffness. They are however associated with potential side effects of gastric irritation, nephrotoxicity and increased cardiovascular risk making their long-term use undesirable. Similarly, simple pain killers such a paracetamol alone or in combination with weak opioids can be useful for some patients. However, these medications do not have any effects on the progression of erosive disease. Therefore, a conventional disease-modifying antirheumatic drug (DMARD) therapy is recommended in treatment of naive patients with early and established RA. If disease activity remains high despite conventional DMARD therapy, then combination conventional DMARD therapy or a biological treatment is recommended. In case of flare, short-term glucocorticoid therapy could be added (Singh et al. 2015). Conventional DMARDs and biological therapy are shown in Table 8.3.

Intervention with early combination DMARD therapy has been shown to have beneficial effects on disease progression independent of treatment in later years, suggesting that there is a "window of opportunity" in which the disease process can be altered (Boers et al. 1997; Möttönen et al. 1999). In addition, it has been calculated that the risk of undertreatment is 5–6 times the risk of overtreatment if all patients are treated aggressively from the outset, providing justification for an aggressive approach in early RA (de Vries-Bouwstra et al. 2006). Recommendations from EULAR have advised that the combined efficacy of methotrexate (MTX) and

Table 8.3 Disease-modifying anti-rheumatic drugs currently available for the management of RA

	Typical dosing	Proposed mechanism of action	Common side effects
Conventional DMARDs			
Methotrexate (MTX)	o, s/c max 30 mg/week	Inhibition of folate metabolism Inhibition of Adenosine release	Myelotoxicity, hepatotoxicity
Sulphasalazine (SSZ)	o, 2000–3000 mg daily	Uncertain	Myelotoxicity, hepatotoxicity
Hydroxychloroquine (HCQ)	o, 200–400 mg daily	Alteration of lysosomal pH	Ocular toxicity
Azathioprine (AZA)	o, 150–200 mg daily	Anti-metabolite	Myelotoxicity, hepatotoxicity
Gold (IMG)	im, 50 mg 3–4 weeks	Uncertain	Myelotoxicity, nephrotoxicity
Cyclosporine A (CSA)	o, 200–400 mg daily	Inhibition of IL-2 signal transduction	Nephrotoxicity
Leflunomide (LEF)	o, 10–20 mg daily	Inhibition of pyrimidine synthesis	Myelotoxicity, hepatotoxicity, gastrointestinal upset
Biological DMARDs			
Etanercept	s/c, 50 mg weekly	Soluble TNF receptor	
Adalimumab	s/c, 40 mg every 2 weeks	Humanised anti-TNF antibody	Infection TB reactivation, drug-induced lupus
Infliximab	iv, 3 mg/kg 0, 2, 6 then 8 weekly	Chimeric anti-TNF antibody	
Golimumab	s/c, 50 mg once monthly	Fully human anti-TNF antibody	
Rituximab	iv, 1000 mg on day 0 and 14 then 6 monthly	B-cell depletion, targets CD-20	Infusion reactions
Anakinra	s/c, 100 mg daily	IL-1 receptor antagonist	Injection site reactions
Abatacept	iv, weeks 0,2,4 then 4 weekly	CTLA4-Ig, blocks co-stimulation	Headache, nausea, infection
Tocilizumab (TCZ)	iv, 8 mg/kg infused monthly	Monoclonal antibody against IL-6	Heptotoxicity, neutropenia, abnormal lipid profiles

Certolizumab, s/c, week 0/2/4: 400 mg; after 200 mg every other week; PEGylated Fab' of monoclonal antibody

o oral, *s/c* subcutaneous, *im* intramuscular, *iv* intravenous leflunomide appears superior to other conventional DMARDs, but this combination may also be associated with increased hepatotoxicity (Gaujoux-Viala et al. 2010).

The evolution of biological therapies has provided insight into the pathogenesis of RA as well as ushering in a new era in its treatment. Currently, there are five anti-TNFs (adalimumab, etanercept, golimumab, infliximab, certolizumab), one interleukin-1 receptor antagonist (anakinra), one T-cell selective co-stimulation modulator (abatacept), one chimeric monoclonal CD20 antibody (rituximab) and one antibody against the IL-6 receptor (tocilizumab) available for the treatment of RA (Tvete et al. 2015). All biological drugs are effective when compared to placebo as well as conventional DMARDs and are generally more effective when given together with conventional DMARDs. Higher doses of biological agents are associated with higher effect compared to lower doses (Tvete et al. 2015).

Anti-TNFs have positive effects on disease activity, well-being and radiographic progression. The efficacy, onset of action and side effect profile are comparable for all anti-TNFs. Lately, certolizumab pegol, a PEGylated anti-TNF, has shown promising results in the treatment of RA. The long-term efficacy over almost 5 years in combination with MTX in patients with active RA was recently reported for certolizumab (Smolen et al. 2015).

The anti-interleukin-6 receptor antibody tocilizumab achieved significantly greater improvement in radiographic progression and physical function after 52 weeks compared to patients treated with conventional DMARDs. Tocilizumab is effective in combination with MTX and also as monotherapy for the treatment of early RA (Burmester et al. 2015). The effectiveness of tocilizumab is similar to anti-TNF with respect to treatment response as measured on the ACR20 and ACR50 (Bergman et al. 2010).

A new class of orally available small molecules, the Janus kinase (JAK) inhibitors, have recently been introduced. Tofacitinib is an inhibitor of JAK 1 and 3 and shows similar efficacy when compared to the anti-TNF agents (van Vollenhoven et al. 2012).

These new agents are costly but efficacious. Their efficacy is usually enhanced by using them in combination with MTX or leflunomide (Nam et al. 2010). Recent evidence suggests that the addition of a biologic agent to those patients who have had an inadequate response to MTX at 3 months is superior to combination conventional DMARD therapy. This may be through synergistic action or because the conventional DMARD prevents formation of neutralising antibodies against the biologic agent. Before the initiation of a biological agent, it is mandatory to screen for past exposure to *Mycobacterium tuberculosis* as the risk of reactivation of latent infection is high. This is usually with a combination of chest radiograph, Mantoux and interferon gamma release assay. Rates of infection with other organisms appear to be increased, particularly in the first 12 months. Most authorities would suggest the use of anti-TNF therapy as initial biological treatment unless contraindicated as a consequence of recent malignancy, active sepsis or hypersensitivity.

The impact of disease in RA can be assessed by looking at the three domains of disease activity, disability and structural damage. There are a number of

Table 8.4 Measures of disease activity in RA

	Formula	Moderate activity	Low activity	Re mission
DAS	$0.54 \bullet \sqrt{RAI} + 0.065 \bullet SJC + 0.33 \bullet \ln(ESR) + 0.0072 \bullet \text{VAS}$	3.7	2.4	1.6
DAS28	$0.56 \bullet \sqrt{TJC} + 0.039 \bullet SJC + 0.72 \bullet \ln(ESR) + 0.0013 \bullet VAS$	5.1	3.2	2.6
SDAI	TJC + *SJC* + VAS + PhysicianVAS + CRP	26	11	3.3
CDAI	TJC + *SJC* + VAS + PhysicianVAS	22	10	2.8

well-validated tools in routine use to facilitate this process including the DAS, DAS28, CDAI, SDAI (Table 8.4) and ACR response criteria for disease activity, the Health Assessment Questionnaire (HAQ) and SF36 for disability and the Sharp-van der Heidje method for assessing joint space narrowing and erosion on plain radiograph. Full details of these measures are beyond the scope of this text. However, important targets are presented in Table 8.4.

8.3 Spondyloarthritis (SpA)

The spondylarthritides (SpA) are a group of inflammatory disorders of the spine and peripheral joints where the hallmark pathologic feature is enthesitis. The classification system for SpA has recently been revised highlighting the distinction between axial and peripheral disease which includes psoriatic arthritis (PsA), reactive arthritis, enteropathic arthritis, the enthesitis-related subtype of juvenile idiopathic arthritis and undifferentiated spondyloarthritis (Rudwaleit et al. 2011). It has been clear for many years that like RA, SpA is influenced by genetic, environmental and immunologic factors. Despite substantial progress into our understanding of these areas, it is not yet possible to link them into a unified theory of pathogenesis. In this chapter radiographic axial SpA (r-axSpA), non-radiographic axial SpA (nr-axSpA) and psoriatic arthritis (PsA) are discussed in detail.

8.3.1 Axial Spondyloarthritis

8.3.1.1 Epidemiology

The term axial SpA includes radiographic axial SpA (previously known as ankylosing spondylitis (AS) or Bechterew's disease) and non-radiographic axial spondyloarthritis according to the Assessment of Spondyloarthritis International Society (ASAS) criteria. The typical clinical feature of axial SpA is the presence of inflammatory back pain ≥3 months, age of onset below the age of 45 years, response to NSAIDs, a positive family history and HLA-B27 positivity. Associated inflammatory diseases such as inflammatory bowel disease (IBD) or uveitis are common features.

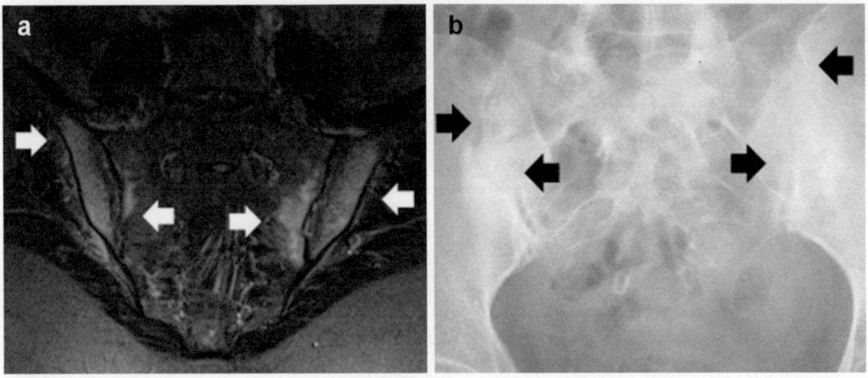

Fig. 8.2 (**a**) MRI (STIR sequence) showing signs of disease activity. *White arrows*: oedema in the bone marrow. (**b**) Pelvic X-ray film showing bilateral sacroiliitis (*black arrows*), New York Criteria 3–4

In r-axSpA, typical radiographic changes can be found with sclerosis, erosion and fusion of the sacroiliac joints as the most important findings. In contrast, nr-axSpA is distinguished by the presence of MRI changes in the absence of radiographic disease (Rudwaleit and Sieper 2012). See Fig. 8.2.

Several studies in different populations have found an annual incidence of 7 cases per 100,000 population for axial SpA. This appears to have remained constant over the last 55 years, with males affected approximately three times more frequently than females (Gabriel and Michaud 2009). Although the mean age of onset is 24–26 years, approximately 15 % experience disease onset in childhood; onset after age 45 is rare.

The incidence of axial SpA follows closely that of the HLA-B27 allele around the world and, as such, is highest in Native North Americans and lowest in Australian Aborigines and Africans (Lau et al. 1998). In white European populations, the frequency of the HLA-B27 gene ranges from 26 % in the Lapps of northern Norway to 4 % in Southern Europeans.

8.3.1.2 Pathology and Cellular Mechanisms

The pathology of SpA differs from RA in two key factors. Firstly, in SpA the enthesis is the site of major histologic changes which progress in an orderly sequence, beginning with a destructive enthesopathy followed by a healing process with new bone formation linking deeper bone to the ligament and ultimately resulting in bony ankylosis (Ball 1971). Vertebral changes typically begin with an erosive lesion at the anterior corner of the annulus fibrosus. Secondly, the healing process results in increased bone formation which is laid down initially as cancellous bone which is then remodelled into mature lamellar bone creating the typical syndesmophytes that are seen on plain radiography of the spine.

The pattern of joint involvement is different from that seen in RA with sacroiliitis, presenting as buttock pain, being a hallmark clinical feature of axial

SpA. Changes seen on plain radiology of the sacroiliac joints are most prominent towards the inferior aspects of the joints. Once disease has been longstanding, there is encasement of the joints as a result of ossification of the capsule, often surrounding small islands of intact articular cartilage. It is important to bear in mind that even in the population without axial SpA, there can be progressive fusion of the sacroiliac joint as a result of osteoarthritis; changes are more marked on the iliac side and towards the superior aspect of the joints in osteoarthritis.

Histological specimens from the hip, zygapophyseal and sacroiliac joints of patients with active disease have identified infiltrates of T-cells, B-cells, macrophages and osteoclasts as well as cells involved in angiogenesis. Synovitis is less common than in RA and can be distinguished by the greater proportion of M2 regulatory CD163+ macrophages. Expression of TNF-α and TGF-β mRNA is increased, and recent advances in treatment with therapies blocking TNF-α have shown that this cytokine is responsible for the pain, fatigue, swelling and stiffness (Francois et al. 2006). Nevertheless, neither the cell stimulating TNF-α production nor its target cell has yet been identified. Since a proportion of patients fail to achieve a complete remission with anti-TNF-α therapy, it is clear that there are still questions to be answered in this area.

Additionally, the target antigen driving the immune response has yet to be identified. However, a T-cell response against a proteoglycan link protein has been demonstrated in humans with axial SpA (Mikecz et al. 1988). In the Balb/c mouse, a clinical picture similar to axial SpA can be induced by immunisation with fetal human cartilage creating a humoral and cellular response against aggrecan (Glant et al. 1987).

8.3.1.3 Genetic and Immunologic Factors

It has long been appreciated that axial SpA are heritable diseases with genetic factors being responsible for ~90 % of the susceptibility. Evidence has come from twin studies: if a condition were entirely genetically determined, one would anticipate 100 % concordance between monozygotic twins; however, for axial SpA, only 63 % concordance is seen suggesting involvement of non-genetic factors to be discussed shortly (Jarvinen 1995).

The discovery of the association of HLA-B27 with axial SpA was made contemporaneously by two groups (Schlosstein et al. 1973; Brewerton et al. 1973). Development of the B27 transgenic rat in the 1990s confirmed the direct involvement of this molecule in the disease process (Taurog and Hammer 1995). To date, approximately 60 different subtypes of HLA-B27 have been identified that have most likely evolved from the B*2705 allele. The most common alleles, B*2702, B*2704, B*2705 and B*2707, are associated with axial disease. Interestingly, in the Sardinian population, B*2709, and, in the Southeast Asian population, B*2706 are much less common in those with axial disease (Khan 2000).

There are currently three hypotheses explaining the possible role of HLA-B27 in the pathogenesis of axial SpA. Firstly, it is known that the HLA-B27 molecule does not behave like other HLA class I molecules in that it is capable of homo-dimerising in the absence of β2-microglobulin – a process termed misfolding. It is postulated that this stimulates NK (natural killer) and T-cells through interaction with cell surface receptors in the leukocyte immunoglobulin-like receptor (LILR) family (Kollnberger et al. 2004). Secondly, the unfolded protein response hypothesis holds that a reduced rate of folding of the HLA-B27 molecule in the endoplasmic reticulum triggers an intracellular signalling response that may in turn lead to IL-23 release. Evidence supporting this unfolded protein response has been found in synovial biopsies from patients with axial SpA, and it is known that the ERAP1 gene product is important in processing the antigenic peptide required during the folding process (Dong et al. 2008). Finally, the molecular mimicry hypothesis suggests that HLA-B27 preferentially binds to a self-antigen that resembles a microbial peptide. This is supported by the presence of T-cells that recognise self-antigens in vivo (Atagunduz et al. 2005).

The finding that the risk of developing axial SpA was 16-fold higher in first-degree relatives of those who are HLA-B27 positive compared to those who are HLA-B27 positive in the general population suggested that although HLA-B27 was important, there were likely to be additional genetic factors to be identified (van der Linden et al. 1984).

Attempts to identify other genetic factors based on a candidate gene approach have proven largely unsuccessful. However, work has progressed with the advent of whole genome scanning techniques, and, to date, there have been three published genome scans in axial SpA including one in the Han Chinese population (Laval et al. 2001; Brown et al. 2003; Gu et al. 2004; Zhang et al. 2004). Seven loci have been identified and validated by this procedure; these include HLA-B27, ERAP-1 (ARTS-1), IL-23R, IL1R2 and ANTXR2 (CMG2); the other two loci do not encode gene sequences. The population attributable risks for HLA-B27, ARTS-1 and IL23R are 90 %, 26 % and 1 %, respectively (Burton et al. 2007). The association with IL-23R has been replicated in Spanish and Canadian cohorts. Other candidate genes are TNF-α and CYP2D6. ERAP-1 is known to be involved in the processing of HLA-B27 molecules as well as in the shedding of receptors for TNF-α, IL-1 and IL-6 from the cell surface. It is therefore conceivable that defective ERAP-1 function could lead to defective cytokine regulation. IL-23R is expressed on Th17 cells, which in turn secrete IL-17, a potent pro-inflammatory cytokine that leads to the release of IL-1, TNF-α. This cell type is being subject to increasing scrutiny in spondyloarthritis as well as in RA and may offer a new therapeutic target in the future. The role of ANTXR2 is still unknown.

8.3.1.4 Environmental Factors

Since the presence of the genetic factors identified to date is neither necessary nor sufficient to initiate disease, investigators have been searching for an environmental trigger. The observation that a syndrome of arthritis, anterior uveitis and urethritis can develop after specific infections (*Campylobacter, Shigella, Salmonella, Yersinia* and *Chlamydia* species) is evidence that infection can be an initiating event for these disorders. Further work with the HLA-B27 transgenic mouse has shown that it only develops axial SpA if it is exposed to environmental pathogens in the animal house.

Indeed, intestinal colonisation with *Bacteroides* species appears to be sufficient to initiate disease in the susceptible host (Rath et al. 1996).

Antibodies against *Klebsiella* have shown an association with human axial SpA, suggesting that there may be a role for this agent in initiating or maintaining disease activity (Rashid and Ebringer 2007). The association is attractive in that it seeks to explain the association between the gastrointestinal tract and axial SpA. Whilst a small open label trial of moxifloxacin has shown promise, more work is clearly required in this area.

A final factor that has been investigated is mechanical stress at the enthesis; it is hypothesised that this leads to downstream events that ultimately result in inflammation, erosion and bone formation (Benjamin et al. 2007).

8.3.1.5 Mechanisms of Bone Erosion and Formation in Axial SpA

The analysis of histological specimens from inflamed tissues shows significant expression of the enzymes cathepsin K and matrix metalloproteinase-1 (MMP1) by the invading mononuclear cells in those with axial SpA (Neidhart et al. 2009). In contrast, those with RA demonstrate an overexpression of RANKL and MMP3 suggesting that the pathways involved in bone metabolism are different.

Evidence from animal modelling suggests involvement of the RANKL-OPG axis in axial SpA (Rauner et al. 2009). The principal mechanism of bone formation at the enthesis appears to be through endochondral ossification. Members of the bone morphogenic protein (BMP) family of growth factors influence chondrocyte development and are important early in the process. Reduced expression of the negative regulator, sclerostin, has been shown in axial SpA suggesting that this pathway may be important in vivo (Appel et al. 2009). Later stages of bone development are influenced by the Wnt signalling pathway through intracellular accumulation of β-catenin which ultimately leads to osteoblast differentiation. Reduced levels of DKK-1, which inhibit this pathway, have been found in axial SpA, and it is postulated that this may be important in syndesmophyte formation (Daoussis et al. 2010).

8.3.1.6 Diagnostic and Classification Criteria

The diagnostic criteria for axial SpA were developed in Rome in 1960 and subsequently underwent modification in New York in 1966 and again in 1984 (Goie The et al. 1985). The current ASAS criteria for the classification of SpA are shown in Fig. 8.3 (Rudwaleit 2009). The ASAS criteria should be applied to patients with chronic back pain and age of onset under 45 years old.

For the individual patient, diagnosis is often delayed because of late presentation; some patients present with features other than axial disease which can mislead the unwary, and plain radiographic changes of sacroiliitis take several years to become evident (Kidd and Cawley 1988). It is for this reason that MRI is becoming increasingly important in the early diagnosis of inflammatory spinal disease where changes are evident at a much earlier stage.

8.3.1.7 Presentation

The classical feature of axial SpA is inflammatory spinal pain. Usually felt in the lower back or buttocks, the pain is typified by its rhythm which is worst overnight

Sacroiliitis on imaging and ≥ 2 SpA typical feature	or	HLA-B-27 and ≥ 2 SpA typical features

Inflammatory back pain Arthritis Enthesitis of the heel Uveitis Dactylitis Psoriasis Inflammatory bowel diseases Good response to NSAIDs Positive family history for SpA HLA-B-27 Elevated CRP	Sacroiliitis on imaging Active inflammation on MRI Radiographic sacroiliitis

Fig. 8.3 The ASAS criteria for classification of axial spondyloarthritis for patients with chronic back pain and age onset under 45 years. Regarding sacroiliitis on imaging: either (i) unilateral sacroiliitis grade 3–4 or bilateral sacroiliitis grade 2–4 by X-rays or, (ii) active inflammatory lesions of sacroiliac joints with definite bone marrow oedema reflecting sacroiliitis by MRI (Modified by Rudwaleit et al. 2009)

or early in the morning. Many patients report prolonged stiffness making it difficult for them to dress. The stiffness and pain tend to improve over the course of the day and almost always with exercise or anti-inflammatory medications. Some patients notice that their symptoms wax and wane over time, with flare-ups followed by periods of little or no symptoms.

If peripheral arthritis is present, it is typically found in the large joints in the lower limbs, in an asymmetrical manner. Enthesitis is a frequent feature, and most often affects the Achilles tendon, plantar fascia or common extensor origin at the lateral humeral epicondyle. Dactylitis may also be seen as the swelling of an entire digit so that it takes on a sausage shape.

Extra-articular features include anterior uveitis, which can affect up to 25–40 % of patients with SpA and is typically unilateral. In adults, this presents as a painful red eye and requires urgent ophthalmological review and management to reduce the chances of complications such at cataract, posterior synechiae and vision loss. The majority of cases in children are asymptomatic, and regular screening is warranted in this population.

Colonoscopic evaluation has found subclinical lesions in 20–70 % of cases, and it was this association that led to the routine use of sulphasalazine for the treatment of axial spondyloarthritis. The corollary of this is that approximately 30 % of patients with inflammatory bowel disease have axial disease; 10–20 % having sacroiliitis alone, 7–12 % having spondylitis and 10 % having features indistinguishable from classical axial SpA.

Other features seldom seen in practice are secondary amyloidosis due to deposition of serum amyloid A protein, apical lung fibrosis, aortitis and cardiac conduction defects due to fibrosis of the conduction system. Rarely, in longstanding axial disease, cauda equina syndrome can develop, and arachnoid diverticulae are described.

It should be evident from the preceding discussion of clinical features that a careful examination can reward the examiner with plentiful signs. Of particular importance is assessment of spinal motion, and this is carried out to best advantage with a physiotherapist so that intervention can be directed appropriately. The modified Schober's test is performed to assess the range of lumbar flexion; lateral lumbar bending is assessed by finger-floor distance and the extent of cervical lordosis by a tragus-wall distance. This latter measure being preferred over the occiput-wall distance as the tragus lies closer to the centre of rotation of the skull on the atlas and is relatively unaffected by the degree of neck flexion and extension. Chest expansion and hip range of motion should also be included in routine measurement.

Imaging has always been an important factor in making the diagnosis, reflecting the presence of sacroiliitis in the diagnostic and classification criteria. As mentioned above, the characteristic findings are within the axial skeleton where early changes include sacroiliitis which can be detected on MRI or CT in the early stages or visualised on plain radiograph once disease has been present for several years. Family studies have suggested that a mean of 9 years is required for plain radiographic changes to become apparent after changes are first detected on MRI. "Bright corners" or Romanus lesions may be seen on plain radiography or MRI in the antero-superior or antero-inferior corners of the vertebral bodies reflecting marginal erosion with reactive sclerotic change. Eventually, this leads to squaring of the vertebral body and finally to ossification of the superficial layers of the annulus fibrosus and longitudinal ligaments forming the typical, end-stage bamboo spine of axial SpA. Some may develop single or multilevel spondylodiscitis (Andersson lesions), which can be mistaken for septic discitis or osteomyelitis.

With the implementation of MRI in routine clinical practice, sacroiliitis can be observed even in the early stages of disease. This has led to the concept of nr-axSpA. Other than the higher proportion of female subjects in nr-axSpA studies, no major differences in patient demographics have been observed between r-axSpA and nr-axSpA. The rate of progression from nr-axSpA to r-axSpA was reported to be about 12 % after 2 years of follow-up (Calin et al. 1994). Given the similarity in the level of symptoms and the response to treatment, nr-axSpA seems to be an early stage of radiographic disease (Corli et al. 2015).

Factors predicting poor prognosis include hip involvement, ESR > 30 mm/h, lack of response to NSAIDs, reduced spinal motion, dactylitis, oligoarticular arthritis, age of onset less than 16 and the presence of inflammatory bowel disease, psoriasis or urethritis (Amor et al. 1994).

8.3.1.8 Assessment and Monitoring of Disease Activity

It is an unfortunate fact that about 40 % of patients with clinically and radiologically active axial disease do not reflect this with a rise in their serum inflammatory markers. Nevertheless, the ESR and CRP are commonly monitored as they have prognostic implications, and in some countries, a sufficient rise in these markers is required for funding of anti-TNF therapy.

Numerous tools have been developed to facilitate both the assessment of activity at one point in time and the change in disease activity over time. The Bath Ankylosing Spondylitis Disease Activity Index (BASDAI) is a validated six-item questionnaire that asks patients to mark the severity of their pain, stiffness and fatigue on visual analogue scales. A score of 4 or greater is consistent with active diseases, and a change of 50 % is considered significant for assessing the efficacy of interventions. Functional limitation can be quantified with numerous composite measures such as the BASFI (Bath Ankylosing Spondylitis Functional Index), HAQ or WOMAC scores. In practice, the BASFI is used most commonly and consists of a ten-item questionnaire on activities of daily living.

Primarily for use in clinical trials, there are response criteria to assess the efficacy of interventions in AS. The Ankylosing Spondylitis Assessment Score (ASAS 20) response criteria are fulfilled if there is a 20 % reduction on patient global assessment, function (as measured on the BASFI), pain and inflammation (Anderson et al. 2001).

The Ankylosing Spondylitis Disease Activity Score (ASDAS) is a composite index to assess disease activity in axial SpA. ASDAS includes back pain, duration of morning stiffness, patient global assessment, peripheral pain/swelling as well as CRP or ESR and is able to discriminate between high and low disease activity in early SpA (Fernández-Espartero et al. 2014).

8.3.1.9 Treatment of Axial SpA

Comparable to the treat-to-target (T2T) concept in RA, the primary treatment goal in axial SpA is defined as remission, with a secondary objective defined as low disease activity (Smolen et al. 2014). The ASAS/EULAR recommend maximising long-term health-related quality by controlling symptoms and inflammation. Moreover, an additional objective of therapy is the prevention of progressive structural damage. The optimal therapy in axial SpA therefore requires a combination of pharmacological and non-pharmacological treatments.

The wide array of potential symptoms and sites of disease involvement means that the treatment has to be tailored to the individual with the overall aims being to reduce pain, maintain mobility and function and possibly modify the underlying disease process.

Patient education is a principal component of any effective management strategy. Group education programmes are offered in some centres, and although comparison of these is not possible because of the heterogeneous nature of the different groups, many believe that they promote coping with the emotional consequences of the disease and help patients participate in the management. The role of education is acknowledged in the EULAR practice recommendations as a cornerstone to effective management (Zochling et al. 2006).

The importance of early and regular physical therapy in axial SpA is of the utmost importance; for many years, it was the sole management for the condition. There is evidence from a well-conducted but small multicentre trial that physical therapy in addition to medical management resulted in better spinal mobility, chest expansion and work capacity than medical management alone (Ince et al. 2006).

The best form of physical therapy is still debated; however, multimodal programmes encouraging stretching and aerobic exercise seem appropriate for most, and others with specific issues may merit tailored programmes.

For axial disease, NSAIDs and physical therapy should be considered first-line interventions. Approximately 70–80 % of patients with axial SpA will derive benefit from NSAIDs compared to only 15 % of those with non-inflammatory lower back pain. Analysis of the data from randomised trials of NSAID used in axial SpA has suggested that all patients received maximum benefit within 2 weeks, and this would therefore seem an appropriate length of treatment trial. Approximately 10–60 % will develop minor gastrointestinal side effects including epigastric pain and nausea, but serious side effects such a gastrointestinal bleeding occur in 2–4 % treated for 12 months. The risk of serious adverse events rises with age, comorbidity and concomitant use of other anti-inflammatory agents. Agents selective for the COX-2 isoenzyme have less gastrointestinal toxicity; however, this has to be balanced with the well-documented rise in cardiovascular events with these agents. It has even been suggested that regular use of NSAIDs – taken daily rather than as needed – can retard radiographic progression. The above measures will control symptoms adequately in approximately 70 % of patients; however, in those with ongoing symptoms, high inflammatory markers, high score on the BASDAI or evidence from imaging that disease is progressing, further treatment is required.

There is now widespread recognition that conventional DMARDs such as leflunomide or methotrexate are not effective for the control of axial disease. Similarly, there is a very limited role for systemic corticosteroids in axial SpA, but intra-articular steroids, given either into a peripheral joint or in to the sacroiliac joint, can provide substantial symptomatic benefit. As a consequence, TNF inhibitors should be considered in the event of high disease activity with or without prior conventional DMARD therapy in patients with axial disease (Braun et al. 2011).

Randomised controlled trials have been conducted showing that improvement can be seen as early as 2 weeks and is maximal after 12 weeks. This can be sustained over several years (van den Bosch et al. 2001; Braun et al. 2002; Gorman et al. 2002; Brandt et al. 2003; Davis et al. 2003; Calin et al. 2004; van der Heijde et al. 2005, 2006; Lambert et al. 2007). In addition, these agents are associated with improvement in quality of life, reduced sick leave, improved work productivity and reduction in inflammation on MRI. Even patients with complete spinal ankylosis report symptomatic benefit from these agents.

Early remission under anti-TNF therapy was reported to be the strongest predictor for achieving remission for up to 5 years in axial SpA. In the ATLAS trial, patients receiving adalimumab and achieving remission after 12 weeks were more than ten times more likely to be in remission after 5 years compared to patients who did not reach remission after 12 weeks (Sieper et al. 2012). In the ABILITY-1 trial, effective disease control has also been reported in nr-axSpA patients treated with adalimumab. Decreased inflammation and improved quality of life were also shown in this study. Patients with age <40, symptom duration <5 years or high baseline C-reactive protein demonstrated a better response at week 12 of treatment (Sieper et al. 2013).

Similar results after 1 year of etanercept treatment were shown in r-ax-SpA and nr-ax-SpA. The ankylosing spondylitis disease activity scores and BASDAI were decreased in r-ax-SpA and nr-ax-SpA, and the response rate to etanercept was similar in the two groups (Song et al. 2013).

In patients with nr-axSpA who had an insufficient response to NSAIDs, etanercept treatment showed significant benefits with regard to function, disease activity and inflammation as shown on MRI at week 12 when compared to placebo (Dougados et al. 2013). Recently, the efficacy and safety of certolizumab pegol, a PEGylated Fc-free anti-TNF, were shown in the RAPID-axSpA trial patients with r-ax-SpA and nr-axSpA. Response rates measured by ASAS-20 were significantly higher in certolizumab when compared with placebo. The treatment response was similar for certolizumab in r-ax-SpA and nr-axSpA, and both subgroups improved in BASDAI (Landewé et al. 2014).

Despite the quick response to anti-TNFs, the improvement in disease-related quality of life, augmented physical mobility and reduced pain, the response rate after discontinuation is high in r-ax-SpA and nr-ax-SpA. A relapse rate of approximately 70 % was reported in patients with r-ax-SpA and nr-ax-SpA (Haibel et al. 2013). These data suggest that discontinuation of biologics is associated with an increased risk of relapse in axial SpA, and discontinuation should be carefully considered.

A positive effect regarding inhibition of radiographic progression in patients with axial SpA treated with anti-TNFs has not been shown in most trials, suggesting that osteoproliferation in axial SpA is independent of anti-TNF treatment (van der Heijde et al. 2009). However, in patients treated with infliximab, bony spinal lesions progressed more slowly (Baraliakos et al. 2005).

Biologic agents appear to be of similar efficacy for treating axial disease with best responses seen in younger patients with high inflammatory markers and active disease on MRI. Anecdotally however, adalimumab and etanercept are given in fixed dosage regimens and for larger patients, and those with very active disease, response may be improved with infliximab as the dose can be adjusted to body weight. In addition, anterior uveitis can develop in those receiving etanercept, and so this agent may not be preferred in patients where this is a prominent feature.

For peripheral disease, the anti-TNF agents show excellent efficacy. Prior to their introduction, studies had focussed on the use of sulphasalazine as a consequence of the link with inflammatory bowel disease. To date, ten randomised, double-blind studies including two multicentre studies have been conducted assessing the efficacy of this agent in peripheral arthritis and finding it to be of modest efficacy (Dougados et al. 1995; Clegg et al. 1996). A subsequent meta-analysis of four trials concluded that the duration and severity of morning stiffness, pain severity and patient global assessment of disease activity reached statistical significance.

Approximately 30 % of patients with axial SpA will not respond to treatment with anti-TNF therapy or will experience side effects leading to its discontinuation. Trials are underway using anti-IL-6, anti-IL-23, anti-17A and targeted B-cell therapy. The IL-6-inhibition by tocilizumab did not demonstrate efficacy in patients with axial SpA (Sieper et al. 2014). However, in a pilot study on the IL-17A-inhibitor secukinumab, treatment up to 2 years was associated to clinical improvement accompanied by regression of spinal inflammation (Baraliakos et al. 2015). There is evidence from one small study that the intravenous bisphosphonate, pamidronate, given in a dose of 60 mg once monthly improved symptoms (Maksymowych et al. 2002). Although it is not helpful in the presence of peripheral joint synovitis, it also has the advantage of treating the well-documented association of AS with osteoporosis.

8.4 Psoriatic Arthritis (PsA)

8.4.1 Epidemiology

Psoriatic arthritis (PsA) is a chronic inflammatory joint disorder with a prevalence of 30 % in patients with skin psoriasis. PsA is characterised by local bone loss and new bone formation. Bone erosions are as frequent as in RA, although have a different morphology (Finzel et al. 2011). PsA is associated with nail disease (which has recently been shown to be a form of enthesitis), uveitis and inflammatory bowel disease (Aydin et al. 2012; Veale 2013). In 75 % of cases, skin involvement precedes the onset of joint disease, which can be delayed by up to 10 years (Duarte 2012). Although the pathophysiology is not fully understood to date, genetic and environmental factors have been discussed (see Sect. 8.3.1).

8.4.2 Diagnostic and Classification Criteria

Based on the ASAS criteria, peripheral arthritis, enthesitis or dactylitis and SpA-typical parameters such as uveitis, HLA-B27-positivity and skin psoriasis lead to the diagnosis of PsA with a sensitivity of 77.8 % and a specificity 82.9 % (Rudwaleit et al. 2011). The CASPAR classification criteria (*C*l*AS*sification criteria for the diagnosis of *P*soriatic *AR*thritis) include inflammatory joint, spine or entheseal disease plus at least three points: (i) symptomatic psoriasis (two points), family history for psoriasis (one point) and psoriasis in the past (one point); (ii) psoriatic nail disease (one point); (iii) rheumatoid factor negativity (one point); (iv) symptomatic dactylitis (one point) and dactylitis in the past (one point); and (v) periarticular bone formation (one point) (Taylor et al. 2006).

8.4.3 Treatment of PsA

EULAR recommends NSAIDs as first-line therapy for the management of PsA. To prevent joint damage, synthetic DMARDs in combination with local injections of glucocorticoids should be considered in patients with active disease, even at an early stage of the disease (Gossec et al. 2012). Systemic glucocorticoids are not recommended for the chronic use due to the potential risk of a withdrawal flare of psoriasis.

In contrast to RA, MTX and other conventional DMARDs are less effective in PsA, as has been shown previously. Low-dose oral MTX does not seem to improve synovitis in active PsA (Kingsley et al. 2012). Only parenteral high-dose MTX and salazopyrin had well-demonstrated effects in a Cochrane analysis of PsA patients (Jones et al. 2000).

TNF-α inhibitors should be considered in patients with active arthritis and an inadequate response to at least one conventional DMARD and in patients with active enthesitis or dactylitis and an insufficient response to local steroid injection and NSAIDs. For very active PsA characterised by structural joint damage, numerous swollen joints and extensive skin disease, biologics should also be considered in the treatment of naive patients. Moreover, biological drugs are especially beneficial in those with mainly axial disease and insufficient response to NSAIDs.

Positive effects for anti-TNFs on disease activity, enthesitis, dactylitis, skin psoriasis and psoriatic nail disease have been reported (Ritchlin et al. 2009; Mease et al. 2000; Kavanaugh et al. 2009). A reduced rate of radiographic progression was shown in PsA patients treated with anti-TNFs compared to conventional DMARDs (Glintborg et al. 2011; Eder et al. 2014). Currently six biological agents are available for the treatment of PsA (etanercept, adalimumab, infliximab, golimumab, certolizumab and ustekinumab). A similar efficacy regarding peripheral arthritis, ACR-50 response after 24 weeks of therapy, the reduction of radiographic progression and the improvement of skin disease was shown for the anti-TNFs (Fénix-Caballero et al. 2013; Ritchlin et al. 2009).

The IL-12/IL-23-antibody ustekinumab was recently approved for the treatment of PsA as well as plaque psoriasis. Ustekinumab treatment reduced joint disease activity and improved quality of life in patients with PsA (Gottlieb and Narang 2013). In the phase 3 trials PSUMMIT I and PSUMMIT II, a significant reduction in radiographic progression of joint damage in patients with active PsA and ustekinumab therapy compared to placebo was shown (Kavanaugh et al. 2014a; 2015).

Another emerging therapeutic area is the inhibition of IL-17A. Secukinumab, a fully human monoclonal antibody, is a promising biological agent in the treatment of PsA and skin psoriasis (McInnes et al. 2014). PsA patients treated with secukinumab showed a significantly higher response rate after 24 weeks and less structural damage compared to those receiving placebo in a phase 2 trial. The infection rate was higher in the secukinumab group than in the placebo group (Mease et al. 2015).

The oral phosphodiesterase-4-inhibitor apremilast showed significant benefits regarding disease activity, skin psoriasis and physical function in phase III studies, with an acceptable safety profile (Kavanaugh et al. 2015b). Apremilast has recently been approved in the USA and Europe.

8.4.4 Systemic Bone Loss in RA and SpA

It has been appreciated for some time that the quality of bone in RA is adversely influenced by the inflammatory response. This is evident as periarticular osteopenia and well as generalised osteoporosis. Therefore, RA is an independent risk factor for secondary osteoporosis. Both vertebral and non-vertebral fracture risk is increased in patients with

RA (Vis et al. 2011; Kim et al. 2010; Dirven et al. 2012; Wright et al. 2011). The increased risk of fracture is most marked in the spine (RR 2.4) and at the hip (RR 1.8). Furthermore, the fracture risk is increased substantially by the use of glucocorticoids (van Staa et al. 2006). Recently, microstructural deteriorations were reported in RA. A decreased trabecular bone volume caused by a decrease in trabecular width and a low trabecular number in female and male RA patients was reported. In addition, cortical thinning with an increase in the cortical perimeter, reflecting a compensatory mechanism to restore bone strength, was found (Kocijan et al. 2014a). Changes of the volumetric BMD, bone volume and inhomogeneity of the trabecular network as well as higher cortical porosity were found in Asian patients with RA (Zhu et al. 2013, 2014).

Besides inflammation, autoimmunity seems to play a major role not only in local but also in systemic bone loss in RA. Patients with seropositive RA (ACPA or rheumatoid factor positive) suffered from more severe systemic bone loss, characterised by deterioration of both trabecular and cortical bone (Kocijan et al. 2014b). The diagnosis of RA was suggested as an independent risk factor for osteoporosis and low-traumatic fracture risk and therefore included in the fracture risk assessment tool (FRAX) (McCloskey et al. 2012).

The relationship between SpA and spinal osteoporosis has been well documented. Spinal fracture may account for some of the exaggerated kyphosis seen in many patients with longstanding disease (Ghozlani et al. 2009). Fracture of the rigid spine can occur with minimal trauma and is usually longitudinal across a syndesmophyte rather than across a vertebral body. As in RA, bone quality is reduced in SpA. The incidence of spinal osteoporotic fractures is increased (Magrey and Khan 2010). Data regarding systemic bone loss in PsA are conflicting. Low, normal and high area bone mineral (BMD) were reported in PsA (Frediani et al. 2001; Attia et al. 2011; Millard et al. 2001; Nolla et al. 1999; Borman et al. 2008; Riesco et al. 2013). Using high-resolution peripheral quantitative computed tomography (HR-pQCT), significant alterations in trabecular, but not cortical bone structure, were observed (see Fig. 8.4, Kocijan et al. 2015). The duration of skin psoriasis was an independent risk factor for patients with PsA. Moreover, significant associations between mean psoriatic disease duration and BMD alterations (D'Epiro et al. 2014) as well as the presence of non-vertebral fractures (Del Puente et al. 2015) were reported.

Fig. 8.4 Bone microstructure in psoriatic arthritis (*PsA*) and healthy control (*CTRL*). Reconstruction of high-resolution peripheral quantitative computed tomography scans; 3D reconstruction, 60 slices, axial/sagittal view

References

Abu-Amer Y, Erdmann J, Alexopoulou L, Kollias G, Ross FP, Teitelbaum SL (2000) Tumor necrosis factor receptors types 1 and 2 differentially regulate osteoclastogenesis. J Biol Chem 275:27307–27310

Aceves-Avila FJ, Baez-Molgado S, Medina F, Fraga A (1998) Paleopathology in osseous remains from the 16th century. A survey of rheumatic diseases. J Rheumatol 25:776–782

Alamanos Y, Voulgari PV, Drosos AA (2006) Incidence and prevalence of rheumatoid arthritis, based on the 1987 American College of rheumatology criteria: a systematic review. Semin Arthritis Rheum 36:182–188

Aletaha D, Neogi T, Silman AJ, Funovits J, Felson DT, Bingham CO et al (2010) Rheumatoid arthritis classification criteria: an American College of Rheumatology/European League Against Rheumatism collaborative initiative. Arthritis Rheum (9):2569–2581

Amor B, Santos RS, Nahal R, Listrat V, Dougados M (1994) Predictive factors for the long-term outcome of spondyloarthropathies. J Rheumatol 21:1883–1887

Anderson JJ, Baron G, van der Heijde D, Felson DT, Dougados M (2001) Ankylosing spondylitis assessment group preliminary definition of short-term improvement in ankylosing spondylitis. Arthritis Rheum 44:1876–1886

Appel H, Ruiz-Heiland G, Listing J, Zwerina J, Herrmann M, Mueller R, Haibel H, Baraliakos X, Hempfing A, Rudwaleit M, Sieper J, Schett G (2009) Altered skeletal expression of sclerostin and its link to radiographic progression in ankylosing spondylitis. Arthritis Rheum 60:3257–3262

Appelboom T, de Boelpaepe C, Ehrlich GE, Famaey J-P (1981) Rubens and the question of antiquity of rheumatoid arthritis. JAMA 245:483–486

Atagunduz P, Appel H, Kuon W, Wu P, Thiel A, Kloetzel PM, Sieper J (2005) Hla-b27-restricted cd8+ t cell response to cartilage-derived self peptides in ankylosing spondylitis. Arthritis Rheum 52:892–901

Attia EA, Khafagy A, Abdel-Raheem S, Fathi S, Saad AA (2011) Assessment of osteoporosis in psoriasis with and without arthritis: correlation with disease severity. Int J Dermatol 50:30–35

Aydin SZ, Ash ZR, Tinazzi I, Castillo-Gallego C, Kwok C, Wilson C (2012) The link between enthesitis and arthritis in psoriatic arthritis: a switch to a vascular phenotype at insertions may play a role in arthritis development 72(6):992–995. doi: 10.1136/annrheumdis-2012-201617. Epub 2012 Aug 3

Ball J (1971) Enthesopathy of rheumatoid and ankylosing spondylitis. Ann Rheum Dis 30:213–223

Baraliakos X, Listing J, Rudwaleit M et al (2005) Radiographic progression in patients with ankylosing spondylitis after 2 years of treatment with the tumour necrosis factor alpha antibody infliximab. Ann Rheum Dis 64:1462–1466

Baraliakos X, Borah B, Braun J, Baeten D, Laurent D, Sieper J, Emery P, McInnes IB, van Laar JM, Wordsworth P, Wollenhaupt J, Kellner H, Colin L, Vandenhende F, Radford K, Hueber W (2015) Long-term effects of secukinumab on MRI findings in relation to clinical efficacy in subjects with active ankylosing spondylitis: an observational study. Ann Rheum Dis 75:408–412

Baron R, Rawadi G, Roman-Roman S (2006) Wnt signaling: a key regulator of bone mass. Curr Top Dev Biol 76:103–127

Benjamin M, Toumi H, Suzuki D, Redman S, Emery P, McGonagle D (2007) Microdamage and altered vascularity at the enthesis-bone interface provides an anatomic explanation for bone involvement in the hla-b27-associated spondylarthritides and allied disorders. Arthritis Rheum 56:224–233

Bergman GJ, Hochberg MC, Boers M, Wintfeld N, Kielhorn A, Jansen JP (2010) Indirect comparison of tocilizumab and other biologic agents in patients with rheumatoid arthritis and inadequate response to disease-modifying antirheumatic drugs. Semin Arthritis Rheum 39(6):425–441

Boers M, Verhoeven AC, Markusse HM, van de Laar MAFJ, Westhovens R, van Denderen JC, van Zeben D, Dijkmans BAC, Peeters AJ, Jacobs P, van den Brink HR, Schouten HJA, van der Heijde DMFM, Boonen A, van der Linden S (1997) Randomised comparison of combined step-down prednisolone, methotrexate and sulphasalazine with sulphasalazine alone in early rheumatoid arthritis. Lancet 350:309–318

Borman P, Babaoglu S, Gur G, Bingol S, Bodur H (2008) Bone mineral density and bone turnover in patients with psoriatic arthritis. Clin Rheumatol 27:443–447

Bottini N, Vang T, Cucca F, Mustelin T (2006) Role of ptpn22 in type 1 diabetes and other autoimmune diseases. Semin Immunol 18:207–213

Brandt J, Khariouzov A, Listing J, Haibel H, Sorensen H, Grassnickel L, Rudwaleit M, Sieper J, Braun J (2003) Six-month results of a double-blind, placebo-controlled trial of etanercept treatment in patients with active ankylosing spondylitis. Arthritis Rheum 48:1667–1675

Braun J, Brandt J, Listing J, Zink A, Alten R, Golder W, Gromnica-Ihle E, Kellner H, Krause A, Schneider M, Sorensen H, Zeidler H, Thriene W, Sieper J (2002) Treatment of active ankylosing spondylitis with infliximab: a randomised controlled multicentre trial. Lancet 359:1187–1193

Braun J, van den Berg R, Baraliakos X et al (2011) 2010 update of the ASAS/EULAR recommendations for the management of ankylosing spondylitis. Ann Rheum Dis 70:896–904

Brewerton DA, Hart FD, Nicholls A, Caffrey M, James DCO, Sturrock RD (1973) Ankylosing spondylitis and hl-a 27. Lancet 301:904–907

Brown MA, Brophy S, Bradbury L, Hamersma J, Timms A, Laval S, Cardon L, Calin A, Wordsworth BP (2003) Identification of major loci controlling clinical manifestations of ankylosing spondylitis. Arthritis Rheum 48:2234–2239

Burmester GR, Rigby WF, van Vollenhoven RF, Kay J, Rubbert-Roth A, Kelman A, Dimonaco S, Mitchell N (2015) Tocilizumab in early progressive rheumatoid arthritis: FUNCTION, a randomised controlled trial. Ann Rheum Dis. 2016;75(6):1081–1091

Burton PR, Clayton DG, Cardon LR, Craddock N, Celoukas P, Duncanson A (2007) Association scan of 14,500 nonsynonymous snps in four diseases identifies autoimmunity variants. Nat Genet 39:1329–1337

Calin A, Garrett S, Whitelock H et al (1994) A new approach to defining functional ability in ankylosing spondylitis: the development of the bath ankylosing spondylitis functional index. J Rheumatol 21:2281–2285

Calin A, Dijkmans BA, Emery P, Hakala M, Kalden J, Leirisalo-Repo M, Mola EM, Salvarani C, Sanmarti R, Sany J, Sibilia J, Sieper J, van der Linden S, Veys E, Appel AM, Fatenejad S (2004) Outcomes of a multicentre randomised clinical trial of etanercept to treat ankylosing spondylitis. Ann Rheum Dis 63:1594–1600

Chen XX, Baum W, Dwyer D, Stock M, Schwabe K, Ke HZ, Stolina M, Schett G, Bozec A (2013) Sclerostin inhibition reverses systemic, periarticular and local bone loss in arthritis. Ann Rheum Dis 72(10):1732–1736

Clegg DO, Reda DJ, Weisman MH et al (1996) Comparison of sulfasalazine and placebo in the treatment of ankylosing spondylitis. A department of veterans affairs cooperative study. Arthritis Rheum 39:2004–2012

Cloutier JF, Veillette A (1999) Cooperative inhibition of t-cell antigen receptor signaling by a complex between a kinase and a phosphatase. J Exp Med 189:111–121

Cohen SB, Dore RK, Lane NE, Ory PA, Peterfy CG, Sharp JT, van der Heijde D, Zhou L, Tsuji W, Newmark R (2008) Denosumab treatment effects on structural damage, bone mineral density, and bone turnover in rheumatoid arthritis: a twelve-month, multicenter, randomized, double-blind, placebo-controlled, phase ii clinical trial. Arthritis Rheum 58:1299–1309

Colmegna I, Diaz-Borjon A, Fujii H, Schaefer L, Goronzy JJ, Weyand CM (2008) Defective proliferative capacity and accelerated telomeric loss of hematopoietic progenitor cells in rheumatoid arthritis. Arthritis Rheum 58:990–1000

Corli J, Flipo RM, Philippe P, Bera-Louville A, Béhal H, Wibaux C, Paccou J (2015) Tumor necrosis factor-α inhibition in ankylosing spondylitis and nonradiographic axial spondyloarthritis: treatment response, drug survival, and patient outcome. J Rheumatol 42:2376–2382

Cutolo M, Lahita RG (2005) Estrogens and arthritis. Rheum Dis Clin North Am 31:19–27

D'Epiro S, Marocco C, Salvi M, Mattozzi C, Luci C, Macaluso L, Giancristoforo S, Campoli M, Scarnò M, Migliaccio S, Calvieri S, Richetta A (2014) Psoriasis and bone mineral density: implications for long-term patients. J Dermatol 41(9):783–787

Daoussis D, Liossis SN, Solomou EE, Tsanaktsi A, Bounia K, Karampetsou M, Yiannopoulos G, Andonopoulos AP (2010) Evidence that dkk-1 is dysfunctional in ankylosing spondylitis. Arthritis Rheum 62:150–158

David JP, Schett G (2010) Tnf and bone. Curr Dir Autoimmun 11:135–144

Davis JC Jr, Van Der Heijde D, Braun J, Dougados M, Cush J, Clegg DO, Kivitz A, Fleischmann R, Inman R, Tsuji W (2003) Recombinant human tumor necrosis factor receptor (etanercept) for treating ankylosing spondylitis: a randomized, controlled trial. Arthritis Rheum 48:3230–3236

de Vries-Bouwstra J, Le Cessie S, Allaart C, Breedveld F, Huizinga T (2006) Using predicted disease outcome to provide differentiated treatment of early rheumatoid arthritis. J Rheumatol 33:1747–1753

Del Puente A, Esposito A, Costa L, Benigno C, Del Puente A, Foglia F, Oriente A, Bottiglieri P, Caso F, Scarpa R (2015) Fragility fractures in patients with psoriatic arthritis. J Rheumatol Suppl 93:36–39

Duarte GV, Faillace C, Freire de Carvalho J (2012) Psoriatic arthritis. Best Pract Res Clin Rheumatol 26(1):147–156. doi: 10.1016/j.berh.2012.01.003

Diarra D, Stolina M, Polzer K, Zwerina J, Ominsky MS, Dwyer D, Korb A, Smolen J, Hoffmann M, Scheinecker C, van der Heide D, Landewe R, Lacey D, Richards WG, Schett G (2007) Dick- kopf-1 is a master regulator of joint remodeling. Nat Med 13:156–163

Dirven L, van den Broek M, van Groenendael JH et al (2012) Prevalence of vertebral fractures in a disease activity steered cohort of patients with early active rheumatoid arthritis. BMC Musculoskelet Disord 13:125

Dong W, Zhang Y, Yan M, Liu H, Chen Z, Zhu P (2008) Upregulation of 78-kda glucose-regulated protein in macrophages in peripheral joints of active ankylosing spondylitis. Scand J Rheumatol 37:427–434

Dougados M, vam der Linden S, Leirisalo-Repo M, Huitfeldt B, Juhlin R, Veys E, Zeidler H, Kvien TK, Olivieri I, Dijkmans B et al (1995) Sulfasalazine in the treatment of spondylarthropa- thy. A randomized, multicenter, double-blind, placebo-controlled study. Arthritis Rheum 38:618–627

Dougados M, van der Heijde D, Sieper J et al (2013) Clinical and imaging efficacy of etanercept in early non-radiographic axial spondyloarthritis: a 12-week, randomized, double-blind, placebo-controlled trial. Ann Rheum Dis 72(Suppl 3):A87–A88

Eder L, Thavaneswaran A, Chandran V et al (2014) Tumour necrosis factor α blockers are more effective than methotrexate in the inhibition of radiographic joint damage progression among patients with psoriatic arthritis. Ann Rheum Dis 73:1007–1011

Fénix-Caballero S, Alegre-del Rey EJ, Castaño-Lara R et al (2013) Direct and indirect comparison of the efficacy and safety of adalimumab, etanercept, infliximab and golimumab in psoriatic arthritis. J Clin Pharm Ther 38:286–293

Fernández-Espartero C, de Miguel E, Loza E et al (2014) Validity of the ankylosing spondylitis disease activity score (ASDAS) in patients with early spondyloarthritis from the Esperanza programme. Ann Rheum Dis 73:1350–1355

Finzel S, Englbrecht M, Engelke K, Stach C, Schett G (2011) A comparative study of periarticular bone lesions in rheumatoid arthritis and psoriatic arthritis. Ann Rheum Dis 70(1):122–127

François RJ, Neure L, Sieper J, Braun J (2006) Immunohistological examination of open sacroiliac biopsies of patients with ankylosing spondylitis: detection of tumour necrosis factor α in two patients with early disease and transforming growth factor β in three more advanced cases. Ann Rheum Dis 65:713–720

Franklin EC, Molman HR, Muller-Eberhard HJ, Kunkel HB (1957) An unusual protein component of high molecular weight in the serum of certain patients with rheumatoid arthritis. J Exp Med 105:425–435

Frediani B, Allegri A, Falsetti P et al (2001) Bone mineral density in patients with psoriatic arthritis. J Rheumatol 28:138–143

Furuzawa-Carballeda J, Vargas-Rojas MI, Cabral AR (2007) Autoimmune inflammation from the th17 perspective. Autoimmun Rev 6:169–175

Gabriel S, Michaud K (2009) Epidemiological studies in incidence, prevalence, mortality, and comorbidity of the rheumatic diseases. Arthritis Res Ther 11:229. doi:10.1186/ar2669

Gaujoux-Viala C, Smolen JS, Landewé R, Dougados M, Kvien TK, Mola EM, Scholte-Voshaar M, van Riel P, Gossec L (2010) Current evidence for the management of rheumatoid arthritis with synthetic disease-modifying antirheumatic drugs: a systematic literature review informing the eular recommendations for the management of rheumatoid arthritis. Ann Rheum Dis 69:1004–1009

Ghozlani I, Ghazi M, Nouijai A, Mounach A, Rezqi A, Achemlal L, Bezza A, Maghraoui AE (2009) Prevalence and risk factors of osteoporosis and vertebral fractures in patients with ankylosing spondylitis. Bone 44:772–776

Girbal-Neuhauser E, Durieux J-J, Arnaud M, Dalbon P, Sebbag M, Vincent C, Simon M, Senshu T, Masson-Bessiere C, Jolivet-Reynaud C, Jolivet M, Serre G (1999) The epitopes targeted by the rheumatoid arthritis-associated antifilaggrin autoantibodies are posttranslationally generated on various sites of (pro)filaggrin by deimination of arginine residues. J Immunol 162:585–594

Glant TT, Mikecz K, Arzoumanian A, Poole AR (1987) Proteoglycan-induced arthritis in balb/c mice. Clinical features and histopathology. Arthritis Rheum 30:201–212

Glintborg B, Østergaard M, Dreyer L et al (2011) Treatment response, drug survival, and predictors thereof in 764 patients with psoriatic arthritis treated with anti-tumor necrosis factor α therapy: results from the nationwide Danish DANBIO registry. Arthritis Rheum 63:382–390

Goie The HS, Steven MM, van der Linden SM, Cats A (1985) Evaluation of diagnostic criteria for ankylosing spondylitis: a comparison of the Rome, New York and modified New York criteria in patients with a positive clinical history screening test for ankylosing spondylitis. Br J Rheumatol 24:242–249

Gorman JD, Sack KE, Davis JC Jr (2002) Treatment of ankylosing spondylitis by inhibition of tumor necrosis factor alpha. N Engl J Med 346:1349–1356

Gossec L, Smolen JS, Gaujoux-Viala C et al (2012) European league against rheumatism recommendations for the management of psoriatic arthritis with pharmacological therapies. Ann Rheum Dis 71:4–12

Gottlieb A, Narang K (2013) Ustekinumab in the treatment of psoriatic arthritis: latest findings and clinical potential. Ther Adv Musculoskelet Dis 5:277–285

Gregersen PK, Silver J, Winchester RJ (1987) The shared epitope hypothesis. An approach to understanding the molecular genetics of susceptibility to rheumatoid arthritis. Arthritis Rheum 30:1205–1213

Gu MM, Yuan WT, Yang JQ, Zhang J, Xiong XY, Yao FJ, Lu ZY, Wang ZG, Huang W, Fan LA (2004) A genomewide scan for the susceptibility gene loci to ankylosing spondylitis in Chinese han population. Yi Chuan Xue Bao 31:217–220

Haibel H, Heldmann F, Braun J et al (2013) Long-term efficacy of adalimumab after drug withdrawal and retreatment in patients with active non-radiographically evident axial spondyloarthritis who experience a flare. Arthritis Rheum 65:2211–2213

Harre U, Georgess D, Bang H, Bozec A, Axmann R, Ossipova E, Jakobsson PJ, Baum W, Nimmerjahn F, Szarka E, Sarmay G, Krumbholz G, Neumann E, Toes R, Scherer HU, Catrina AI, Klareskog L, Jurdic P, Schett G (2012) Induction of osteoclastogenesis and bone loss by human autoantibodies against citrullinated vimentin. J Clin Invest 122(5):1791–1802

Harre U, Lang SC, Pfeifle R, Rombouts Y, Frühbeißer S, Amara K, Bang H, Lux A, Koeleman CA, Baum W, Dietel K, Gröhn F, Malmström V, Klareskog L, Krönke G, Kocijan R, Nimmerjahn F, Toes RE, Herrmann M, Scherer HU, Schett G (2015) Glycosylation of immunoglobulin G determines osteoclast differentiation and bone loss. Nat Commun 6:6651

Harrison MJ, Tricker KJ, Davies L, Hassell A, Dawes P, Scott DL, Knight S, Davis M, Mulherin D, Symmons DPM (2005) The relationship between social deprivation, disease outcome measures, and response to treatment in patients with stable, long-standing rheumatoid arthritis. J Rheumatol 32:2330–2336

Haschka J, Englbrecht M, Hueber AJ, Manger B, Kleyer A, Reiser M et al (2016) Relapse rates in patients with rheumatoid arthritis in stable remission tapering or stopping antirheumatic therapy: interim results from the prospective randomised controlled RETRO study. Ann Rheum Dis 75:45–51

Hetland ML, Ejbjerg B, Horslev-Petersen K et al (2009) Mri bone oedema is the strongest predictor of subsequent radiographic progression in early rheumatoid arthritis. Results from a 2-year randomised controlled trial (cimestra). Ann Rheum Dis 68:384–390

Ince G, Sarpel T, Durgun B, Erdogan S (2006) Effects of a multimodal exercise program for people with ankylosing spondylitis. Phys Ther 86:924–935

Jarvinen P (1995) Occurrence of ankylosing spondylitis in a nationwide series of twins. Arthritis Rheum 38:381–383

Jones G, Crotty M, Brooks P (2000) Interventions for psoriatic arthritis. Cochrane Database Syst Rev (3):CD000212

Jørgensen KT, Wiik A, Pedersen M, Hedegaard CJ, Vestergaard BF, Gislefoss RE, Kvien TK, Wohl- fahrt J, Bendtzen K, Frisch M (2008) Cytokines, autoantibodies and viral antibodies in premorbid and postdiagnostic sera from patients with rheumatoid arthritis: case–control study nested in a cohort of Norwegian blood donors. Ann Rheum Dis 67:860–866

Kaltenhauser S, Pierer M, Arnold S, Kamprad M, Baerwald C, Hantzschel H, Wagner U (2007) Antibodies against cyclic citrullinated peptide are associated with the drb1 shared epitope and predict joint erosion in rheumatoid arthritis. Rheumatology 46:100–104

Kavanaugh A, McInnes I, Mease P et al (2009) Golimumab, a new human tumor necrosis factor alpha antibody, administered every four weeks as a subcutaneous injection in psoriatic arthritis: twenty-four-week efficacy and safety results of a randomized, placebo-controlled study. Arthritis Rheum 60:976–986

Kavanaugh A, Ritchlin C, Rahman P et al (2014a) Ustekinumab, an anti-IL-12/23 p40 monoclonal antibody, inhibits radiographic progression in patients with active psoriatic arthritis: results of an integrated analysis of radiographic data from the phase 3, multicentre, randomised, double-blind, placebo-controlled PSUMMIT-1 and PSUMMIT-2 trials. Ann Rheum Dis 73:1000–1006

Kavanaugh A, Mease PJ, Gomez-Reino JJ et al (2014b) Treatment of psoriatic arthritis in a phase 3 randomised, placebo-controlled trial with apremilast, an oral phosphodiesterase 4 inhibitor. Ann Rheum Dis 73:1020–1026

Kavanaugh A, Puig L, Gottlieb AB, Ritchlin C, Li S, Wang Y, Mendelsohn AM, Song M, Zhu Y, Rahman P, McInnes IB, PSUMMIT I Study Group (2015) Maintenance of clinical efficacy and radiographic benefit through 2 years of ustekinumab therapy in patients with active psoriatic arthritis: results from the PSUMMIT 1 trial. Arthritis Care Res (Hoboken) 67:1739–1749

Khan MA (2000) Update: the twenty subtypes of hla-b27. Curr Opin Rheumatol 12:235–238

Kidd BL, Cawley MI (1988) Delay in diagnosis of spondarthritis. Br J Rheumatol 27:230–232

Kim N, Kadono Y, Takami M, Lee J, Lee SH, Okada F, Kim JH, Kobayashi T, Odgren PR, Nakano H, Yeh WC, Lee SK, Lorenzo JA, Choi Y (2005) Osteoclast differentiation independent of the trance-rank-traf6 axis. J Exp Med 202:589–595

Kim SY, Schneeweiss S, Liu J et al (2010) Risk of osteoporotic fracture in a large population based cohort of patients with rheumatoid arthritis. Arthritis Res Ther 12(4):R154

Kingsley GH, Kowalczyk A, Taylor H et al (2012) A randomized placebo-controlled trial of methotrexate in psoriatic arthritis. Rheumatology (Oxford) 51:1368–1377

Klareskog L, Padyukov L, Ronnelid J, Alfredsson L (2006) Genes, environment and immunity in the development of rheumatoid arthritis. Curr Opin Rheumatol 18:650–655

Klareskog L, Padyukov L, Alfredsson L (2007) Smoking as a trigger for inflammatory rheumatic diseases. Curr Opin Rheumatol 19:49–54

Kleyer A, Finzel S, Rech J, Manger B, Krieter M, Faustini F, Araujo E, Hueber AJ, Harre U, Engelke K, Schett G (2014) Bone loss before the clinical onset of rheumatoid arthritis in subjects with anticitrullinated protein antibodies. Ann Rheum Dis 73(5):854–860

Kobayashi K, Takahashi N, Jimi E, Udagawa N, Takami M, Kotake S, Nakagawa N, Kinosaki M, Yamaguchi K, Shima N, Yasuda H, Morinaga T, Higashio K, Martin TJ, Suda T (2000) Tumor necrosis factor alpha stimulates osteoclast differentiation by a mechanism independent of the odf/rankl-rank interaction. J Exp Med 191:275–286

Kocijan R, Finzel S, Englbrecht M, Engelke K, Rech J, Schett G (2014a) Decreased quantity and quality of the periarticular and non-periarticular bone in patients with rheumatoid arthritis: a cross-sectional HR-pQCT study. J Bone Miner Res 29(4):1005–1014

Kocijan R, Finzel S, Englbrecht M, Engelke K, Rech J, Schett G (2014b) Differences in bone structure between rheumatoid arthritis and psoriatic arthritis patients relative to autoantibody positivity. Ann Rheum Dis 73(11):2022–2028

Kocijan R, Englbrecht M, Haschka J, Simon D, Kleyer A, Finzel S, Kraus S, Resch H, Muschitz C, Engelke K, Sticherling M, Rech J, Schett G (2015) Quantitative and qualitative changes of bone in psoriasis and psoriatic arthritis patients. J Bone Miner Res 30:1775–1783

Kollnberger S, Bird LA, Roddis M, Hacquard-Bouder C, Kubagawa H, Bodmer HC, Breban M, McMichael AJ, Bowness P (2004) Hla-b27 heavy chain homodimers are expressed in hla-b27 transgenic rodent models of spondyloarthritis and are ligands for paired ig-like receptors. J Immunol 173:1699–1710

Lambert RG, Salonen D, Rahman P, Inman RD, Wong RL, Einstein SG, Thomson GT, Beaulieu A, Choquette D, Maksymowych WP (2007) Adalimumab significantly reduces both spinal and sacroiliac joint inflammation in patients with ankylosing spondylitis: a multicenter, randomized, double-blind, placebo-controlled study. Arthritis Rheum 56:4005–4014

Landewé R, Braun J, Deodhar A et al (2014) Efficacy of certolizumab pegol on signs and symptoms of axial spondyloarthritis including ankylosing spondylitis: 24-week results of a double-blind randomised placebo-controlled phase 3 study. Ann Rheum Dis 73:39–47

Lau CS, Burgos-Vargas R, Louthrenoo W, Mok MY, Wordsworth P, Zeng QY (1998) Features of spondyloarthritis around the world. Rheum Dis Clin North Am 24:753–770

Laval SH, Timms A, Edwards S, Bradbury L, Brophy S, Milicic A, Rubin L, Siminovitch KA, Weeks DE, Calin A, Wordsworth BP, Brown MA (2001) Whole-genome screening in ankylosing spondylitis: evidence of non-mhc genetic-susceptibility loci. Am J Hum Genet 68:918–926

Leadbetter EA, Rifkin IR, Hohlbaum AM, Beaudette BC, Shlomchik MJ, Marshak-Rothstein A (2002) Chromatin-igg complexes activate b cells by dual engagement of igm and toll-like receptors. Nature 416:603–607

Lu X, Gilbert L, He X, Rubin J, Nanes MS (2006) Transcriptional regulation of the osterix (osx, sp7) promoter by tumor necrosis factor identifies disparate effects of mitogen-activated protein kinase and nf kappa b pathways. J Biol Chem 281:6297–6306

Lundy S, Sarkar S, Tesmer L, Fox D (2007) Cells of the synovium in rheumatoid arthritis. T lymphocytes. Arthritis Res Ther 9:202

Magrey M, Khan MA (2010) Osteoporosis in ankylosing spondylitis. Curr Rheumatol Rep 12:332–336

Maksymowych WP, Jhangri GS, Fitzgerald AA, LeClercq S, Chiu P, Yan A, Skeith KJ, Aaron SL, Homik J, Davis P, Sholter D, Russell AS (2002) A six-month randomized, controlled, double-blind, dose–response comparison of intravenous pamidronate (60mg versus 10mg) in the treatment of nonsteroidal antiinflammatory drug-refractory ankylosing spondylitis. Arthritis Rheum 46:766–773

Manfredsdottir VF, Vikingsdottir T, Jonsson T, Geirsson AJ, Kjartansson O, Heimisdottir M, Sigurdardottir SL, Valdimarsson H, Vikingsson A (2006) The effects of tobacco smoking and rheumatoid factor seropositivity on disease activity and joint damage in early rheumatoid arthritis. Rheumatology 45:734–740

McCloskey E, Johansson H, Oden A, Kanis JA (2012) Fracture risk assessment. Clin Biochem 45(12):887–893

McInnes IB, Sieper J, Braun J et al (2014) Efficacy and safety of secukinumab, a fully human anti-interleukin-17A monoclonal antibody, in patients with moderate-to-severe psoriatic arthritis: a 24-week, randomised, double-blind, placebo-controlled, phase II proof-of concept trial. Ann Rheum Dis 73:349–356

Mease PJ, Goffe BS, Metz J et al (2000) Etanercept in the treatment of psoriatic arthritis and psoriasis: a randomised trial. Lancet 356:385–390

Mease PJ, McInnes IB, Kirkham B, Kavanaugh A, Rahman P, van der Heijde D, Landewé R, Nash P, Pricop L, Yuan J, Richards HB, Mpofu S, FUTURE 1 Study Group (2015) Secukinumab Inhibition of Interleukin-17A in patients with psoriatic arthritis. N Engl J Med 373(14):1329–1339

Mikecz K, Glant TT, Baron M, Poole AR (1988) Isolation of proteoglycan-specific t lymphocytes from patients with ankylosing spondylitis. Cell Immunol 112:55–63

Millard TP, Antoniades L, Evans AV, Smith HR, Spector TD, Barker JN (2001) Bone mineral density of patients with chronic plaque psoriasis. Clin Exp Dermatol 26:446–448

Möttönen T, Hannonen P, Leirisalo-Repo M, Nissilä M, Kautiainen H, Korpela M, Laasonen L, Julkunen H, Luukkainen R, Vuori K, Paimela L, Blåfield H, Hakala M, Ilva K, Yli-Kerttula U, Puolakka K, Järvinen P, Hakola M, Piirainen H, Ahonen J, Pälvimäki I, Forsberg S, Koota K, Friman C (1999) Comparison of combination therapy with single-drug therapy in early rheumatoid arthritis: a randomised trial. Lancet 353:1568–1573

Nakamura I, Kadono Y, Takayanagi H, Jimi E, Miyazaki T, Oda H, Nakamura K, Tanaka S, Rodan GA, le Duong T (2002) Il-1 regulates cytoskeletal organization in osteoclasts via tnf receptor-associated factor 6/c-src complex. J Immunol 168:5103–5109

Nam JL, Winthrop KL, van Vollenhoven RF, Pavelka K, Valesini G, Hensor EMA, Worthy G, Landewé R, Smolen JS, Emery P, Buch MH (2010) Current evidence for the management of rheumatoid arthritis with biological disease-modifying antirheumatic drugs: a systematic literature review informing the eular recommendations for the management of RA. Ann Rheum Dis 2010;69(6):976–986. doi:10.1136/ard.2009.126573

Neidhart M, Baraliakos X, Seemayer C, Zelder C, Gay RE, Michel BA, Boehm H, Gay S, Braun J (2009) Expression of cathepsin k and matrix metalloproteinase 1 indicate persistent osteode- structive activity in long-standing ankylosing spondylitis. Ann Rheum Dis 68:1334–1339

Nepom GT (1989) Hla genes associated with rheumatoid arthritis: identification of susceptibility alleles using specific oligonucleotide probes. Arthritis Rheum 32:15–21

Nolla JM, Fiter J, Rozadilla A et al (1999) Bone mineral density in patients with peripheral psoriatic arthritis. Rev Rhum Engl Ed 66:457–461

Rashid T, Ebringer A (2007) Ankylosing spondylitis is linked to klebsiella – the evidence. Clin Rheumatol 26:858–864

Rath HC, Herfarth HH, Ikeda JS, Grenther WB, Hamm TE, Balish E, Taurog JD, Hammer RE, Wilson KH, Sartor RB (1996) Normal luminal bacteria, especially bacteroides species, mediate chronic colitis, gastritis, and arthritis in hla-b27/human beta2 microglobulin transgenic rats. J Clin Invest 98:945–953

Rauner M, Stupphann D, Haas M, Fert I, Glatigny S, Sipos W, Breben M, Pietschmann P (2009) The hla-b27 transgenic rat, a model of spondyloarthritis, has decreased bone mineral density and increased rankl to osteoprotegerin mrna ratio. J Rheumatol 36:120–126

Raza K, Falciani F, Curnow SJ, Ross E, Lee C-Y, Akbar A, Lord J, Gordon C, Buckley C, Salmon M (2005) Early rheumatoid arthritis is characterized by a distinct and transient synovial fluid cytokine profile of t cell and stromal cell origin. Arthritis Res Ther 7:R784–R795

Riesco M, Manzano F, Font P, Garcia A, Nolla JM (2013) Osteoporosis in psoriatic arthritis: an assessment of densitometry and fragility fractures. Clin Rheumatol 32:1799–1804

Ritchlin CT, Kavanaugh A, Gladman DD et al (2009) Treatment recommendations for psoriatic arthritis. Ann Rheum Dis 68:1387–1394

Rossini M, Viapiana O, Adami S, Fracassi E, Idolazzi L, Dartizio C, Povino MR, Orsolini G, Gatti D (2015) In patients with rheumatoid arthritis, Dickkopf-1 serum levels are correlated with parathyroid hormone, bone erosions and bone mineral density. Clin Exp Rheumatol 33(1):77–83

Rothschild BM, Turner KR, DeLuca MA (1988) Symmetrical erosive peripheral polyarthritis in the late archaic period of alabama. Science 241:1498–1501

Rudwaleit M, Sieper J (2012) Referral strategies for early diagnosis of axial spondyloarthritis. Nat Rev Rheumatol 8:262–268

Rudwaleit M, van der Heijde D, Landewé R, Listing J, Akkoc N, Brandt J et al (2009) The development of assessment of spondyloarthritis international society classification criteria for axial spondyloarthritis (part II): validation and final selection. Ann Rheum Dis 68:777–783. doi: 10.1136/ard.2009.108233. Epub 2009 Mar 17

Rudwaleit M, van der Heijde D, Landewé R et al (2011) TheAssessment of SpondyloArthritis International Society classification criteria for peripheral spondyloarthritis and for spondyloarthritis in general. Ann Rheum Dis 70:25–31

Sanmartí R, Gómez-Centeno A, Ercilla G, Larrosa M, Viñas O, Vazquez I, Gómez-Puerta J, Gratacós J, Salvador G, Cañete J (2007) Prognostic factors of radiographic progression in early rheumatoid arthritis: a two year prospective study after a structured therapeutic strategy using dmards and very low doses of glucocorticoids. Clin Rheumatol 26:1111–1118

Schlosstein L, Terasaki PI, Bluestone R, Pearson CM (1973) High association of an hl-a antigen, w27, with ankylosing spondylitis. N Engl J Med 288:704–706

Sieper J, van der Heijde D, Dougados M et al (2012) Early response to adalimumab predicts long-term remission through 5 years of treatment in patients with ankylosing spondylitis. Ann Rheum Dis 71:700–706

Sieper J, van der Heijde D, Dougados M et al (2013) Efficacy and safety of adalimumab in patients with non-radiographic axial spondyloarthritis: results of a randomised placebo-controlled trial (ABILITY-1). Ann Rheum Dis 72:815–822

Sieper J, Porter-Brown B, Thompson L et al (2014) Assessment of short-term symptomatic efficacy of tocilizumab in ankylosing spondylitis: results of randomised, placebo-controlled trials. Ann Rheum Dis 73:95–100

Singh JA, Saag KG, Bridges SL Jr, Akl EA, Bannuru RR, Sullivan MC, Vaysbrot E et al (2015) 2015 American College of rheumatology guideline for the treatment of rheumatoid arthritis. Arthritis Rheumatol 68:1–26

Smolen JS, Aletaha D, Bijlsma JW, Breedveld FC, Boumpas D, Burmester G, Combe B, Cutolo M, de Wit M, Dougados M, Emery P, Gibofsky A, Gomez Reino JJ (2010) Treating rheumatoid arthritis to target: recommendations of an international task force. Ann Rheum Dis 4:631–637

Smolen J, Braun J, Dougados M et al (2014) Treating spondyloarthritis, including ankylosing spondylitis and psoriatic arthritis, to target: recommendations of an international task force. Ann Rheum Dis 73:6–16

Smolen JS, van Vollenhoven R, Kavanaugh A et al (2015) Certolizumab pegol plus methotrexate 5-year results from the rheumatoid arthritis prevention of structural damage (RAPID) 2 randomized controlled trial and long-term extension in rheumatoid arthritis patients. Arthritis Res Ther 17:245

Smolen JS, Breedveld FC, Burmester GR, Bykerk V, Dougados M, Emery P, Kvien TK et al (2016) Treating rheumatoid arthritis to target: 2014 update of the recommendations of an international task force. Ann Rheum Dis 75(1):3–15

Song IH, Weiß A, Hermann KG et al (2013) Similar response rates in patients with ankylosing spondylitis and non-radiographic axial spondyloarthritis after 1 year of treatment with etanercept: results from the ESTHER trial. Ann Rheum Dis 72:823–825

Stamp LK, O'Donnell JL, Chapman PT, Zhang M, Frampton C, James J, Barclay ML (2009) Determinants of red blood cell methotrexate polyglutamate concentrations in rheumatoid arthritis patients receiving long-term methotrexate treatment. Arthritis Rheum 60:2248–2256

Taurog JD, Hammer RE (1995) Experimental spondyloarthropathy in hla-b27 transgenic rats. Clin Rheumatol 15(Suppl 1):22–27

Taylor W, Gladman D, Helliwell P, Marchesoni A, Mease P, Mielants H, CASPAR Study Group (2006) Classification criteria for psoriatic arthritis: development of new criteria from a large international study. Arthritis Rheum 54(8):2665–2673

Turesson C, Matteson EL (2009) Vasculitis in rheumatoid arthritis. Curr Opin Rheumatol 21:35–40

Tvete IF, Natvig B, Gåsemyr J, Meland N, Røine M, Klemp M (2015) Comparing effects of biologic agents in treating patients with rheumatoid arthritis: a multiple treatment comparison regression analysis. PLoS One 10(9):e0137258. doi:10.1371/journal.pone.0137258. eCollection
van den Bosch F, Baeten D, Kruithof E, De Keyser F, Mielants H, Veys EM (2001) Treatment of active spondyloarthropathy with infliximab, the chimeric monoclonal antibody to tumour necrosis factor alpha. Ann Rheum Dis 60(Suppl 3):iii33–iii36
van der Heijde D, Dijkmans B, Geusens P, Sieper J, DeWoody K, Williamson P, Braun J (2005) Efficacy and safety of infliximab in patients with ankylosing spondylitis: results of a randomized, placebo-controlled trial (assert). Arthritis Rheum 52:582–591
van der Heijde D, Kivitz A, Schiff MH, Sieper J, Dijkmans BA, Braun J, Dougados M, Reveille JD, Wong RL, Kupper H, Davis JC Jr (2006) Efficacy and safety of adalimumab in patients with ankylosing spondylitis: results of a multicenter, randomized, double-blind, placebo- controlled trial. Arthritis Rheum 54:2136–2146
van der Heijde D, Salonen D, Weissman BN, Landewe R, Maksymowych WP, Kupper H, Ballal S, Gibson E, Wong R (2009) Assessment of radiographic progression in the spines of patients with ankylosing spondylitis treated with adalimumab for up to 2 years. Arthritis Res Ther 11:R127. doi:10.1186/ar2794
van der Helm-van Mil AH, Huizinga TW, Schreuder GM, Breedveld FC, de Vries RR, Toes RE (2005) An independent role of protective hla class ii alleles in rheumatoid arthritis severity and susceptibility. Arthritis Rheum 52:2637–2644
van der Helm-van Mil AHM, Huizinga TWJ, de Vries RRP, Toes REM (2007) Emerging patterns of risk factor make-up enable subclassification of rheumatoid arthritis. Arthritis Rheum 56:1728–1735
van der Linden SM, Valkenburg HA, de Jongh BM, Cats A (1984) The risk of developing ankylosing spondylitis in hla-b27 positive individuals. A comparison of relatives of spondylitis patients with the general population. Arthritis Rheum 27:241–249
van Staa TP, Geusens P, Bijlsma JW, Leufkens HG, Cooper C (2006) Clinical assessment of the long- term risk of fracture in patients with rheumatoid arthritis. Arthritis Rheum 54:3104–3112
van Vollenhoven RF, Fleischmann R, Cohen S, Lee EB, García Meijide JA, Wagner S, Forejtova S, Zwillich SH, Gruben D, Koncz T, Wallenstein GV, Krishnaswami S, Bradley JD, Wilkinson B, ORAL Standard Investigators (2012) Tofacitinib or adalimumab versus placebo in rheumatoid arthritis. N Engl J Med 367:508–519
Veale DJ (2013) Psoriatic arthritis: recent progress in pathophysiology and drug development. Arthritis Res Ther 15:224
Vis M, Haavardsholm EA, Boyesen P et al (2011) High incidence of vertebral and non vertebral fractures in the OSTRA cohort study: a 5 year follow up study in postmenopausal women with rheumatoid arthritis. Osteoporos Int 22(9):2413–2419
Wagner UG, Koetz K, Weyand CM, Goronzy JJ (1998) Perturbation of the t cell repertoire in rheumatoid arthritis. Proc Natl Acad Sci U S A 95:14447–14452
Wesoly J, van der Helm-van Mil AHM, Toes REM, Chokkalingam AP, Carlton VEH, Begovich AB, Huizinga TWJ (2005) Association of the ptpn22 c1858t single- nucleotide polymorphism with rheumatoid arthritis phenotypes in an inception cohort. Arthritis Rheum 52:2948–2950
Weyand CM, Hicok KC, Conn DL, Goronzy JJ (1992) The influence of hla-drb1 genes on disease severity in rheumatoid arthritis. Ann Intern Med 117:801–803
Wright NC, Lisse JR, Walitt BT, Eaton CB, Chen Z (2011) Arthritis increases the risk for fractures—results from the Women's Health Initiative. J Rheumatol 38:1680–1688
Zhang G, Luo J, Bruckel J, Weisman MA, Schumacher HR, Khan MA, Inman RD, Mahowald M, Maksymowych WP, Martin TM, Yu DTY, Stone M, Rosenbaum JT, Newman P, Lee J, McClain JA, West OC, Jin L, Reveille JD (2004) Genetic studies in familial ankylosing spondylitis susceptibility. Arthritis Rheum 50:2246–2254

Zhu TY, Griffith JF, Qin L, Hung VW, Fong TN, Au SK, Tang XL, Kwok AW, Leung PC, Li EK, Tam LS (2013) Structure and strength of the distal radius in female patients with rheumatoid arthritis: a case–control study. J Bone Miner Res 28(4):794–806

Zhu TY, Griffith JF, Qin L, Hung VW, Fong TN, Au SK, Li M, Lam YY, Wong CK, Kwok AW, Leung PC, Li EK, Tam LS (2014) Alterations of bone density, microstructure, and strength of the distal radius in male patients with rheumatoid arthritis: a case–control study with HR-pQCT. J Bone Miner Res 29(9):2118–2129

Zochling J, van der Heijde D, Dougados M, Braun J (2006) Current evidence for the management of ankylosing spondylitis: a systematic literature review for the asas/eular management recommendations in ankylosing spondylitis. Ann Rheum Dis 65:423–432

Zvaifler NJ (1973) The immunopathology of joint inflammation in rheumatoid arthritis. Adv Immunol 16:265–336

9 Antibodies for the Treatment of Bone Diseases: Preclinical Data

Wolfgang Sipos

9.1 Animal Models of Osteoporosis

9.1.1 General Aspects

Osteoporosis is a complex systemic disease (Pietschmann et al. 2008). Therefore, researchers have to rely on rodent and large animal models for the development of new antiosteoporotic therapeutics. Rodent models are well established and have been widely used in osteoporosis research. However, the US Food and Drug Administration (FDA), besides rats, also demands the use of large animal models in preclinical testing of anti-osteoporotic substances with an experimental time frame of 12 months when using rats and 16 months when using larger species. According to FDA regulations, valid animal models have to develop an osteoporotic phenotype either spontaneously or after ovariectomy (OVX) (FDA Guidelines 1994). In order to give an idea of the different biomechanical forces on the skeletons of small vs large animal species, Fig. 9.1 shows the skeletons of a rat, a sheep, and a minipig. There exist several ways of inducing osteopenia or osteoporotic phenotypes in mammals, such as *OPG* gene knockout, systemic RANKL administration, prolonged glucocorticoid administration, age-related osteoporosis, dietary calcium shortage, OVX, and combinations. All these manipulations have specific advantages and disadvantages, and the physiological relevance has to be proven for each model. Although calcium shortage itself may induce osteopenia to some extent, it is usually combined with OVX, which may be considered the main model of postmenopausal and thus estrogen deficiency-induced osteoporosis.

In this context it is essential to be aware of species-specific sexual endocrinology. Whereas women (and macaques) have a sexual cycle of 28 days, sows experience spontaneous ovulations as well but with a cycle length of 21 days. With regard

W. Sipos, DVM
Clinical Department for Farm Animals, University of Veterinary Medicine Vienna, Vienna 1210, Austria
e-mail: Wolfgang.Sipos@vetmeduni.ac.at

P. Pietschmann (ed.), *Principles of Osteoimmunology*,
DOI 10.1007/978-3-319-34238-2_9

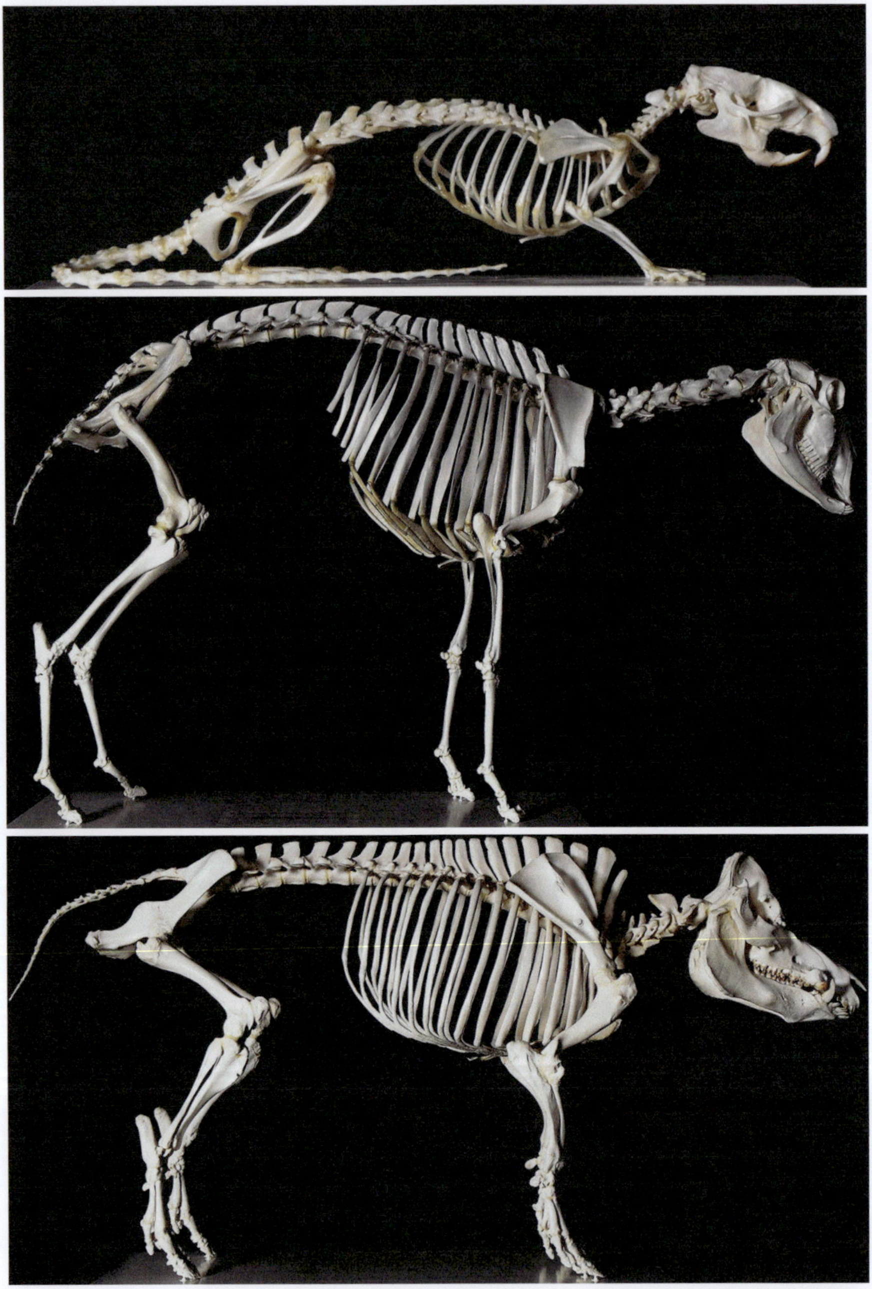

Fig. 9.1 Skeletons of a rat, a sheep, and a minipig. Not only due to large differences concerning body weight but also due to the different position of extremities in relation to the body and different functions of the vertebral column with regard to ensuring the stability of thorax and especially the abdomen between the shown species, interspecies differences concerning the mechanical forces exerting an impact on the axial and the appendicular skeletons are the logical consequence (Photos with kind permission of G. Weissengruber, Vienna)

to sheep, there are seasonal and aseasonal breeds with a mean cycle length of 17 days. The seasonal rhythm is regulated by the epiphyseal melatonin secretion, which is significantly higher in winter ("short day breeders"). The mainly used merinos are an aseasonal breed. Another feature, which has to be kept in mind, is extragonadal estrogen synthesis, which may cause a rebound effect in ovariectomized sheep in terms of lowering bone mineral density (BMD) change differences when compared to sham animals starting from 4 months after OVX (Sigrist et al. 2007). Moreover, ovariectomized sows experience extragonadal estrogen synthesis to some extent (Sipos et al. 2011a). Female dogs are seasonal monestric, usually in March and September. Heat includes proestrus, estrus, and early metestrus and takes 3 weeks. Estradiol increases only during this very short period to titers between 18 and 50 pg/ml during proestrus and decreases already after the start of estrus to baseline levels of 5 pg/ml (Sipos 1997). Rats have sexual cycles of 4–5 days, whereas mice experience inducible ovulations. Furthermore, the occurrence of a natural menopause has not been proven in all model animals, except for macaques.

Nevertheless, estrogen deficiency induced by OVX leads to an osteoporotic phenotype in rodents as well as diverse ungulates and can be aggravated by additional glucocorticoid administration or calcium shortage. However, it has to be kept in mind that osteoporosis as defined for humans (i.e., osteopenia plus fragility fractures due to disturbed cortical and trabecular microarchitecture) does not naturally occur in our domestic mammals. Therefore, all models have their weaknesses. In this chapter, we will focus on rodent and artiodactyl models, but largely spare canine as well as primate models, as these are used less frequently due to ethical reasons, although especially primate models most closely reflect the situation in humans.

9.1.2 Rodent Models

In contrast to large animal models, there is a broad range of genetically modified rodent models. *OPG*$^{-/-}$ mice suffer from severe osteoporosis associated with a high incidence of fractures and vertebral deformities but lack unwanted side effects on immune function, which can be observed in *RANKL*$^{-/-}$ mice (Bucay et al. 1998; Kong et al. 1999; Yun et al. 2001). Systemic RANKL administration also leads to increased osteoclast numbers and to an osteoporotic phenotype in mice (Lacey et al. 1998). Prolonged glucocorticoid exposure by subcutaneously implanted slow-release prednisolone pellets (2.1 mg/kg/day) is another reasonable means of inducing osteoporosis in rodents (Weinstein et al. 1998). Wang et al. (2001) demonstrated an age-related decline of bone mass in the axial and appendicular skeleton and a decrease in bone formation in the spine in aged male Sprague–Dawley rats. Pietschmann et al. (2007) described similar changes, i.e., markedly reduced cancellous bone mineral density (BMD), bone volume (BV), and trabecular number (Tb.N), in the proximal tibiae of aged male rats. Additionally, bone formation as well as osteoclastogenesis appeared decreased, which was substantiated by decreased insulin-like growth factor 1 (IGF-1), osteocalcin, and RANKL serum levels. Bone density and bone strength are reduced in young growing rats by dietary calcium restriction in the range of 0.3–2.5 g Ca^{2+}/kg diet (Thomas et al. 1988; Persson et al. 1993; Talbott et al. 1998).

In the rat, rapid loss of cancellous bone mass and strength occurs following OVX, which after a while reaches a steady-state phase of bone mass with an increase in the rate of bone turnover (Wronski et al. 1986, 1988, 1989, 1990; Wronski and Yen 1992; Kalu 1991). The proximal tibial metaphysis, the lumbar vertebral bodies, and the femoral neck are affected by cancellous bone loss within 1 month after OVX, whereas OVX-induced bone loss does not occur in the trabecular bone of long bone epiphyses, the distal tibial metaphysis, and the caudal vertebrae (Ma et al. 1994; Li et al. 1996; Westerlind et al. 1997; Miyakoshi et al. 1999). The situation of the cortical bone in the ovariectomized rat is complex, as increased bone resorption taking place at the endosteum of the diaphysis of long bones is antagonized by stimulated periosteal bone growth (Turner et al. 1987; Miller et al. 1991; Aerssens et al. 1996). Thus, enlargement of the marrow cavity is the most sensitive index of cortical bone loss in this species. As the rat, like most other experimental animal models, does not experience naturally occurring pathological fractures, bone strength has to be tested mechanically. Significant loss of vertebral and femoral neck bone strength occurs 3 months after OVX (Mosekilde et al. 1991; Mosekilde et al. 1993a; Peng et al. 1994; Sogaard et al. 1994; Jiang et al. 1997; Yoshitake et al. 1999). Reduction of dietary calcium from 0.4 to 0.2 % in ovariectomized rats results in a further decrease in bone density and mechanical properties (Shahnazari et al. 2009).

Alternative rodent osteoporosis models include the Botox-induced and the LPS-induced bone loss models (Grimston et al. 2007; Ochi et al. 2010). In the former, mice are injected intramuscularly with 2.0 U Botox per 100 g body weight in the quadriceps, hamstrings, and posterior calf muscles of one hind limb. Maximum limb dysfunction occurs by days 2–3 after Botox injection without gender differences. By 3–4 weeks postinjection, full activity is restored. Bone loss in the injected limb is rapid and profound with the difference to the noninjected limb being significant by week 2. At 12 weeks trabecular BV in the injected limb is substantially reduced, and cortical thickness is lowered. In the endotoxin model, male adult mice are subcutaneously injected with 20 mg/kg LPS. LPS treatment reduces BV/TV for approximately 40 % in the proximal region of tibial bones by 48 h posttreatment. This LPS-induced enhancement of osteoclastogenesis can be blocked by OPG administration giving evidence of the importance of RANKL signaling in this setting, but LPS or OPG does not affect osteoclastogenesis in $TNFR1^{-/-}$ mice. Interestingly, osteoblast surface is remarkably reduced in these mice as a result of enhanced osteoblast apoptosis due to TRAIL-mediated signaling, which triggers apoptosis of primary osteoblasts only when the TNFR1 signal is ablated in vitro.

9.1.3 Ovine Models

There is also a bulk of literature dealing with the ovine species and its suitability as a postmenopausal osteoporosis model. The compact bone of growing sheep is predominantly plexiform. A well-developed Haversian system consisting of secondary osteons is not developed until an age of 7–9 years (Turner 2002; Pearce et al. 2007). Seasonal fluctuations in bone metabolism complicate the interpretation of data. It should also be mentioned that biochemical bone metabolism markers are regarded as having only limited informative value in the ovine model (Gerlach 2002). Due to anatomical peculiarities of ovine vertebrae, biomechanical parameters fail to correlate with BMD of these bone sites, but do so with femoral BMD. Anatomical features of the iliac crest seem to be close to the corresponding human bone site (Ito et al. 1998). Although sheep do not experience a natural menopause, decreasing bone mass in aging sheep might prove this species a potential model for age-related osteoporosis (Turner et al. 1993). OVX alone is not as efficient as in other species, whereas glucocorticoid treatment induces an osteoporotic phenotype comparable to the one in other species including humans (Chavassieux et al. 1993; Hornby et al. 1995). BMD of L5 and distal radius as evaluated by DXA is significantly changed 6 months and the one of L4 1 year after OVX, whereas the proximal parts of the femur, humerus, and tibia do not exhibit changes to that extent (Turner et al. 1995). However, MacLeay et al. (2004) were not able to detect areal BMD changes in lumbar vertebrae in ovariectomized sheep 3 months after surgery. Another study showed a significantly decreased femoral but not lumbar vertebral BMD as well as significant effects on cortical bone parameters by 6 months after OVX (Chavassieux et al. 2001). Interesting and seemingly inconsistent with the likelihood of extragonadal estrogen synthesis is the fact that sheep experience significant microarchitectural changes in the vertebral cancellous bone (decreased BV/TV by approximately 30 %, trabecular thickness (Tb.Th) by 13 %, and increased trabecular separation (Tb.Sp) by 46 %) 2 years after OVX and show histomorphometric changes (i.e., significantly increased osteoclast numbers) already 3 months after surgery (Giavaresi et al. 2001; Pogoda et al. 2006).

Combining OVX with profound dietary calcium restriction (1.5 g Ca^{2+} and 100 IU $VitD_3$/day instead of the physiological daily need for 4–5 g Ca^{2+} and 1.000 IU $VitD_3$/day) and weekly doses of 120–200 mg methylprednisolone in 7–9 years old ewes for 7 months led to an approximately 35–40 % reduction of BMD of the spongiosa of the radius and tibia without affecting the corticalis (Lill et al. 2002). Lumbar vertebral bodies exhibited a decrease in BMD of 13 %. Also, all analyzed histomorphological parameters of the iliac crest as well as BV/TV of L4, and Tb.Th, BV/TV, and bone surface (BS/BV) of femoral head were significantly altered and corresponded well with biomechanical data. Although the profound decrease of BMD of approximately 4 SD is remarkable in this model, the increased susceptibility for opportunistic infections and the drawbacks concerning osteoimmunological analyses due to the prolonged exposure to glucocorticoids have to be considered. Another study showed the advantages of combining OVX and glucocorticoid administration over combining OVX and calcium restriction with a higher decrease of BMD of the distal radius, distal tibia, and calcaneus (the spongiosa by 25 % and the corticalis by 17 % in the former group and 10 and 5 % in the latter, respectively) (Lill et al. 2000). Combining all three measures led to the most pronounced reductions (60 and 25 %). Age dependency of osteoporosis inducibility in sheep is still a matter of debate, but there is some consensus that inducing measures should last at least 7 months, and the time span between OVX and lactation should be over a year due to the increased intestinal calcium resorption (Hornby et al. 1995; Lill et al. 2002). Chavassieux et al. (1993) induced osteoporosis in adult ewes by daily intramuscular (IM) injections of 30 mg methylprednisolone for 2 months followed by 15 mg for 1 month. Mineral apposition rate in the iliac crest was decreased by 63 %, and bone formation rate was decreased by 84 %.

9.1.4 Porcine Models

Pigs seem to have some advantages over sheep concerning their suitability as biomedical large model species because of several anatomical and physiological reasons. Although there are a lot of similarities of diverse porcine organ systems to their human orthologs, the pig's usefulness as an osteological model species is still not entirely clear. However, the porcine femoral compact bone is predominantly plexiform, but is converted to well-developed osteonal bone earlier than in sheep (Mori et al. 2005). Peak bone mass is obtained with an age of 2–3 years. Ovariectomized and calcium-restricted (0.3 % Ca^{2+}) multiparous conventional sows seem not to be ideal models for osteoporosis research, most probably due to their extremely high mean areal BMD of 1.5 g/cm^2 as measured at the femoral neck, which seemingly reduces bone plasticity (Sipos et al. 2011a). Contrary, growing pigs aged 2 months have a mean areal BMD of 0.64 g/cm^2 as measured by DXA (Sipos et al. 2011b). Therefore, the main body of investigations in this area of research was performed using growing minipigs, which, however, might not appropriately reflect the situation of the postmenopausal osteoporotic woman due to their juvenile age. On the other hand, minipigs achieve sexual maturity earlier than conventional pigs, and thus OVX may induce the desired phenotype earlier than in conventional sows. Mosekilde et al. (1993b) successfully established a minipig bone loss model. OVX in 10-month-old minipigs resulted in a 6 % decrease in BMD, 15 % in BV, and 13 % in Tb.N, and an increase of 15 % in Tb.Sp after 6 months, whereas OVX in combination with a mild nutritive calcium shortage (0.75 % Ca^{2+}), which had already been started at an age of 4 months, led to a 10 % reduction in vertebral BMD and significant increases in final erosion depth and vertebral marrow star volume. In the growing pig model, calcium restriction (0.1–0.4 %) for 1 month leads to increased plasma PTH, calcitriol, alkaline phosphatase (ALP), propeptide of type 1 procollagen (P1CP), and hydroxyproline titers, which are associated with osteoporotic changes of metacarpals (Eklou-Kalonji et al. 1999). A study investigating multiparous sows being fed a standard diet (1.5 Ca^{2+}) showed a significant increase in plasma PTH, calcitriol, and AP levels over a time span of 1 year after OVX (Scholz-Ahrens et al. 1996). This could not be reproduced by our group in a similar setting (Sipos et al. 2011a). However, the former group also did not observe significant bone morphometric changes.

Whereas glucocorticoid treatment alone has been shown not to be sufficient to induce osteoporosis in the ovine model, this does not apply to the porcine model (Scholz-Ahrens et al. 2007). In adult (30 months old) primiparous Göttingen minipigs, an osteoporotic phenotype could be induced by daily oral treatment with 1 mg/

kg of prednisolone for 2 months and a reduction to 0.5 mg/kg thereafter until the end of the experiment, which was after 8 months in the short-term group and 15 months in the long-term group. In the short term, glucocorticoids reduced BMD at the lumbar spine by 48 mg/cm^3 from baseline, whereas in the control group the reduction was 12 mg/cm^3. These changes were also evidenced by plasma BAP levels, which decreased significantly in the glucocorticoid group. In the long term, the loss of BMD became more pronounced and bone mineral content, Tb.Th, and mechanical stability tended to be lower compared to the control group. There was a negative association between the cumulative dose of glucocorticoids and BMD, which could be traced back to impaired osteoblastogenesis. Other authors used growing minipigs (Ikeda et al. 2003). They treated 8-month-old Göttingen minipigs subcutaneously with prednisolone at a dosage of 0.5 mg/kg, 5 days/week for 26 weeks. Glucocorticoid treatment significantly reduced bone turnover marker (serum osteocalcin, urinary type-1 collagen N-telopeptide) levels at 13 weeks and thereafter also serum BAP levels relative to baseline. At 26 weeks, the longitudinal axis of the lumbar vertebral bone and length of the femur were smaller in the glucocorticoid group compared to the control group. The same applied to the BMD of the femur, but not L2, as measured by DXA. Age-dependent increases in trabecular bone structure were also reduced by glucocorticoids. L2 and femora of these animals were also tested mechanically, and prednisolone was shown to significantly reduce the ultimate load and maximum absorption energy of both sites. Further regression analyses revealed that bone minerals, bone structure, and chemical markers correlated with mechanical properties of L2 and mid-femur. It was concluded that prednisolone reduced systemic bone formation and resorption and suppressed the age-dependent increases in bone minerals, structure, and mechanical properties of L2 and mid-femur.

9.2 Animal Models of Rheumatoid Arthritis

Like osteoporosis, rheumatoid arthritis (RA) is a very complex pathological condition. Whereas this has not been proven for osteoporosis, rheumatoid arthritis is clearly identified as an autoimmune disease with a chronic inflammatory character. Although the responsible autoantigens are still speculative (most seemingly cyclic citrullinated peptides), there are several models, which reflect the pathophysiological situation satisfactorily on the focal (including redness, joint swelling, cartilage, and bone destruction) as well as the systemic level with upregulated proinflammatory cytokines. In the following, the most frequently used models will be presented in short: adjuvant-induced arthritis (AIA), collagen-induced arthritis (CIA), and the TNF-α transgenic mouse. Contrary to biomedical osteoporosis research, in rheumatoid arthritis research only rodent models are used.

Classically, AIA is induced by an intradermal injection of heat-killed *Mycobacterium tuberculosis* in paraffin oil at the base of the tail followed by a consecutive injection into one knee joint. The clinical onset is 9 days following the second injection as indicated by hind paw swelling and locomotory difficulties

(Feige et al. 2000). In a modified protocol, mice are immunized intradermally at the base of the tail and four footpads with 100 mg of methylated BSA (mBSA) emulsified in an equal volume of complete Freund's adjuvant (Ohshima et al. 1998). Additionally, mice are intraperitoneally injected with *Bordetella pertussis*. This procedure is repeated 7 days later. On day 21, 100 mg of mBSA in 10 ml of saline is injected into one knee joint. As a control, the same volume of saline is injected into the contralateral one. The acute phase of AIA lasts for 1 week starting after the booster injection and is followed by a chronic phase. Mice usually are sacrificed 35 days after the first immunization.

CIA is elicited by an intradermal injection or intravenous infusion of heterologous type II collagen emulsified 1:1 with incomplete Freund's adjuvant (Cremer et al. 1983; Stolina et al. 2009a). Following other regimens, CIA is induced after intradermal immunization with collagen emulsified in complete Freund's adjuvant, followed by a booster dose of collagen emulsified in incomplete Freund's adjuvant 3 weeks later. Disease susceptibility is strongly linked to the MHC class II haplotype. CIA-susceptible mice such as DBA/1, B10.Q, and B10.III are the most commonly used strains for the CIA model. Mice should be young (approx. 8 weeks) and healthy, and in many settings male mice are preferred because they develop disease earlier than females. CIA is consistently induced 16–35 days after immunization in 90–100 % of male mice and in 60–100 % of females using bovine collagen. Notably, progression of the two models, AIA and CIA, is mediated by distinct immunopathogenic mechanisms (Cremer et al. 1983). With respect to the dominant pro-arthritic cytokines, AIA is driven mainly by TNF-α (Stolina et al. 2009b), while CIA is provoked mainly by IL-1 (Stolina et al. 2008). Of importance for a systemic proinflammatory disease is that both CIA and AIA are characterized by a systemic upregulation of acute-phase proteins, IL-1β, IL-8, CCL2, and RANKL, whereas TNF-α, IL-17, and PGE_2 are elevated exclusively in clinical AIA. In contrast, Sarkar et al. (2009) found in their CIA model that joint inflammation was associated with a higher ratio of systemic IL-17/IFN-γ. Interestingly, neutralization of IFN-γ accelerated the course of CIA and was associated with increased IL-17 levels in serum and joints. The authors concluded that the absolute level of IL-17 is not the only determinant of joint inflammation. Instead, the balance of Th1, Th2, and Th17 cytokines is suggested to control the immune events leading to joint inflammation.

More recently, the serum transfer arthritis model has been generated. This model is based on T cells expressing a single autoreactive TCR recognizing glucose-6-phosphate isomerase (G6PI). These cells escape negative selection in mice bearing a specific MHC class II allele, IAg7. In the periphery, these T cells promote a breach in B cell tolerance, and high levels of anti-G6PI antibodies are produced, leading to a destructive and erosive arthritis similar to that seen in human RA. The adoptive transfer of serum from these mice results in peripheral joint swelling in most recipient strains. Early events that trigger paw swelling in the serum transfer model include signaling through FcRs. Further analysis demonstrated that $Fc\gamma RII^{-/-}$ mice manifested accelerated arthritis whereas $Fc\gamma RIII^{-/-}$ mice experienced a more slowly progressing arthritis. In the K/BxN serum transfer model of arthritis, there is a clinically apparent acute phase, which is modulated by FcγRII and FcγRIII, and a subacute component, which results in bone erosion, even in the absence of FcγR signaling (Corr and Crain 2002).

The TNF-α transgenic mouse (Keffer et al. 1991) allows deregulated human *TNF-α* gene expression. These mice develop a chronic inflammatory and destructive polyarthritis within 6 weeks after birth (Redlich et al. 2002).

9.3 Animal Models of Cancer-Associated Osteolytic Lesions

Many aggressively growing malign entities metastasize into the bone causing osteolytic lesions, which relies on the RANKL-mediated osteoclastogenesis-promoting nature of these tumor cells. Among the clinically most relevant bone-affecting tumors are the sexual steroid-controlled mammary and prostate cancers as well as multiple myeloma. In rodent mammary cancer models, usually the human breast cancer cell line MDA-231 is injected into the left ventricle of female athymic BALB/c nude mice aged 4–8 weeks under general anesthesia (Mbalaviele et al. 1996; Morony et al. 2001). Transplanted mice develop a profound cachexia. Multiple osteolytic lesions with highly active osteoclasts are evident in the bones of the proximal and distal extremities within 1 month after tumor inoculation.

More sophisticated data than those produced by radiography can be acquired by means of in vivo whole-body bioluminescence imaging (BLI). This technique demands the application of luciferase (*Luc*)-gene transduced cells of interest. Visibility is induced by a preceding intraperitoneal injection of luciferin (Canon et al. 2008). For example, BLI has been applied in the 4 T1 model. The 4 T1 orthotopic breast cancer model has been extensively utilized to examine the efficacy of a series of bisphosphonate compounds for the treatment of breast cancer bone metastases (Yoneda et al. 2000). This model is characterized by the occurrence of bone metastases in nearly 100 % of animals. Histological examination reveals the occurrence of profound osteoclastic bone resorption, and luciferase activity assays confirm tumor burden (Reinholz et al. 2010). In the 4 T1/Luc model, washed 4 T1 luc-transduced mouse mammary cancer cells are suspended in sterile PBS and subcutaneously injected into mammary fat pads of 4–5 week old syngeneic female BALB/c mice. Primary mammary tumors form approximately 1 week after cell inoculation, and metastases to the lung and liver develop within 2 weeks. Metastases to the bone, adrenals, kidneys, spleen, and heart occur by 3 weeks post-inoculation. Mice typically succumb by 4 weeks after tumor cell injection.

Prostate cancer (PC) bone metastases can also be induced by intracardial tumor cell infusions into male nude mice (Miller et al. 2008). *Luc*-transduced cells were found to develop a pattern of bioluminescence consistent with tumor metastatic foci in the bone with the highest concentrations in hind limbs and mandible as early as 3–5 days after intracardial injection. Another route for administration of PC cells is the intratibial injection (Ignatoski et al. 2008). Herein, the hind limb is shaved, the knee cap is located, and cells are injected in a volume of 50 μl into the tibia percutaneously via the tibial crest into the marrow cavity.

Xenograft models are also utilized for inducing multiple myeloma-associated bone metastases. The KAS-6/1-MIP-1α mouse model may serve as an example (Reinholz et al. 2010). Herein, genetically engineered KAS-6/1 myeloma cells carrying the osteoclast activating factor MIP-1α are injected into female SCID mice. Bone loss occurs within 2 weeks, hind limb paralysis occurs within 2 months, and mice typically die 1 week later.

Other tumor cell types used for generating osteolytic bone metastases include human A431 epidermoid carcinoma cells and murine colon adenocarcinoma-26 cells. A431/Luc cells were found to cause osteolytic lesions in the hind limbs after 20 days, and Colon-26 cells colonize the skeleton and cause significant localized bone destruction in syngeneic 7–8-week-old BALB/c DBA/2 mice within 12 days after intracardial tumor inoculation (Morony et al. 2001; Canon et al. 2010).

9.4 Immunotherapy of Osteoporosis in Animal Models

9.4.1 Osteoprotegerin as Surrogate for Denosumab in Rodent Models

The following two subchapters focus on RANKL blockade in rodent and non-primate large animal models. First, the effects of RANKL blockade have been investigated in models of postmenopausal osteoporosis. Fewer preclinical data are available for this treatment option of male osteoporosis. Min et al. (2000) demonstrated the protective and osteoporosis-reversing effects of the treatment of $OPG^{-/-}$ 8-week-old mice with high intravenous doses of 50 mg/kg of recombinant human (rh) OPG three times per week for 4 weeks. The advantage of rhOPG, which is fused to an Fc fragment, is its sustained serum half-life enabling a prolonged antiresorptive activity. In a next step, Capparelli et al. (2003) treated male, 10-week-old Sprague–Dawley rats with a single intravenous bolus injection of 5 mg/kg rhOPG. Maximum rhOPG concentrations were seen within 12 h after injection and coincided with significant elevations of serum PTH levels, which normalized 24 h later. Although a remarkable decline in rhOPG serum levels started at day 10, rhOPG serum concentrations remained at measurable levels throughout the 30-day study. The suppression of osteoclastic bone resorption started within 24 h after treatment. Significant gains in tibial cancellous BV were evident within 5 days. Femoral BMD increased between days 10 and 20. The significant decrease of osteoclast surface of 95 % in the rhOPG group was paralleled by a 35 % decrease in the serum bone resorption marker TRAP5b. Repeated rhOPG treatment (5 times 2 mg/kg within 2 weeks) further led to an increase of bone fracture strength at the femur mid-diaphysis in three-point bending by 30 % without affecting elastic or maximum strength in young male Sprague–Dawley rats (Ross et al. 2001). At the femoral neck, rhOPG significantly increased elastic (45 %), maximum (15 %), and fracture (35 %) strengths. Additionally, rhOPG treatment significantly increased whole bone dry mass (25 %), mineral mass (30 %), organic mass (17 %), and percent mineralization (4 %). Overall, rhOPG augments mineralization and strength indices in the rat femur with its effects on strength being more pronounced in the femoral neck than at the mid-diaphysis.

Next, effects of RANKL blockade have been tested in diverse osteoporosis models. Ominsky et al. (2008) treated ovariectomized rats aged 3 months with 10 mg/kg rhOPG twice weekly. OVX was associated with significantly higher serum RANKL titers, increased osteoclast surface, and reduced areal and volumetric BMD. Recombinant human OPG markedly reduced osteoclast surface and serum TRAP5b while completely preventing OVX-associated bone loss in the lumbar vertebrae, distal femur, and femoral neck. μCT analyses showed that trabecular compartments in rhOPG-treated OVX rats had a significantly greater BV fraction, volumetric BMD, bone area, Tb.Th, and Tb.N. Additionally, rhOPG improved the cortical area in the lumbar vertebrae and femoral neck to levels that were significantly greater than in OVX or sham controls. Also, bone strength was increased in OVX rats by rhOPG treatment.

Clinical development of rhOPG was discontinued in favor of denosumab. Thus, preclinical experiments demanded a rodent model in which this fully humanized monoclonal antibody that specifically inhibits primate and human RANKL would be effective. Preliminary experiments showed that denosumab did not suppress bone resorption in normal mice or rats but prevented the resorptive response in mice challenged with a human RANKL fragment encoded primarily by the fifth exon of the RANKL gene. Therefore, exon 5 from murine RANKL was replaced by its human ortholog. The resulting huRANKL mice exclusively express chimeric (human/murine) RANKL that maintains bone resorption at slightly reduced levels when compared to wild type controls (Kostenuik et al. 2009). In these mice denosumab reduced bone resorption, increased cortical and cancellous bone mass, and improved trabecular microarchitecture. Hofbauer et al. (2009) analyzed the bone protective effects of denosumab (10 mg/kg subcutaneously twice weekly over 4 weeks) in glucocorticoid-treated male, 8-month-old homozygous huRANKL knock-in mice. They showed that prednisolone treatment induced the loss of vertebral and femoral volumetric BMD, which was associated with suppressed vertebral bone formation and increased bone resorption as evidenced by increases in the number of osteoclasts, TRAP5b protein in bone extracts, serum levels of TRAP5b, and urinary excretion of deoxypyridinoline. More detailed analysis showed that glucocorticoid-induced bone loss was most pronounced in the cortical and subcortical compartment in the distal femoral metaphysis whereas trabecular BMD remained unchanged by prednisolone treatment. Denosumab prevented prednisolone-induced loss of total BMD at the spine and the distal femur. Additionally, biomechanical compression tests of lumbar vertebrae revealed a detrimental effect of prednisolone on bone strength that could also be prevented by denosumab.

In analogy to rhOPG and denosumab as RANKL antagonists, Kim et al. (2009) developed a cell-permeable inhibitor termed RANK receptor inhibitor (RRI), which targets a cytoplasmic motif of RANK. The RRI peptide blocked RANKL-induced osteoclast formation from murine bone marrow-derived macrophages. Furthermore, RRI inhibited the resorptive function of osteoclasts, induced osteoclast apoptosis, and protected against OVX-induced bone loss in mice. As RANK blockade may theoretically impair the immune system, the authors also determined whether the RRI peptide interferes with phagocytosis or dendritic cell differentiation; both immune functions were not affected.

Li et al. (2009a) tested the effects of rhOPG treatment on bone biology of orchiectomized (ORX) rats. Whereas serum testosterone declined within 2 weeks after surgery, no changes in serum RANKL could be observed. In contrast, there was an increase of RANKL in bone marrow plasma, which correlated positively with marrow plasma TRAP5b. RANKL inhibition was induced by treating ORX rats twice weekly for 6 weeks subcutaneously with 10 mg/kg rhOPG. Whereas vehicle-treated ORX rats showed significant deficits in BMD of the femur and tibia as well as lower trabecular BV in the distal femur, rhOPG treatment increased femoral and tibial BMD and trabecular BV to levels that significantly exceeded values for ORX or sham controls. Histologically, rhOPG treatment reduced trabecular osteoclast surfaces in ORX rats by 99 %. μCT of lumbar vertebrae from rhOPG-treated ORX rats demonstrated significantly greater cortical and trabecular bone volume and density versus ORX-vehicle controls. These data substantiate RANKL inhibition as a strategy for preventing bone loss associated with androgen ablation or deficiency.

9.4.2 Osteoprotegerin as Surrogate for Denosumab in Pigs

Data on the effects of rhOPG treatment in large animal models are limited. The homology between a porcine RANKL-specific sequence and the corresponding human RANKL sequence was found to be 79 %. Also, RANKL is upregulated during in vitro osteoclastogenesis and expressed by a variety of different cell types including immunocytes in pigs (Sipos et al. 2005). Sipos et al. (2011b) analyzed the effects of a single intravenous rhOPG bolus of 5 mg/kg in pigs aged 2 months. Serum rhOPG levels peaked at day 5 and coincided with significantly decreased calcium, phosphate, and bone turnover markers. TRACP5b, P1CP, and BAP levels significantly decreased by 40–70 % relative to vehicle controls in the rhOPG group between days 5 and 10, indicating that pharmacologic concentration of rhOPG led to systemic concomitant inhibition of bone formation and resorption. At termination of the experiment at day 20, μCT analysis showed a significantly higher connectivity density in the proximal femur, proximal tibia, and L4 as well as BMD of the femur, hip, and tibia. Interestingly, Tb.Th (femur, hip, L4) was significantly lower in the rhOPG-treated pigs, but Tb.N (femur) was higher, and a trend toward lower Tb.Sp was evident in the tibia. In summary, rhOPG treatment inhibited osteoclast function as evidenced by TRAP5b decrease and led to a more arborescent architecture and higher mineralisation in the weight-bearing skeleton. These data show that not only anatomical but also microstructural bone parameters as well as the responsiveness to RANKL inhibition may differ in detail between large mammal species such as humans and pigs and also between these species and rodents with their distinct biomechanical forces on the axial as well as the appendicular skeleton.

9.4.3 Romosozumab

Romosozumab is a humanized monoclonal antibody that targets sclerostin for the treatment of osteoporosis. Sclerostin is a potent inhibitor of the canonical Wnt signaling pathway, which is essential for osteoblastogenesis. Mutations in the sclerostin-encoding gene SOST lead to sclerostosis, which is the result of progressively increasing bone formation. Thus, romosozumab acts by supporting osteoblastogenesis, which is different from the final outcome achieved by denosumab, which exerts its effects by suppressing osteoclastogenesis.

In preclinical studies, the therapeutic potential of a neutralizing sclerostin antibody was tested in ovariectomized rats (Li et al. 2009b). This therapeutic intervention not only resulted in complete prevention of osteoporosis but had also strong osteoanabolic effects. In gonad-intact female cynomolgus macaques, the administration of a humanized neutralizing anti-sclerostin monoclonal antibody over 2 months had clear osteoanabolic effects with marked dose-dependent increases in bone formation on trabecular, periosteal, endocortical, and intracortical surfaces. Significant increases in trabecular thickness and bone strength were found at the lumbar vertebrae in the highest-dose group (Ominsky et al. 2010).

9.4.4 Secukinumab

Secukinumab is a human monoclonal antibody targeting IL-17A and is approved for the treatment of psoriasis. The therapeutic spectrum may soon be expanded to the treatment of uveitis, rheumatoid arthritis, ankylosing spondylitis, and psoriatic arthritis. Basis for the therapeutic application is the disturbance of the Th17/Treg balance in a large spectrum of diverse autoimmune diseases. Notably, also for osteoporosis the therapeutic use of anti-IL-17 antibodies has now been proposed, as a reduction of proinflammatory cytokines, an increase in regulatory T cell number, a positive effect on cortical as well as trabecular bone and bone mechanical parameters, and an increased osteoblastogenesis could be demonstrated in ovariectomized mice, which was even superior to anti-RANKL and anti-TNF-α antibodies (Tyagi et al. 2014). These effects may be considered as another hint toward an involvement of a dysregulation of the immune system in the pathogenesis of osteoporosis.

9.5 Immunotherapy of Rheumatoid Arthritis in Animal Models

Antirheumatic drugs exhibit pronounced adverse side effects. Therefore, efforts aim to establish more specific anti-inflammatory and osteoclastogenesis-inhibitory therapeutic regimens including RANKL blockade. First, rhOPG monotherapy was tested for its efficacy to ameliorate or abolish bone destructive processes of RA. In their study investigating male Lewis AIA rats treated with subcutaneously administered rhOPG, Campagnuolo et al. (2002) found that rhOPG provided dose- and schedule-dependent preservation of BMD and periarticular bone while essentially eliminating intralesional osteoclasts. Dosages >2.5 mg/kg/day preserved or enhanced BMD and essentially prevented all erosions. OPG treatment was also successful in preventing the loss of cartilage matrix proteoglycans and was shown to be most effective when initiated early in the course of the disease. However, signs of inflammation could not be affected by rhOPG treatment. In another study based on TNF-α transgenic mice, the anti-inflammatory as well as anti-osteoclastogenic potential of infliximab, an anti-TNF-α monoclonal antibody, was compared to rhOPG and pamidronate treatment (Redlich et al. 2002). As may be delineated from the preceding study, clinical improvement was achieved only in the infliximab group. However, radiographic analyses revealed a significant retardation of joint damage in animals treated with rhOPG (55 % reduction of erosions), pamidronate (50 % reduction), a combination therapy of rhOPG and pamidronate (64 % reduction), and with infliximab (66 % reduction). These data show that rhOPG alone or in combination with bisphosphonates may be an effective therapeutic tool for the prevention of inflammatory bone destruction. Stolina et al. (2009a, b) also investigated Lewis rats with established AIA or CIA. Rats were treated with pegsunercept (a TNF-α inhibitor), anakinra (an IL-1 receptor antagonist), or rhOPG. Anti-TNF-α treatment ameliorated paw swelling in both models and reduced ankle BMD loss in AIA rats. Anti-IL-1 treatment decreased paw swelling in CIA rats and reduced ankle BMD loss in both models. Both anti-TNF-α and anti-IL-1 applications reduced systemic markers of inflammation as well as, at least in part, systemic RANKL, but failed to prevent vertebral BMD loss in either model. OPG reduced TRAP5b by over 90 % and consequently also BMD loss in ankles and vertebrae in both models, but as anticipated had no effect on paw swelling.

The former studies demonstrated that rat AIA and CIA feature bone loss and systemic increases in TNF-α, IL-1β, and RANKL. Anti-cytokine therapies targeting inflammatory cytokines consistently reduce inflammation in these models, but systemic bone loss often persists. On the other hand, RANKL inhibition consistently prevents bone loss without reducing joint inflammation. Logically, RANKL inhibition has to be further combined with anti-inflammatory, most preferably site-specific, drugs. Oelzner et al. (2010) treated AIA rats with dexamethasone (0.25 mg/kg/day, i.p.), rhOPG (2.5 mg/kg/day, i.p.), or a combination of both at regular intervals for 3 weeks. As expected, dexamethasone monotherapy substantially suppressed joint swelling without inhibiting bone loss of the secondary spongiosa, whereas rhOPG monotherapy showed no anti-inflammatory effect. Interestingly, rhOPG monotherapy failed to inhibit AIA-induced bone loss, whereas the combination of dexamethasone and rhOPG produced an anti-inflammatory effect and resulted in inhibition of periarticular and axial bone loss. Thus, the principle of combining an anti-inflammatory drug with RANKL inhibition may prove an effective bone-saving therapy in RA.

More recently, the disturbance of the Th17/Treg balance in the case of autoimmune diseases has come into focus of (pre)clinical research. Not only are these diseases characterized by a bias toward a Th17 activation, but it has also been shown that these cells can even transdifferentiate out of regulatory T cells. The conversion of Foxp3⁺CD4⁺ T cells to Th17 cells was mediated by synovial fibroblast-derived IL-6 (Komatsu et al. 2014). The reestablishment of a preferable balance of Th17 cells to regulatory T cells can be achieved by different means, such as anti-IL17 antibodies or STAT3-inhibitors (Li et al. 2014; Park et al. 2014). Of notice, anti-IL17 treatment has the additional effect of indirectly suppressing also the expression of other proinflammatory cytokines, such as IL-1β and TNF-α (Li et al. 2014).

9.6 Immunotherapy of Cancer-Associated Osteolytic Lesions in Animal Models

RANKL is one of the key mediators of malignant bone resorption. Thus, RANKL inhibition prevents primary tumor or metastasis-induced osteolysis and decreases skeletal tumor burden. Morony et al. (2001) were among the first, who tested the ability of rhOPG to inhibit tumor-induced osteoclastogenesis, osteolysis, and skeletal tumor burden in two animal models. In one model they induced lytic bone lesions by transfecting mice with mouse colon adenocarcinoma (Colon-26) cells and found that treatment with rhOPG dose dependently decreased the number and area of radiographically evident lytic lesions with high efficacy. This therapeutic approach was also effective in nude mice transplanted with human MDA-231 breast cancer cells, completely preventing radiographic osteolytic lesions. Histologically, rhOPG decreased skeletal tumor burden by 75 % and completely eradicated MDA-231 tumor-associated osteoclasts. In both models, rhOPG had no effect on tumor metastases in soft tissue organs. The MDA-231 model for investigating the effects of RANKL inhibition in order to reduce bone tumor burden was also used by other authors, who confirmed the above findings. Investigating this model, Canon et al. (2008) found that RANKL protein levels were significantly higher in tumor-bearing bones than in tumor-free animals. They monitored the antitumor efficacy of RANKL inhibition by rhOPG on MDA-231 cells in a temporal manner using bioluminescence imaging. One mechanism by which RANKL inhibition reduced tumor burden appeared to be indirect through increasing tumor cell apoptosis as measured by active caspase-3. In this model, treatment with rhOPG resulted in an overall improvement in survival. Zheng et al. (2007) combined the antitumor effects of rhOPG and ibandronate, again using the MDA-231 mouse model. Ten days after intratibial tumor cell injection (when the tumors were evident radiologically), mice were treated with rhOPG (1 mg/kg/day), ibandronate (160 μg/kg/day), or a combination for 1 week, and the effects of each treatment on lytic lesions, tumor cell growth, cell apoptosis, and proliferation were measured. Compared to vehicle controls, treatment with all regimens prevented the expansion of osteolytic bone lesions at a similar rate (2.3 % increase in the mean vs 232.5 % increase in sham animals). Treatment with all regimens produced similar reductions in tumor area (mean 51.3 %) as well as a similar increase in cancer cell apoptosis (339.7 %) and decrease in cancer cell proliferation (59.7 %).

The inhibition of prostate cancer bone metastases is also subject of rodent model-based research. For treatment of metastatic hormone-refractory prostate cancer, docetaxel is a well-established medication. However, the side effects associated with docetaxel treatment can be severe, resulting in the discontinuation of therapy. Thus, efforts are ongoing to identify an adjuvant therapy to allow lower doses of docetaxel. As advanced prostate carcinomas are typically accompanied by skeletal metastases, targeting RANKL is a reasonable option. A combined treatment regimen based on RANKL inhibition and docetaxel decreased the establishment and progression of prostate carcinoma growth in the bone in murine models (Ignatoski et al. 2008). Fortunately, the combination of RANKL inhibition and docetaxel reduced tumor burden in the bone greater than either treatment alone and increased median survival time by 16.7 % (Miller et al. 2008).

Another study investigated whether the reduction of osteolysis by RANKL inhibition could enhance the antitumor effects of an anti-EGFR antibody (panitumumab) in a novel murine model of human A431 epidermoid carcinoma bone metastasis by bioluminescence imaging (Canon et al. 2010). As shown by earlier studies, these authors also found that RANKL inhibition by rhOPG treatment resulted in a reduction in tumor progression in bone sites and in tumor-induced osteolysis. The antitumor efficacy of panitumumab could be increased by rhOPG. The combination completely blocked tumor-induced bone breakdown. These studies demonstrate an additive effect of RANKL inhibition to the first-line agent in various murine bone metastasis settings and may lead to sanguine novel therapeutic options in the fight against tumor-associated bone diseases.

References

Aerssens J, van Audekercke R, Talalaj M, Geusens P, Bramm E, Dequeker J (1996) Effect of 1-alpha-vitamin D3 and estrogen on cortical bone mechanical properties in the ovariectomized rat model. Endocrinology 137:1358–1364

Bucay N, Sarosi I, Dunstan CR, Morony S, Tarpley J, Capparelli C, Scully S, Tan HL, Xu WL, Lacey DL, Boyle WJ, Simonet WS (1998) Osteoprotegerin-deficient mice develop early onset osteoporosis and arterial calcification. Genes Dev 12:1260–1268

Campagnuolo G, Bolon B, Feige U (2002) Kinetics of bone protection by recombinant osteoprotegerin therapy in Lewis rats with adjuvant arthritis. Arthritis Rheum 46:1926–1936

Canon JR, Roudier M, Bryant R, Morony S, Stolina M, Kostenuik PJ, Dougall WC (2008) Inhibition of RANKL blocks skeletal tumor progression and improves survival in a mouse model of breast cancer bone metastasis. Clin Exp Metastasis 25:119–129

Canon J, Bryant R, Roudier M, Osgood T, Jones J, Miller R, Coxon A, Radinsky R, Dougall WC (2010) Inhibition of RANKL increases the anti-tumor effect of the EGFR inhibitor panitumumab in a murine model of bone metastasis. Bone 46:1613–1619

Capparelli C, Morony S, Warmington K, Adamu S, Lacey D, Dunstan CR, Stouch B, Martin S, Kostenuik PJ (2003) Sustained antiresorptive effects after a single treatment with human recombinant osteoprotegerin (OPG): a pharmacodynamic and pharmacokinetic analysis in rats. J Bone Miner Res 18:852–858

Chavassieux P, Pastoureau P, Chapuy C, Delmas PD, Meunier PJ (1993) Glucocorticoid-induced inhibition of osteoblastic bone formation in ewes: a biomechanical and histomorphometric study. Osteoporos Int 3:97–102

Chavassieux P, Garnero P, Duboeuf F, Vergnaud O, Brunner-Ferber F, Delmas PD, Meunier PJ (2001) Effects of a new selective estrogen receptor modulator (MDL 103,323) on cancellous and cortical bone in ovariectomized ewes: A biochemical, histomorphometric, and densitometric study. J Bone Miner Res 16:89–96

Corr M, Crain B (2002) The role of FcγR signaling in the K/BxN serum transfer model of arthritis. J Immunol 169:6604–6609

Cremer MA, Hernandez AD, Townes AS, Stuart JM, Kang AH (1983) Collagen-induced arthritis in rats: antigen-specific suppression of arthritis and immunity by intravenously injected native type II collagen. J Immunol 131:2995–3000

Eklou-Kalonji E, Zerath E, Colin C, Lacroix C, Holy X, Denis I, Pointillart A (1999) Calcium-regulating hormones, bone mineral content, breaking load and trabecular remodeling are altered in growing pigs fed calcium-deficient diets. J Nutr 129:188–193

FDA Guidelines (1994) Guidelines for Preclinical and Clinical Evaluation of Agents Used for the prevention or Treatment of Postmenopausal osteoporosis, Division of Metabolism and Endocrine Drug Products, Food and Drug Administration. Chapter: Preclinical Studies, 2–6

Feige U, Hu YL, Gasser J, Campagnuolo G, Munyakazi L, Bolon B (2000) Anti-interleukin-1 and anti-tumor necrosis factor-alpha synergistically inhibit adjuvant arthritis in Lewis rats. Cell Mol Life Sci 57:1457–1470

Gerlach UV (2002) Tierexperimentelles Modell zur Untersuchung der Frakturbehandlung beim osteoporotischen Knochen. Thesis, Justus-Liebig-Universität Giessen

Giavaresi G, Fini M, Torricelli P, Giardino R (2001) The ovariectomized ewe model in the evaluation of biomaterials for prosthetic devices in spinal fixation. Int J Artif Organs 24:814–820

Grimston SK, Silva MJ, Civitelli R (2007) Bone loss after temporarily induced muscle paralysis by Botox is not fully recovered after 12 weeks. Ann NY Acad Sci 1116:444–460

Hofbauer LC, Zeitz U, Schoppet M, Skalicky M, Schüler C, Stolina M, Kostenuik PJ, Erben RG (2009) Prevention of glucocorticoid-induced bone loss in mice by inhibition of RANKL. Arthritis Rheum 60:1427–1437

Hornby SB, Ford SL, Mase CA, Evans GP (1995) Skeletal changes in the ovariectomized ewe and subsequent response to treatment with 17β-estradiol. Bone 17:387–394

Ignatoski KM, Escara-Wilke JF, Dai JL, Lui A, Dougall W, Daignault S, Yao Z, Zhang J, Day ML, Sargent EE, Keller ET (2008) RANKL inhibition is an effective adjuvant for docetaxel in a prostate cancer bone metastases model. Prostate 68:820–829

Ikeda S, Morishita Y, Tsutsumi H, Ito M, Shiraishi A, Arita S, Akahoshi S, Narusawa K, Nakamura T (2003) Reductions in bone turnover, mineral, and structure associated with mechanical properties of lumbar vertebra and femur in glucocorticoid-treated growing minipigs. Bone 33:779–787

Ito M, Nakamura T, Matsumoto T, Tsurusaki K, Hayashi K (1998) Analysis of trabecular microarchitecture of human iliac bone using microcomputed tomography in patients with hip arthrosis with or without vertebral fracture. Bone 23:163–169

Jiang YB, Zhao J, Genant HK, Dequeker J, Geusens P (1997) Long-term changes in bone mineral and biomechanical properties of vertebrae and femur in aging, dietary calcium restricted and/or estrogen-deprived/-replaced rats. J Bone Miner Res 12:820–831

Kalu DN (1991) The ovariectomized rat as a model of postmenopausal osteopenia. Bone Miner 15:175–191

Keffer J, Probert L, Cazlaris H, Georgopoulos S, Kaslaris E, Kioussis D, Kollias G (1991) Transgenic mice expressing human tumour necrosis factor: a predictive genetic model of arthritis. EMBO J 10:4025–4031

Kim H, Choi HK, Shin JH, Kim KH, Huh JY, Lee SA, Ko CY, Kim HS, Shin HI, Lee HJ, Jeong D, Kim N, Choi Y, Lee SY (2009) Selective inhibition of RANK blocks osteoclast maturation and function and prevents bone loss in mice. J Clin Invest 119:813–825

Komatsu N, Okamoto K, Sawa S, Nakashima T, Oh-hora M, Kodama T, Tanaka S, Bluestone JA, Takayanagi H (2014) Pathogenic conversion of Foxp3+ T cells into TH17 cells in autoimmune arthritis. Nat Med 20:62–68

Kong YY, Yoshida H, Sarosi I, Tan HL, Timms E, Capparelli C, Morony S, Oliveira-dos-Santos AJ, Van G, Itie A, Khoo W, Wakeham A, Dunstan CR, Lacey DL, Mak TW, Boyle WJ, Penninger JM (1999) OPGL is a key regulator of osteoclastogenesis, lymphocyte development and lymph-node organogenesis. Nature 397:315–323

Kostenuik PJ, Nguyen HQ, McCabe J, Warmington KS, Kurahara C, Sun N, Chen C, Li L, Cattley RC, Van G, Scully S, Elliott R, Grisanti M, Morony S, Tan HL, Asuncion F, Li X, Ominsky MS, Stolina M, Dwyer D, Dougall WC, Hawkins N, Boyle WJ, Simonet WS, Sullivan JK (2009) Denosumab, a fully human monoclonal antibody to RANKL, inhibits bone resorption and increases BMD in knock-in mice that express chimeric (murine/human) RANKL. J Bone Miner Res 24:182–195

Lacey DL, Timms E, Tan HL, Kelley MJ, Dunstan CR, Burgess T, Elliott R, Colombero A, Elliott G, Scully S, Hsu H, Sullivan J, Hawkins N, Davy E, Capparelli C, Eli A, Qian YX, Kaufman S, Sarosi I, Shalhoub V, Senaldi G, Guo J, Delaney J, Boyle WJ (1998) Osteoprotegerin ligand is a cytokine that regulates osteoclast differentiation and activation. Cell 93:165–176

Li M, Shen Y, Qi H, Wronski TJ (1996) Comparison study of skeletal response to estrogen depletion at red and yellow marrow sites in rats. Anat Rec 245:472–480

Li X, Ominsky MS, Stolina M, Warmington KS, Geng Z, Niu QT, Asuncion FJ, Tan HL, Grisanti M, Dwyer D, Adamu S, Ke HZ, Simonet WS, Kostenuik PJ (2009a) Increased RANK ligand in bone marrow of orchiectomized rats and prevention of their bone loss by the RANK ligand inhibitor osteoprotegerin. Bone 45:669–676

Li X, Ominsky MS, Warmington KS, Morony S, Gong J, Cao J, Gao Y, Shalhoub V, Tipton B, Haldankar R, Chen Q, Winters A, Boone T, Geng Z, Niu QT, Ke HZ, Kostenuik PJ, Simonet WS, Lacey DL, Paszty C (2009b) Sclerostin antibody treatment increases bone formation, bone mass, and bone strength in a rat model of postmenopausal osteoporosis. J Bone Miner Res 24:578–588

Li Q, Ren G, Xu L, Wang Q, Qi J, Wang W, Zhou B, Han X, Sun C, Wu Q, Yu Y, Peng Z, Zheng S, Li D (2014) Therapeutic efficacy of three bispecific antibodies on collagen-induced arthritis mouse model. Int Immunopharmacol 21:119–127

Lill CA, Fluegel AK, Schneider E (2000) Sheep model for fracture treatment in osteoporotic bone: a pilot study about different induction regimens. J Orthop Trauma 14:559–566

Lill CA, Gerlach UV, Eckhardt C, Goldhahn J, Schneider E (2002) Bone changes due to glucocorticoid application in an ovariectomized animal model for fracture treatment in osteoporosis. Osteoporos Int 13:407–414

Ma YF, Ke HZ, Jee WSS (1994) Prostaglandin E2 adds bone to a cancellous bone site with a closed growth plate and low bone turnover in ovariectomized rats. Bone 15:137–146

MacLeay JM, Olson JD, Enns RM, Les CM, Toth CA, Wheeler DL, Turner AS (2004) Dietary-induced metabolic acidosis decreases bone mineral density in mature ovariectomized ewes. Calcif Tissue Int 75:431–437

Mbalaviele G, Dunstan CR, Sasaki A, Williams PJ, Mundy GR, Yoneda T (1996) E-cadherin expression in human breast cancer cells suppresses the development of osteolytic bone metastases in an experimental metastasis model. Cancer Res 56:4063–4070

Miller SC, Bowman BM, Miller MA, Bagi CM (1991) Calcium absorption and osseous organ-, tissue-, and envelope-specific changes following ovariectomy in rats. Bone 12:439–446

Miller RE, Roudier M, Jones J, Armstrong A, Canon J, Dougall WC (2008) RANK ligand inhibition plus docetaxel improves survival and reduces tumor burden in a murine model of prostate cancer bone metastasis. Mol Cancer Ther 7:2160–2169

Min H, Morony S, Sarosi I, Dunstan CR, Capparelli C, Scully S, Van G, Kaufman S, Kostenuik PJ, Lacey DL, Boyle WJ, Simonet WS (2000) Osteoprotegerin reverses osteoporosis by inhibiting endosteal osteoclasts and prevents vascular calcification by blocking a process resembling osteoclastogenesis. J Exp Med 192:463–474

Miyakoshi N, Sato K, Tsuchida T, Tamura Y, Kudo T (1999) Histomorphometric evaluation of the effects of ovariectomy on bone turnover in rat caudal vertebrae. Calcif Tissue Int 64:318–324

Mori R, Kodaka T, Soeta S, Sato J (2005) Preliminary study of histological comparison on the growth patterned of long-bone cortex in young calf, pig, and sheep. J Vet Med Sci 67: 1223–1229

Morony S, Capparelli C, Sarosi I, Lacey DL, Dunstan CR, Kostenuik PJ (2001) Osteoprotegerin inhibits osteolysis and decreases skeletal tumor burden in syngeneic and nude mouse models of experimental bone metastasis. Cancer Res 61:4432–4436

Mosekilde L, Sogaard CH, Danielson CC, Torring O, Nilsson MHL (1991) The anabolic effects of human parathyroid hormone (hPTH) on rat vertebral body mass are also reflected in the quality of bone, assessed by biomechanical testing: A comparison study between hPTH-(1–34) and hPTH- (1–84). Endocrinology 129:421–428

Mosekilde L, Danielsen DD, Knudsen UB (1993a) The effect of aging and ovariectomy on the vertebral bone mass and biomechanical properties of mature rats. Bone 14:1–6

Mosekilde L, Weisbrode SE, Safron JA, Stills HF, Jankowsky ML, Ebert DC, Danielsen CC, Sogaard CH, Franks F, Stevens ML, Paddock CL, Boyce RW (1993b) Evaluation of the skeletal effects of combined mild dietary calcium restriction and ovariectomy in Sinclair S-1 minipigs: a pilot study. J Bone Miner Res 8:1311–1321

Ochi H, Hara Y, Tagawa M, Shinomiya K, Asou Y (2010) The roles of TNFR1 in lipopolysaccharide-induced bone loss: dual effects of TNFR1 on bone metabolism via osteoclastogenesis and osteoblast survival. J Orthop Res 28:657–663

Oelzner P, Fleissner-Richter S, Bräuer R, Hein G, Wolf G, Neumann T (2010) Combination therapy with dexamethasone and osteoprotegerin protects against arthritis-induced bone alterations in antigen-induced arthritis of the rat. Inflamm Res 59:731–741

Ohshima S, Saeki Y, Mima T, Sasai M, Nishioka K, Nomura S, Kopf M, Katada Y, Tanaka T, Suemura M, Kishimoto T (1998) Interleukin 6 plays a key role in the development of antigen-induced arthritis. Proc Natl Acad Sci U S A 95:8222–8226

Ominsky MS, Li X, Asuncion FJ, Barrero M, Warmington KS, Dwyer D, Stolina M, Geng Z, Grisanti M, Tan HL, Corbin T, McCabe J, Simonet WS, Ke HZ, Kostenuik PJ (2008) RANKL inhibition with osteoprotegerin increases bone strength by improving cortical and trabecular bone architecture in ovariectomized rats. J Bone Miner Res 23:672–682

Ominsky MS, Vlasseros F, Jolette J, Smith SY, Stouch B, Doellgast G, Gong J, Gao Y, Cao J, Graham K, Tipton B, Cai J, Deshpande R, Zhou L, Hale MD, Lightwood DJ, Henry AJ, Popplewell AG, Moore AR, Robinson MK, Lacey DL, Simonet WS, Paszty C (2010) Two doses of sclerostin antibody in cynomolgus monkeys increases bone formation, bone mineral density, and bone strength. J Bone Miner Res 25:948–959

Park JS, Kwok SK, Lim MA, Kim EK, Ryu JG, Kim SM, Oh HJ, Ju JH, Park SH, Kim HY, Cho ML (2014) STA-21, a promising STAT-3 inhibitor that reciprocally regulates Th17 and Treg cells, inhibits osteoclastogenesis in mice and humans and alleviates autoimmune inflammation in an experimental model of rheumatoid arthritis. Arthritis Rheumatol 66:918–929

Pearce AI, Richards RG, Milz S, Schneider E, Pearce SG (2007) Animal models from implant biomaterial research in bone: a review. Eur Cell Mater 13:1–10

Peng Z, Tuukkanen J, Vaananen HK (1994) Exercise can provide protection against bone loss and prevent the decrease in mechanical strength of femoral neck in ovariectomized rats. J Bone Miner Res 9:1559–1564

Persson P, Gagnemo-Persson R, Hakanson R (1993) The effect of high or low dietary calcium on bone and calcium homeostasis in young male rats. Calcif Tissue Int 52:460–464

Pietschmann P, Skalicky M, Kneissel M, Rauner M, Hofbauer G, Stupphann D, Viidik A (2007) Bone structure and metabolism in a rodent model of male senile osteoporosis. Exp Gerontol 42:1099–1108

Pietschmann P, Rauner M, Sipos W (2008) Osteoporosis: an age-related and gender-specific disease – a mini-review. Gerontology 55:3–12

Pogoda P, Egermann M, Schnell JC, Priemel M, Schilling AF, Alini M, Schinke T, Rueger JM, Schneider E, Clarke I, Amling M (2006) Leptin inhibits bone formation not only in rodents but also in sheep. J Bone Miner Res 21:1591–1599

Redlich K, Hayer S, Maier A, Dunstan CR, Tohidast-Akrad M, Lang S, Türk B, Pietschmann P, Woloszczuk W, Haralambous S, Kollias G, Steiner G, Smolen JS, Schett G (2002) Tumor necrosis factor alpha-mediated joint destruction is inhibited by targeting osteoclasts with osteoprotegerin. Arthritis Rheum 46:785–792

Reinholz MM, Zinnen SP, Dueck AC, Dingli D, Reinholz GG, Jonart LA, Kitzmann KA, Bruzek AK, Negron V, Abdalla AK, Arendt BK, Croatt AJ, Sanchez-Perez L, Sebesta DP, Lönnberg H, Yoneda T, Nath KA, Jelinek DF, Russell SJ, Ingle JN, Spelsberg TC, Dixon HB, Karpeisky A, Lingle WL (2010) A promising approach for treatment of tumor-induced bone diseases: utilizing bisphosphonate derivatives of nucleoside antimetabolites. Bone 47:12–22

Ross AB, Bateman TA, Kostenuik PJ, Ferguson VL, Lacey DL, Dunstein CR, Simske SJ (2001) The effects of osteoprotegerin on the mechanical properties of rat bone. J Materials Sci 12:583–588

Sarkar S, Cooney LA, White P, Dunlop DB, Endres J, Jorns JM, Wasco MJ, Fox DA (2009) Regulation of pathogenic IL-17 responses in collagen-induced arthritis: roles of endogenous interferon-gamma and IL-4. Arthritis Res Ther 11:R158

Scholz-Ahrens KE, Delling G, Jungblut PW, Kallweit E, Barth CA (1996) Effect of ovariectomy on bone histology and plasma parameters of bone metabolism in nulliparous and multiparous sows. Z Ernahrungswiss 35:13–21

Scholz-Ahrens KE, Delling G, Stampa B, Helfenstein A, Hahne HJ, Açil Y, Timm W, Barkmann R, Hassenpflug J, Schrezenmeir J, Glüer CC (2007) Glucocorticosteroid-induced osteoporosis in adult primiparous Göttingen miniature pigs: effects on bone mineral and mineral metabolism. Am J Physiol Endocrinol Metab 293:E385–E395

Shahnazari M, Martin BR, Legette LL, Lachcik PJ, Welch J, Weaver CM (2009) Diet calcium level but not calcium supplement particle size affects bone density and mechanical properties in ovariectomized rats. J Nutr 139:1308–1314

Sigrist HM, Gerhardt C, Alini M, Schneider E, Egermann M (2007) The long-term effects of ovariectomy on bone metabolism in sheep. J Bone Miner Metab 25:28–35

Sipos W (1997) In-vitro Maturation und In-vitro Fertilisation canincer Oozyten unter besonderer Berücksichtigung der Supplementierung der In-vitro Maturationsmedien mit 17β-Östradiol und Progesteron. Thesis, University of Veterinary Medicine Vienna

Sipos W, Duvigneau JC, Hofbauer G, Schmoll F, Baravalle G, Exel B, Hartl R, Dobretsberger M, Pietschmann P (2005) Characterization of the cytokine pattern of porcine bone marrow-derived cells treated with 1α,25$(OH)_2D_3$. J Vet Med A 52:382–387

Sipos W, Kralicek E, Rauner M, Duvigneau JC, Worliczek HL, Schamall D, Hartl RT, Sommerfeld-Stur I, Dall'Ara E, Varga P, Resch H, Schwendenwein I, Zysset P, Pietschmann P (2011a) Bone and cellular immune system of multiparous sows are insensitive to ovariectomy and nutritive calcium shortage. Horm Metab Res 43:404–409

Sipos W, Zysset P, Kostenuik P, Mayrhofer E, Bogdan C, Rauner M, Stolina M, Dwyer D, Sommerfeld-Stur I, Pendl G, Resch H, Dall'Ara E, Varga P, Pietschmann P (2011b) OPG-Fc treatment in growing pigs leads to rapid reductions in bone resorption markers, serum calcium, and bone formation markers. Horm Metab Res 43:944–949

Sogaard CH, Danielsen CC, Thorling EB, Mosekilde L (1994) Long-term exercise of young and adult female rats: effect on femoral neck biomechanical competence and bone structure. J Bone Miner Res 9:409–416

Stolina M, Bolon B, Dwyer D, Middleton S, Duryea D, Kostenuik PJ, Feige U, Zack DJ (2008) The evolving systemic and local biomarker milieu at different stages of disease progression in rat collagen-induced arthritis. Biomarkers 13:692–712

Stolina M, Schett G, Dwyer D, Vonderfecht S, Middleton S, Duryea D, Pacheco E, Van G, Bolon B, Feige U, Zack D, Kostenuik P (2009a) RANKL inhibition by osteoprotegerin prevents bone loss without affecting local or systemic inflammation parameters in two rat arthritis models: comparison with anti-TNFalpha or anti-IL-1 therapies. Arthritis Res Ther 11:R187

Stolina M, Bolon B, Middleton S, Dwyer D, Brown H, Duryea D, Zhu L, Rohner A, Pretorius J, Kostenuik P, Feige U, Zack D (2009b) The evolving systemic and local biomarker milieu at different stages of disease progression in rat adjuvant-induced arthritis. J Clin Immunol 29:158–174

Talbott SM, Rothkopf MM, Shapses SA (1998) Dietary restriction of energy and calcium alters bone turnover and density in younger and older female rats. J Nutr 128:640–645

Thomas ML, Ibarra MJ, Solcher B, Wetzel S, Simmons DJ (1988) The effect of low dietary calcium and calcium supplements on calcium metabolism and bone in the immature growing rat. Bone Miner 4:73–82

Turner AS (2002) The sheep as a model for osteoporosis in humans. Vet J 163:232–239

Turner RT, Vandersteenhoven JJ, Bell NH (1987) The effects of ovariectomy and 17 beta estradiol on cortical bone histomorphometry in growing rats. J Bone Miner Res 2:115–122

Turner AS, Park RD, Aberman HM, Villanueva AR, Nett TM, Trotter GW, Eckhoff DG (1993) Effects of age and ovariectomy on trabecular bone of the proximal femur and iliac crest in sheep. Trans Orthop Res Soc 18:548

Turner AS, Mallinckrodt CH, Alvis MR, Bryant HU (1995) Dose-response effects of estradiol implants on bone mineral density in ovariectomized ewes. Bone 17:421–427

Tyagi AM, Mansoori MN, Srivastava K, Khan MP, Kureel J, Dixit M, Shukla P, Trivedi R, Chattopadhyay N, Singh D (2014) Enhanced immunoprotective effects by anti-IL-17 antibody translates to improved skeletal parameters under estrogen deficiency compared with anti-RANKL and anti-TNF-α antibodies. J Bone Miner Res 29:1981–1992

Wang L, Banu J, McMahan CA, Kalu DN (2001) Male rodent model of age-related bone loss in men. Bone 29:141–148

Weinstein RS, Jilka RL, Parfitt AM, Manolagas SC (1998) Inhibition of osteoblastogenesis and promotion of apoptosis of osteoblasts and osteocytes by glucocorticoids. J Clin Invest 102:274–282

Westerlind KC, Wronski TJ, Ritman EL, Luo Z-P, An K-N, Bell NH, Turner RT (1997) Estrogen regulates the rate of bone turnover but bone balance in ovariectomized rats is modulated by prevailing mechanical strain. Proc Natl Acad Sci U S A 94:4199–4204

Wronski TJ, Yen CF (1992) The ovariectomized rat as an animal model for postmenopausal bone loss. Cells Mater (Suppl) 1:69–74

Wronski TJ, Walsh CC, Ignaszewski LA (1986) Histologic evidence for osteopenia and increased bone turnover in ovariectomized rats. Bone 7:119–124

Wronski TJ, Cintron M, Dann LM (1988) Temporal relationship between bone loss and increased bone turnover in ovariectomized rats. Calcif Tissue Int 42:179–183

Wronski TJ, Dann LM, Scott KS, Cintron M (1989) Long-term effects of ovariectomy and aging on the rat skeleton. Calcif Tissue Int 45:360–366

Wronski TJ, Dann LM, Horner SL (1990) Time course of vertebral osteopenia in ovariectomized rats. Calcif Tissue Int 46:101–110

Yoneda T, Michigami T, Yi B, Williams PJ, Niewolna M, Hiraga T (2000) Actions of bisphosphonate on bone metastasis in animal models of breast carcinoma. Cancer 88:2979–2988

Yoshitake K, Yokota K, Kasugai Y, Kagawa M, Sukamoto T, Nakamura T (1999) Effects of 16 weeks of treatment with tibolone on bone mass and bone mechanical and histomorphometric indices in mature ovariectomized rats with established osteopenia on a low-calcium diet. Bone 25:311–319

Yun TJ, Tallquist MD, Aicher A, Rafferty KL, Marshall AJ, Moon JJ, Ewings ME, Mohaupt M, Herring SW, Clarck EA (2001) Osteoprotegerin, a crucial regulator of bone metabolism, also regulates B cell development and function. J Immunol 166:1482–1491

Zheng Y, Zhou H, Brennan K, Blair JM, Modzelewski JRK, Seibel MJ, Dunstan CR (2007) Inhibition of bone resorption, rather than direct cytotoxicity, mediates the anti-tumour actions of ibandronate and osteoprotegerin in a murine model of breast cancer bone metastasis. Bone 40:471–478

Antibodies for the Treatment of Bone Diseases: Clinical Data

10

Maria Winzer, Martina Rauner, and Lorenz C. Hofbauer

10.1 Introduction

Poor bone quality with localized or generalized reduced bone strength is associated with an increased incidence of fractures, high health care costs, and increased mortality. An increased need of analgetics, painful joint dysfunction, or insufficient fracture healing is associated with a wide spectrum of secondary disease conditions such as chronic inflammatory or malignant bone diseases (Svedbom et al. 2013). Rapidly expanding knowledge of the molecular changes underlying metabolic bone diseases has revealed novel bone targets that are currently being explored in clinical trials.

The approval of denosumab, a monoclonal antibody directed against receptor activator of nuclear factor kappa-B ligand (RANKL), was just the start for a new era for bone therapy. Among anabolic agents, monoclonal antibodies against the Wnt inhibitors sclerostin (SOST) and dickkopf-1 (DKK-1) are promising in clinical trials.

10.2 Receptor Activator of NF-kB Ligand (RANKL)

Bone turnover is based on osteoclastic bone resorption and osteoblastic bone formation. Due to these processes, a constant rate of bone renewal while preserving bone mass, quality, and integrity is guaranteed. The healthy bone continuously adapts

M. Winzer, MD
Division of Endocrinology, Diabetes and Bone Diseases,
Medical Faculty of the Technische Universität Dresden, Dresden, Germany

Division of Rheumatology, Department of Medicine III,
Medical Faculty of the Technische Universität Dresden, Dresden, Germany

M. Rauner, PhD • L.C. Hofbauer, MD (✉)
Division of Endocrinology, Diabetes and Bone Diseases,
Medical Faculty of the Technische Universität Dresden, Dresden, Germany
e-mail: lorenz.hofbauer@uniklinikum-dresden.de

P. Pietschmann (ed.), *Principles of Osteoimmunology*,
DOI 10.1007/978-3-319-34238-2_10

to changing environmental requirements. In the osteoclast–osteoblast–osteocyte interaction, the receptor activator of NF-kB ligand (RANKL) acts as pacemaker of osteoclast activity. By binding to its receptor RANK, RANKL activates the differentiation, proliferation, and activity of osteoclasts and thus determines the level of osteoclastic bone resorption (Hofbauer and Schoppet 2004). This potent signaling pathway is mitigated by the dimeric glycoprotein osteoprotegerin (OPG). OPG can act as a decoy receptor by binding and neutralizing RANKL (Simonet et al. 1997). Hence, the ratio of RANKL to OPG is the essential determinant at the level of bone turnover. In diseases characterized by excessive osteoclast activity such as osteoporosis, rheumatoid arthritis, and osteolytic bone metastases, bone loss is generally associated with an enhanced RANKL to OPG ratio (Hofbauer and Schoppet 2004). The role of RANKL and OPG as essential mediators of bone disease has since been established in most forms of primary and secondary osteoporosis (Hofbauer and Schoppet 2004). In general, RANKL is upregulated and OPG is suppressed in the presence of substances that are bone catabolic. In contrast, substances such as estrogens have a protective effect on the bone by suppressing RANKL and enhancing OPG (Eghbali-Fatourechi et al. 2003). Additionally, an overexpression of RANKL was shown in a range of osteolytic malignancies such as myeloma, breast, and prostate cancer (Giuliani et al. 2001a, b). Furthermore, RANKL seems to directly attract cancer cells, promoting their migration to the bone (Jones et al. 2006). Even in progestin-driven carcinogenesis, RANKL is involved (Schramek et al. 2010).

The idea of blocking RANKL was first put into practice by developing forms of recombinant OPG. Although animal studies suggested that inhibition of RANK using an OPG-Fc fusion protein is capable of preventing the formation of bone metastases, OPG-Fc was abandoned due to several concerns such as a relatively short half-life, the formation of autoantibodies against OPG-Fc, and the potential of OPG to bind and block tumor necrosis factor (TNF)-related apoptosis-inducing ligand (TRAIL) (Emery et al. 1998). TRAIL is a member of the TNF superfamily that selectively induces apoptosis in malignant cells. As a consequence, a fully human monoclonal antibody directed against RANKL, denosumab, was engineered.

10.3 Clinical Trials on Denosumab

In a phase 1 trial, 49 women received a single subcutaneous injection of either AMG 162 (0.01, 0.03, 0.1, 0.3, 1.0, and 3.0 mg/kg) or placebo. Subjects were followed up to 6 months in all cohorts and 9 months in the three highest-dose cohorts. The bone turnover markers, urinary N-terminal telopeptide (NTX), serum NTX, and serum bone-specific alkaline phosphatase (BALP) were measured. At 6 months, there was a mean change from baseline of −81 % in the 3.0 mg/kg AMG 162 group compared with −10 % in the placebo group. Serum NTX changes were −56 % and 2 %, respectively. BALP levels did not decrease remarkably until after 1 month. Intact parathyroid hormone (PTH) levels increased up to approximately threefold after 4 days in the 3.0 mg/kg dose group but returned toward baseline with follow-up. Shortly, injection of denosumab led to a dose-dependent, rapid, and persistent suppression of urinary N-telopeptide (uNTx), a marker of bone resorption (Bekker et al. 2004).

A total of seven doses and two-dosing schedules of denosumab (application every 3 months and every 6 months) were evaluated at a 1-year primary end point in a phase 2 dose-ranging study. Postmenopausal women with low bone mineral density (BMD) (T score from −1.8 to −4.0 at the lumbar spine or −1.8 to −3.5 at the proximal femur) were randomly assigned to receive denosumab either every 3 months or biannually and compared to open-label oral alendronate weekly or placebo.

Denosumab treatment for 12 months resulted in an increase in bone mineral density at the lumbar spine of 3.0–6.7 % (as compared with an increase of 4.6 % with alendronate and a loss of 0.8 % with placebo), at the total hip of 1.9–3.6 % (as compared with an increase of 2.1 % with alendronate and a loss of 0.6 % with placebo), and at the distal third of the radius of 0.4–1.3 % (as compared with decreases of 0.5 % with alendronate and 2.0 % with placebo). Near-maximal reductions in mean levels of serum C-telopeptide from baseline were evident 3 days after the administration of denosumab. The duration of the suppression of bone turnover appeared to be dose dependent. In postmenopausal women with low bone mass, denosumab increased bone mineral density and decreased bone resorption (McClung et al. 2006).

10.4 Denosumab in Female Osteoporosis and Women with Low Bone Mass

Three-Year Fracture Risk Reduction with Denosumab in Postmenopausal Women with Osteoporosis (The FREEDOM Trial) The aim of osteoporosis treatment is to reduce fractures. Spinal fractures are characteristic of osteoporosis and indicate disease progression and severity. Fractures at peripheral skeletal sites account for most of the patient disability and costs are associated with osteoporosis-related fractures (Budhia et al. 2012). In the crucial efficacy trial known as the FREEDOM trial, the efficacy of denosumab was evaluated in the prevention of both types of fractures compared with placebo over a 3-year period. 7868 subjects were grouped to receive either denosumab 60 mg or placebo subcutaneously biannually. The primary objective was to determine the reduction in the number of new vertebral fractures at 3 years and the secondary objectives assessed times to first nonvertebral and first hip fractures. Subjects with prior treatment with oral bisphosphonates could be included if bisphosphonate therapy was discontinued for 1 year and oral bisphosphonate therapy was received for less than 3 years. The prevalence of vertebral fractures was approximately 24 % for both groups at baseline. In both groups, the average age of subjects was 72.3 years and the average bone mineral density T-scores were −2.8 at the lumbar spine, −1.9 for the total hip, and −2.2 at the femoral neck. Denosumab treatment significantly reduced the risk of new vertebral (−68 %), hip (−40 %), and nonvertebral fractures (−20 %) vs placebo.

There was no significant difference in the overall incidence of adverse events (AEs) between denosumab and placebo-treated subjects, and occurrences of serious AEs or discontinuation of study treatment due to AEs were similar as well. Eighty-two percent of the subjects completed the study and 76 % received all the injections. No significant differences were observed in the overall incidence of malignancy, cardiovascular events, and infections. Eczema occurred in 0.3 % of the denosumab group vs 0.1 % in the placebo group. Other significant differences for denosumab vs placebo were falls not associated with a fracture (4.5 % vs 5.7 %), flatulence (2.2 % vs 1.4 %), and concussion (<0.1 % vs 0.3 %). No subject developed neutralizing antibodies to denosumab and there were no reports of osteonecrosis of the jaw in this study (Cummings et al. 2009).

Denosumab vs Alendronate (The DECIDE Trial) A head-to-head comparison in a randomized, double-blind, international, non-inferiority study with postmenopausal subjects with low BMD was undertaken. All subjects were randomized to receive either denosumab 60 mg subcutaneously every 6 months or oral alendronate 70 mg per week, each with a matching placebo. The primary end point was the percent change from baseline in total hip BMD at 12 months. After 6 months and after 12 months, denosumab significantly increased BMD at the total hip (3.5 %) compared with alendronate (2.6 %), as well as all other measured sites. Denosumab treatment resulted in greater BMD gains at 12 months compared to alendronate at all measured sites.

There was no significant difference in the overall incidence of AEs between denosumab and alendronate-treated subjects. The most common AEs included infections, gastrointestinal disorders, and arthralgia. Serious AEs, including neoplasms and serious infections, were not significantly different among the groups. At month 1, the denosumab group experienced a 2.36 % decrease in the mean albumin-adjusted serum calcium concentration compared with a 0.88 % decrease in the alendronate group. At months 6 and 12, these values were similar for each group and there were no reports of symptomatic hypocalcemia. This study was not powered to compare fracture rates, and fractures were reported as adverse events. Overall, similar numbers of subjects in each treatment group reported >1 on-study fracture (Brown et al. 2009).

Denosumab vs Alendronate (The STAND Trial) In this study postmenopausal women had been receiving alendronate for at least 6 months and were then switched to denosumab or maintained on alendronate. This study reflected a situation that may be more relevant to clinical practice, when patients may have to change therapies. Subjects having switched to denosumab experienced a significant increase in BMD at the total hip and lumbar spine compared with patients continuing on alendronate therapy after 12 months. Significantly greater gains were also observed in BMD for denosumab compared with alendronate at the femoral neck and 1/3 radius. Denosumab treatment resulted in significantly greater increases in BMD, as early as month 6, for the lumbar spine and all measured femoral sites. Additionally, median serum C-telopeptide levels in the denosumab group were significantly decreased vs alendronate at all time points.

There was no significant difference in the overall incidence of (serious) AEs between subjects (Kendler et al. 2010; Bone et al. 2011).

In the DECIDE and the STAND trials, denosumab was found to be superior to alendronate, as demonstrated by the increase in BMD at the total hip at 12 months, which exceeded the primary end point of non-inferiority. Denosumab significantly increased BMD at all measured skeletal sites, including sites that were predominantly cortical, in contrast to alendronate, and significantly reduced bone resorption as measured by levels of serum C-telopeptide. Denosumab increased BMD beyond the steady-state level achieved with long-term alendronate use and at skeletal sites where bisphosphonates showed a weak effect. This effect may be due to its different mechanisms of action to reduce bone resorption. Osteoclasts in the cortical bone may be more sensitive to RANKL inhibition than they are to local deposits of bisphosphonates (Lacey et al. 2012).

Denosumab in Postmenopausal Women with Low Bone Mass (The DEFEND Trial) To assess the effect of denosumab on BMD, 332 postmenopausal women with low bone mass (BMD T-scores between −1.0 and −2.5) were recruited. Subjects received either denosumab 60 mg subcutaneously every 6 months or placebo for 2 years. The primary end point was percent change from baseline in lumbar spine BMD at 24 months compared with placebo. Significant increases in lumbar spine BMD were observed in denosumab subjects vs placebo at 1 month and continued through month 24. The denosumab group exhibited significant increases in cortical BMD, bone mineral content (BMC), and thickness and a significant increase in trabecular BMD compared to placebo (Bone et al. 2008; Genant et al. 2010).

Denosumab and Long-Term Use in Postmenopausal Osteoporosis An open-label, single-arm extension of the 3-year, pivotal phase 3 FREEDOM study is ongoing to further characterize the long-term safety and efficacy of denosumab. Overall, 4550 eligible subjects are signed up in the extension study, 2343 in the long-term group (previously on denosumab), and 2207 in the crossover group (previously on placebo). Subjects in the extension period receive subcutaneous denosumab 60 mg every 6 months with daily calcium and vitamin D supplementation for up to 10 years of continued exposure.

In a pre-specified analysis of BMD changes at year 8 of the extension study, denosumab treatment for up to 8 years in the long-term group resulted in further significant increases in lumbar spine and total hip BMD compared with FREEDOM and extension baseline. These gains in BMD observed in the long-term group were accompanied by progressive increases in the percentage of subjects who experienced improvements from the FREEDOM baseline T-score at both the lumbar spine and total hip over 8 years of denosumab treatment.

The yearly incidence of new vertebral fractures in the long-term group at 8 years was 1.2 %. At year 9, the yearly incidence of new nonvertebral fractures in the long-term and crossover groups was 1.1 % and 1.5 %, respectively. The yearly incidence of hip fracture was <0.1 % and 0.1 % at year 9 for the long-term and crossover groups, respectively. Denosumab administration continued to result in significant reductions in serum C-telopeptide.

Exposure-adjusted AE rates are similar to rates observed during the first 3 years of the FREEDOM study and overtime. Osteonecrosis of the jaw and atypical femoral fracture have been observed in the FREEDOM extension study at a low frequency (<0.3 %).

Data will continue to be collected and monitored in the ongoing clinical trial program (Bone et al. 2013; Papapoulos 2013, 2015; Ferrari et al. 2015).

10.5 Teriparatide and Denosumab, Alone or Combined, in Women with Postmenopausal Osteoporosis: The DATA Study Randomized Trial

Postmenopausal women with osteoporosis were assigned in a 1:1:1 ratio to receive 20 μg teriparatide daily, 60 mg denosumab biannually, or both. BMD was measured at 0, 3, 6, and 12 months. At 12 months, posterior-anterior lumbar spine BMD increased more in the combination group (9.1 %) than in the teriparatide (6.2 %) or denosumab (5.5 %) groups. The gain in femoral-neck BMD was higher in the combination group (4.2 %) than in the teriparatide (0.8 %) and denosumab (2.1 %) groups, as did total hip BMD (combination, 4.9 %; teriparatide, 0.7 %; denosumab, 2.5 %). Combined teriparatide and denosumab increased BMD more than either agent alone, and more than has been reported with approved therapies. Combination treatment might, therefore, be useful to treat patients at high risk of fracture (Tsai et al. 2013).

10.6 Denosumab and Teriparatide Transitions in Postmenopausal Osteoporosis (The DATA-Switch Study): Extension of a Randomized Controlled Trial

Until now osteoporosis treatments are generally limited to a single drug at a fixed dose and frequency. Nonetheless, no approved therapy is able to restore skeletal integrity in most osteoporotic patients, and the long-term use of osteoporosis drugs is controversial. Thus, many patients are treated with the sequential use of two or more therapies. The DATA study showed that combined teriparatide and denosumab increased BMD more than either drug alone. Discontinuing teriparatide and denosumab, however, resulted in rapidly declining BMD. In the DATA-Switch study, changes in BMD in postmenopausal osteoporotic women who transitioned between treatments were assessed. Women who originally assigned to teriparatide received denosumab (teriparatide to denosumab group), those originally assigned to denosumab received teriparatide (denosumab to teriparatide group), and those originally assigned to both received an additional 24 months of denosumab alone (combination to denosumab group). BMD at the spine, hip, and wrist were measured 6, 12, 18, and 24 months after the drug transitions as were biochemical markers of bone turnover. The primary end point was the percent change in posterior-anterior spine BMD over 4 years.

Eighty-three postmenopausal women were enrolled in the extension study and switched from denosumab to teriparatide ($n = 27$), from teriparatide to denosumab ($n = 27$), or from combined teriparatide plus denosumab to continued denosumab for another 24 months ($n = 23$). Interestingly, switching from teriparatide to denosumab resulted in a further increase of BMD at all measured sites at 48 months (e.g., the study's primary end point, mean lumbar spine BMD, increased by 18.3 % over 48 months and by 8.6 % over 24–48 months). Similarly, switching from combination therapy to denosumab was associated with an increase in mean lumbar spine BMD of 16.0 % over 48 months and 3.4 % over 24–48 months. In contrast, the switch from denosumab to teriparatide led to transient bone loss at the spine and hip and progressive bone loss at the distal radius (i.e., a decrease of 5.0 % over 24–48 months). In postmenopausal osteoporotic women switching from teriparatide to denosumab, BMD continued to increase, whereas switching from denosumab to teriparatide results in progressive or transient bone loss. This information becomes valuable when choosing the initial and subsequent management of postmenopausal osteoporotic patients (Leder et al. 2015).

10.7 Denosumab in Male Osteoporosis

The efficacy and safety of denosumab for the treatment of male osteoporosis was evaluated in a multicenter, randomized, double-blind, placebo-controlled, 24-month, phase 3 study that enrolled 242 male subjects between 30 and 85 years of age with a T-score <–2.0 and <3.5 at the lumbar spine or femoral neck or subjects that had a prior major osteoporotic fracture and a T-score <–1.0 and <3.5 at the lumbar spine or femoral neck. Denosumab significantly increased BMD at all skeletal sites measured at 12 months in men with low BMD. These effects were independent of gonadal function level, baseline BMD status, age, estimated fracture risk, and baseline bone turnover level. Overall rates of AEs and serious AEs observed in both treatment groups were similar. Biopsies from a small subset of subjects indicated similar bone quality and reduced turnover in denosumab compared to placebo-treated subjects (Orwoll et al. 2012; Langdahl et al. 2015; Dempster et al. 2013).

10.8 Denosumab in Malignant Diseases

Denosumab Clinical Studies in Patients Receiving Aromatase Inhibitors for Nonmetastatic Breast Cancer Two randomized, double-blind, placebo-controlled phase 3 studies analyzed the effect of denosumab treatment in patients with early-stage nonmetastatic breast cancer receiving adjuvant aromatase inhibitor therapy.

The first study showed increases in BMD at the lumbar spine and all measured areas of the trabecular and cortical bone over the 24-month duration of the study in patients treated with denosumab. In the second study, adjuvant denosumab therapy could prolong the time to first clinical fracture compared to placebo in postmenopausal patients receiving aromatase inhibitor therapy. Increases in BMD in the first study and reduction in fractures in the second study were observed across patient subgroups studied. The most common AEs considered by the investigators to be possibly or probably treatment related were arthralgia, pain in extremity, bone pain, fatigue, and pain (Gnant et al. 2015; Ellis et al. 2009).

10.9 Denosumab Clinical Studies in Patients with Breast Cancer and Bone Metastases

Bone metastases are common in patients with solid tumors, particularly in breast, prostate, and lung cancer (Coleman 2001). Tumor cells that metastasize to the bone cause local bone destruction through a vicious circle involving the secretion of factors that result in RANKL-induced stimulation of osteoclast-mediated bone resorption, which in turn promotes proliferation of metastases (Roodman 2004). The clinical consequences of bone metastases include fractures, spinal cord compression, and pain that often require radiation or surgery to the bone. These debilitating complications in patients with advanced cancer are collectively called skeletal-related events (SREs).

Intravenously administered bisphosphonates are effective in SRE prevention; however, a substantial proportion of patients continue to experience SREs while on bisphosphonate treatment and there are several safety and tolerability concerns associated with these agents (Stopeck et al. 2010; Barrett-Lee et al. 2007). Zoledronic acid is contraindicated in patients with severe renal impairment. Furthermore, some patients treated with zoledronic acid experience acute-phase reactions (flu-like symptoms) that further complicate the clinical picture. The requirement for intravenous access can also be a limiting factor (Lacey et al. 2012; Barrett-Lee et al. 2007). Thus, the efficacy of denosumab was tested in patients with breast cancer and bone metastases.

Treatment with denosumab in patients with breast cancer and bone metastases was investigated in one phase 3 study (Stopeck et al. 2010) and two phase 2 studies (Lipton et al. 2007; Fizazi et al. 2009). The phase 3 study was a randomized, double-blind, double-dummy trial. Patients with breast cancer and bone metastases were randomized to get treatment either with denosumab 120 mg subcutaneously and placebo intravenously every 4 weeks or zoledronic acid 4 mg intravenously and placebo subcutaneously. The primary objective was to evaluate whether denosumab was non-inferior to zoledronic acid with respect to time to first on-study SRE, defined as pathological fracture, radiation to the bone, surgery to the bone, or spinal cord compression. Secondary objectives were to evaluate whether denosumab was superior to zoledronic acid with respect to time to the first on-study SRE, as well as time to first-and-subsequent on-study SRE. Additional objectives included the safety of denosumab compared to zoledronic acid.

Denosumab was superior to zoledronic acid in delaying time to first on-study SRE and time to first-and-subsequent on-study SRE in patients with breast cancer and bone metastases. Overall survival and time to disease progression were balanced between treatment arms (Stopeck et al. 2010; Lipton et al. 2012).

10.9.1 Patients Naïve to Intravenous Bisphosphonate Treatment

A randomized, active-controlled phase 2 study assessed the safety and efficacy of denosumab given at different doses and schedules in IV bisphosphonate-naïve patients with breast cancer and bone metastases. Patients were randomized to receive either blinded subcutaneous denosumab administered every 4 or every 12 weeks or open-label intravenous bisphosphonates. The primary end point was the median percent change in urinary N-telopeptide corrected for creatinine (Cr) from baseline to week 13. Additionally end points included the percent change from baseline to week 25 in uNTx/Cr, the percent of patients with at least one on-study SRE, and the incidence of AE (Lipton et al. 2008).

10.9.2 Patients with Prior Intravenous Bisphosphonate Treatment

A randomized, open-label, active-controlled phase 2 study evaluated the safety and efficacy of denosumab in patients with bone metastases and uNTx levels >50 nmol/L/mM Cr despite >8 weeks of treatment with intravenous bisphosphonates. 111 patients with breast cancer, prostate cancer, multiple myeloma (MM), or other solid malignancies were randomized to receive denosumab or to continue receiving intravenous bisphosphonates or to continue receiving intravenous bisphosphonates for 25 weeks. The primary end point was the percentage of patients with uNTx <50 nmol/L/mM Cr at week 13 (Stopeck et al. 2010).

In both studies, denosumab reduced uNTx levels in patients who were naïve to intravenous bisphosphonate therapy and in patients with elevated uNTx levels despite intravenous bisphosphonate therapy. Adverse events included bone pain, nausea, anemia, vomiting, and diarrhea. One serious AE of hypophosphatemia was considered by the investigators to be possibly related to denosumab. No neutralizing antibodies to denosumab were detected and no cases of osteonecrosis of the jaw were reported (Fizazi et al. 2009; Lipton et al. 2008).

10.10 Denosumab Clinical Studies in Patients with Prostate Cancer and Bone Metastases

One phase 3 and one phase 2 study investigated the effect of denosumab in patients with prostate cancer and bone metastases. Denosumab was better in delaying time to first on-study SRE and symptomatic skeletal events (SSE) as well as time to first-and-subsequent SREs and SSEs compared to zoledronic acid in the phase 3 study. Overall survival and time to disease progression were balanced between treatment arms. Hypocalcemia was reported in 13 % and 6 % of patients in the denosumab and zoledronic acid arms, respectively. No significant difference in the incidence of osteonecrosis of the jaw was reported between treatment arms (Fizazi et al. 2011; Smith et al. 2012).

A randomized, open-label, active-controlled phase 2 study evaluated the safety and efficacy of denosumab in patients with bone metastases and uNTx levels >50 nmol/L/mM creatinine (Cr) despite >8 weeks of ongoing treatment with intravenous bisphosphonates at the time of enrollment. Patients with prostate cancer, breast cancer, multiple myeloma, or other solid malignancies were randomized to receive denosumab or to continue receiving intravenous bisphosphonates. Among patients with elevated uNTx despite ongoing intravenous bisphosphonate therapy, denosumab normalized uNTx levels more frequently than the continuation of intravenous bisphosphonate therapy. Fewer patients receiving denosumab experienced on-study SREs than those receiving intravenous bisphosphonate (Fizazi et al. 2009).

10.11 Phase 3 Study of Denosumab vs Zoledronic Acid in the Treatment of Bone Metastases in Patients with Solid Tumors or Multiple Myeloma

This study compared denosumab with zoledronic acid for delaying or preventing SRE. 1776 patients with bone metastases from advanced cancer (excluding the prostate and breast) or multiple myeloma (MM) were randomized 1:1 to receive either denosumab 120 mg subcutaneously and placebo intravenously every 4 weeks or zoledronic acid 4 mg intravenously and placebo subcutaneously every 4 weeks. Patients were randomly assigned in a double-blind, double-dummy design to receive monthly subcutaneous denosumab 120 mg ($n=886$) or intravenous zoledronic acid 4 mg (dose adjusted for renal impairment; $n=890$). Daily supplemental calcium and vitamin D were strongly recommended. The primary end point was time to first on-study SRE (pathologic fracture, radiation or surgery to the bone, or spinal cord compression). Denosumab was non-inferior to zoledronic acid in delaying time to first on-study SRE. Although directionally favorable, denosumab was not statistically superior to zoledronic acid in delaying time to first on-study SRE or time to first-and-subsequent (multiple) SRE. Overall survival and disease progression were similar between groups. Hypocalcemia occurred more frequently with denosumab. Osteonecrosis of the jaw occurred at similarly low rates in both groups. Acute-phase reactions after the first dose occurred more frequently with zoledronic acid, as did renal adverse events and elevations in serum creatinine based on National Cancer Institute Common Toxicity Criteria for Adverse Events grading. Denosumab was non-inferior (trending to superiority) to zoledronic acid in preventing or delaying first on-study SRE in patients with advanced cancer metastatic to the bone or myeloma. Denosumab represents a potential novel treatment option with the convenience of subcutaneous administration and no requirement for renal monitoring or dose adjustment (Henry et al. 2011).

10.12 Denosumab Use in Patients with Giant Cell Tumor of the Bone

Data-evaluating denosumab in patients with giant cell tumor of the bone (GCTB) is available from two phase 2 studies; an open-label, proof-of-concept phase 2 study evaluated the safety and efficacy of denosumab in patients >18 years with recurrent or unresectable GCTB, and an open-label, phase 2 study evaluated the safety and efficacy of denosumab in adults or skeletally mature adolescents with primary or recurrent GCTB who had surgically unsalvageable disease as determined by the treating surgeon (e.g., sacral or spinal GCTB or multiple GCTB lesions including pulmonary metastases, cohort 1); or a planned surgery that was associated with severe morbidity (e.g., joint resection, limb amputation, or hemipelvectomy, cohort 2); or who transitioned from a previous denosumab GCTB study (cohort 3) and continued denosumab treatment on this study. Denosumab was associated with tumor responses, reduced the need for morbid surgery, and clinically relevant decreases in pain in patients with GCTB. The safety profile of denosumab in patients with GCTB was consistent with the known safety profile of denosumab (Thomas et al. 2010a, b; Chawla et al. 2013; Branstetter et al. 2012).

In the European Union *Prolia* (denosumab), 60 mg has received marketing authorization for the treatment of osteoporosis in postmenopausal women and men at high fracture risk and for the treatment of bone loss due to androgen-deprivation therapy in patients with prostate cancer at high fracture risk.

In the European Union *XGEVA* (denosumab), 120 mg has received marketing authorization for the prevention of skeletal-related events in adults with bone metastases from solid tumors and for the treatment of adults and skeletally mature adolescents with giant cell tumor of bone that is unresectable or where surgical resection is likely to result in severe morbidity.

Besides targeting RANKL as a classical antiresorptive therapy, recent discoveries in the field of bone biology have highlighted novel pathways and targets that may act in a bone-anabolic fashion (see Chap. 1).

Two of them, sclerostin and dickkopf-1 (DKK-1), appear particularly attractive for future therapeutic options for treating osteoporosis. The clinical evaluation of neutralizing antibodies against sclerostin and DKK-1 will be discussed in the following.

10.13 Targeting Sclerostin

Sclerostin is primarily secreted by osteocytes. Mutations of the SOST gene, which encodes for sclerostin, result in progressive pathological bone thickening and clinical conditions named sclerosteosis and Van Buchem's disease (Loots et al. 2005; Balemans et al. 2001). Furthermore, SOST −/− mice have an increased bone formation and high bone mass. Thus, neutralizing monoclonal antibodies directed against sclerostin were generated. These were successfully tested in a rodent model of postmenopausal osteoporosis, where the application of the sclerostin antibody for a period of 5 weeks resulted in pronounced bone-anabolic effects (Li et al. 2009).

Similar positive effects were reported in a study with cynomolgus monkeys (Ominsky et al. 2010). Sclerostin antibodies were first tested in 72 healthy adults in a single-dose phase 1 study and were applied either via subcutaneous or intravenous injections in ascending doses (Padhi et al. 2011). After 85 days a subcutaneous dose of 10 mg/kg had increased the BMD at the lumbar spine (5.3 %) and total hip (2.8 %). An intravenous dose of 5 mg/kg increased the lumbar spine BMD by 5.2 % and BMD at the total hip by 1.1 %. In both treatment groups, sclerostin antibodies dose-dependently increased the bone formation markers total procollagen type 1 N-terminal propeptide (P1NP), BALP, and osteocalcin and were generally well tolerated. One patient who received the highest subcutaneous dosing (10 mg/kg) reported a severe hepatitis 1 day after injection. Of note, six of the 54 patients developed antibodies, two of which were neutralizing.

Romosozumab and Postmenopausal Osteoporosis Romosozumab is an investigational, humanized, monoclonal antibody (IgG2) that binds with high affinity and targets the activity of sclerostin, an inhibitor of osteoblast-mediated bone formation. Romosozumab is currently being investigated for its ability to inhibit binding of sclerostin and stimulate osteoblast-mediated bone formation. Clinical studies are underway to evaluate the safety and efficacy of romosozumab in the treatment of osteoporosis and other diseases associated with bone loss (Padhi et al. 2011, 2014; McClung et al. 2014).

The efficacy, safety, and tolerability of romosozumab for the treatment of postmenopausal women with low BMD was investigated in a randomized, dose-ranging study of 419 women aged 55–85 years with a lumbar spine, total hip, or femoral neck T-score <–2.0 and <3.5. During the first 12 months, women were randomized to one of five regimens of subcutaneous romosozumab or placebo and one of two open-label active comparators (oral alendronate or 20 μg daily subcutaneously teriparatide). The primary end point was percent change from baseline in lumbar spine BMD at month 12 (Padhi et al. 2014; McClung et al. 2014).

All doses of romosozumab significantly increased BMD compared with placebo at each of the bone sites at month 12. In all romosozumab dosing groups, increases in serum levels of the bone formation marker P1NP were observed 1 week after the initial dose was administered and greatest at 1 month. Increases in P1NP were transitory, returning to baseline levels or below between months 2 and 6, depending on the dose. All doses of romosozumab reduced serum levels of the bone resorption marker C-terminal telopeptide (CTX) from baseline by week 1 and, in groups that received monthly doses, remained below baseline levels at month 12. Serum levels of P1NP and CTX were decreased following alendronate treatment and increased following teriparatide treatment.

The proportions of subjects reporting AEs and serious AEs were similar between placebo and total romosozumab groups. No apparent association between dose and AEs was observed. Injection site reactions were reported more frequently in subjects receiving romosozumab (12 %) compared to placebo (4 %) but were generally mild, not dose related, and did not lead to study drug discontinuation or study withdrawal. Adverse events observed in >5 % of subjects treated with romosozumab and occurring more frequently than in placebo-treated groups included nasopharyngitis, arthralgia, pain in the extremity, back pain, gastroenteritis, headache, pain binding antibodies occurred in 20 % of subjects treated with romosozumab. Using a cell-based in vitro biological assay, neutralizing antibodies were detected in 3 % of subjects treated with romosozumab. There were no discernible effects of these binding antibodies on pharmacokinetics, pharmacodynamics, or safety.

After 12 months, subjects in the romosozumab and placebo groups continued their assigned treatments for another 12 months. Romosozumab treatment led to continued BMD increases through 24 months. In romosozumab subjects, both P1NP and CTX levels remained below baseline during year two. AEs continued to be balanced between romosozumab and placebo groups, with the exception of mild injection site reactions. Thus, these studies demonstrate the high efficacy of blocking sclerostin and underline its potential as a novel bone-anabolic therapeutic approach.

10.14 Blosozumab and Postmenopausal Osteoporosis

Blosozumab, a humanized IgG4 monoclonal antibody against sclerostin, was investigated in two clinical phase 1 studies. Healthy postmenopausal women received blosozumab in escalating doses and showed no relevant signs of intolerance or side effects. Dose-dependent responses were observed in sclerostin, P1NP, bone-specific alkaline phosphatase, osteocalcin, CTX, and BMD (McColm et al. 2014). Built up to these promising results, a randomized, double-blind, placebo-controlled multicenter phase 2 clinical trial of blosozumab was conducted with postmenopausal women with low BMD.

Blosozumab treatment caused statistically significant dose-related increases in spine, femoral neck, and total hip BMD as compared with placebo. In the highest dose group, BMD increases from baseline reached 17.7 % at the spine and 6.2 % at the total hip (Recker et al. 2015).

To evaluate the effect of discontinuing blosozumab, women enrolled in this 1-year randomized, placebo controlled phase 2 trial were monitored for an additional year after they completed treatment. At the end of follow-up which was 1 year after discontinuing treatment, lumbar spine and total hip BMD remained significantly greater than placebo in women initially treated with blosozumab (Recknor et al. 2015). These results encourage future studies of blosozumab as an anabolic therapy for treatment of osteoporosis.

10.15 Targeting Dickkopf-1 (BHQ880)

Elevated DKK-1 serum levels have been associated with the presence of bone lesions in patients with multiple myeloma, lung cancer, and breast cancer, and DKK-1 inhibition has been shown to successfully prevent the formation of bone lesions in preclinical models of metastases (Fulciniti et al. 2009).

BHQ880, a human neutralizing IgG1 anti-DKK1 monoclonal antibody, is being investigated for its impact on multiple myeloma-related bone disease and as an agent with potential anti-myeloma activity. A phase 1 study conducted by Iyer et al. showed no dose-limiting toxicity, and a general trend toward increased BMD was observed (Iyer et al. 2014).

Results of phase 2 clinical trials evaluating BHQ880 in patients with multiple myeloma are pending.

The wide-tissue expression of DKK-1 presumably impedes the use of DKK-1 antibodies in osteoporosis.

10.16 Conclusion and Future Prospects

With the development of denosumab, antibody-based therapies have now also finally made it into the therapeutic options for the treatment of bone diseases such as osteoporosis. Its implementation has been most successful, reducing the risk of fractures significantly while at the same time, having a very limited amount of adverse effects. Further, the treatment mode is more convenient than current treatment forms, which will likely improve the adherence, which has traditionally been a major drawback of oral bisphosphonates and teriparatide. Finally, with novel bone-anabolic drugs on the horizon, new treatment concepts will be made available, which will not only be able to halt bone destruction but even add new bone. Thus, such therapies will be extremely useful for the treatment of severe forms of osteoporosis where alleviating bone destruction is not sufficient to prevent fractures.

Taken together, these examples of antibody-based therapies with specific molecular targets demonstrate clearly how basic research findings can translate into potent therapeutics that quickly become an integral part of modern treatment of osteoporosis and other bone-related disorders.

References

Balemans W et al (2001) Increased bone density in sclerosteosis is due to the deficiency of a novel secreted protein (SOST). Hum Mol Genet 10:537–543

Barrett-Lee P et al (2007) An audit to determine the time taken to administer intravenous bisphosphonate infusions in patients diagnosed with metastatic breast cancer to bone in a hospital setting. Curr Med Res Opin 23:1575–1582

Bekker PJ et al (2004) A single-dose placebo-controlled study of AMG 162, a fully human monoclonal antibody to RANKL, in postmenopausal women. J Bone Miner Res 19:1059–1066

Bone HG et al (2008) Effects of denosumab on bone mineral density and bone turnover in postmenopausal women. J Clin Endocrinol Metab 93:2149–2157

Bone HG et al (2011) Effects of denosumab treatment and discontinuation on bone mineral density and bone turnover markers in postmenopausal women with low bone mass. J Clin Endocrinol Metab 96:972–980

Bone HG et al (2013) The effect of three or six years of denosumab exposure in women with postmenopausal osteoporosis: results from the FREEDOM extension. J Clin Endocrinol Metab 98:4483–4492

Branstetter DG et al (2012) Denosumab induces tumor reduction and bone formation in patients with giant-cell tumor of bone. Clin Cancer Res 18:4415–4424

Brown JP et al (2009) Comparison of the effect of denosumab and alendronate on BMD and biochemical markers of bone turnover in postmenopausal women with low bone mass: a randomized, blinded, phase 3 trial. J Bone Miner Res 24:153–161

Budhia S, Mikyas Y, Tang M, Badamgarav E (2012) Osteoporotic fractures: a systematic review of U.S. healthcare costs and resource utilization. Pharmacoeconomics 30:147–170

Chawla S et al (2013) Safety and efficacy of denosumab for adults and skeletally mature adolescents with giant cell tumour of bone: interim analysis of an open-label, parallel-group, phase 2 study. Lancet Oncol 14:901–908

Coleman RE (2001) Metastatic bone disease: clinical features, pathophysiology and treatment strategies. Cancer Treat Rev 27:165–176

Cummings SR et al (2009) Denosumab for prevention of fractures in postmenopausal women with osteoporosis. N Engl J Med 361:756–765

Dempster DW et al (2013) Standardized nomenclature, symbols, and units for bone histomorphometry: a 2012 update of the report of the ASBMR Histomorphometry Nomenclature Committee. J Bone Miner Res 28:2–17

Eghbali-Fatourechi G et al (2003) Role of RANK ligand in mediating increased bone resorption in early postmenopausal women. J Clin Invest 111:1221–1230

Ellis GK et al (2009) Effect of denosumab on bone mineral density in women receiving adjuvant aromatase inhibitors for non-metastatic breast cancer: subgroup analyses of a phase 3 study. Breast Cancer Res Treat 118:81–87

Emery JG et al (1998) Osteoprotegerin is a receptor for the cytotoxic ligand TRAIL. J Biol Chem 273:14363–14367

Ferrari S et al (2015) Further reductions in nonvertebral fracture rate with long-term denosumab treatment in the FREEDOM open-label extension and influence of hip bone mineral density after 3 years. Osteoporos Int 26:2763–2771

Fizazi K et al (2009) Randomized phase II trial of denosumab in patients with bone metastases from prostate cancer, breast cancer, or other neoplasms after intravenous bisphosphonates. J Clin Oncol 27:1564–1571

Fizazi K et al (2011) Denosumab versus zoledronic acid for treatment of bone metastases in men with castration-resistant prostate cancer: a randomised, double-blind study. Lancet 377:813–822

Fulciniti M et al (2009) Anti-DKK1 mAb (BHQ880) as a potential therapeutic agent for multiple myeloma. Blood 114:371–379

Genant HK et al (2010) Denosumab improves density and strength parameters as measured by QCT of the radius in postmenopausal women with low bone mineral density. Bone 47:131–139

Giuliani N, Bataille R, Mancini C, Lazzaretti M, Barille S (2001a) Myeloma cells induce imbalance in the osteoprotegerin/osteoprotegerin ligand system in the human bone marrow environment. Blood 98:3527–3533

Giuliani C, Napolitano G, Bucci I, Montani V, Monaco F (2001b) Nf-kB transcription factor: role in the pathogenesis of inflammatory, autoimmune, and neoplastic diseases and therapy implications. Clin Ter 152:249–253

Gnant M et al (2015) Adjuvant denosumab in breast cancer (ABCSG-18): a multicentre, randomised, double-blind, placebo-controlled trial. Lancet 386:433–443

Henry DH et al (2011) Randomized, double-blind study of denosumab versus zoledronic acid in the treatment of bone metastases in patients with advanced cancer (excluding breast and prostate cancer) or multiple myeloma. J Clin Oncol 29:1125–1132

Hofbauer LC, Schoppet M (2004) Clinical implications of the osteoprotegerin/RANKL/RANK system for bone and vascular diseases. JAMA 292:490–495

Iyer SP et al (2014) A Phase IB multicentre dose-determination study of BHQ880 in combination with anti-myeloma therapy and zoledronic acid in patients with relapsed or refractory multiple myeloma and prior skeletal-related events. Br J Haematol 167:366–375

Jones DH et al (2006) Regulation of cancer cell migration and bone metastasis by RANKL. Nature 440:692–696

Kendler DL et al (2010) Effects of denosumab on bone mineral density and bone turnover in postmenopausal women transitioning from alendronate therapy. J Bone Miner Res 25:72–81

Lacey DL et al (2012) Bench to bedside: elucidation of the OPG-RANK-RANKL pathway and the development of denosumab. Nat Rev Drug Discov 11:401–419

Langdahl BL et al (2015) A 24-month study evaluating the efficacy and safety of denosumab for the treatment of men with low bone mineral density: results from the ADAMO trial. J Clin Endocrinol Metab 100:1335–1342

Leder BZ et al (2015) Denosumab and teriparatide transitions in postmenopausal osteoporosis (the DATA-Switch study): extension of a randomised controlled trial. Lancet 386:1147–1155

Li X et al (2009) Sclerostin antibody treatment increases bone formation, bone mass, and bone strength in a rat model of postmenopausal osteoporosis. J Bone Miner Res 24:578–588

Lipton A et al (2007) Randomized active-controlled phase II study of denosumab efficacy and safety in patients with breast cancer-related bone metastases. J Clin Oncol 25:4431–4437

Lipton A et al (2008) Extended efficacy and safety of denosumab in breast cancer patients with bone metastases not receiving prior bisphosphonate therapy. Clin Cancer Res 14:6690–6696

Lipton A et al (2012) Superiority of denosumab to zoledronic acid for prevention of skeletal-related events: a combined analysis of 3 pivotal, randomised, phase 3 trials. Eur J Cancer 48:3082–3092

Loots GG et al (2005) Genomic deletion of a long-range bone enhancer misregulates sclerostin in Van Buchem disease. Genome Res 15:928–935

McClung MR et al (2006) Denosumab in postmenopausal women with low bone mineral density. N Engl J Med 354:821–831

McClung MR et al (2014) Romosozumab in postmenopausal women with low bone mineral density. N Engl J Med 370:412–420

McColm J, Hu L, Womack T, Tang CC, Chiang AY (2014) Single- and multiple-dose randomized studies of blosozumab, a monoclonal antibody against sclerostin, in healthy postmenopausal women. J Bone Miner Res 29:935–943

Ominsky MS et al (2010) Two doses of sclerostin antibody in cynomolgus monkeys increases bone formation, bone mineral density, and bone strength. J Bone Miner Res 25:948–959

Orwoll E et al (2012) A randomized, placebo-controlled study of the effects of denosumab for the treatment of men with low bone mineral density. J Clin Endocrinol Metab 97:3161–3169

Padhi D, Jang G, Stouch B, Fang L, Posvar E (2011) Single-dose, placebo-controlled, randomized study of AMG 785, a sclerostin monoclonal antibody. J Bone Miner Res 26:19–26

Padhi D et al (2014) Multiple doses of sclerostin antibody romosozumab in healthy men and postmenopausal women with low bone mass: a randomized, double-blind, placebo-controlled study. J Clin Pharmacol 54:168–178

Papapoulos SE (2013) Bone diseases: bisphosphonates in osteoporosis--beyond 5 years. Nat Rev Rheumatol 9:263–264

Papapoulos S et al (2015) The effect of 8 or 5 years of denosumab treatment in postmenopausal women with osteoporosis: results from the FREEDOM Extension study. Osteoporos Int 26:2773–2783

Recker RR et al (2015) A randomized, double-blind phase 2 clinical trial of blosozumab, a sclerostin antibody, in postmenopausal women with low bone mineral density. J Bone Miner Res 30:216–224

Recknor CP et al (2015) The effect of discontinuing treatment with blosozumab: follow-up results of a phase 2 randomized clinical trial in postmenopausal women with low bone mineral density. J Bone Miner Res 30:1717–1725

Roodman GD (2004) Mechanisms of bone metastasis. N Engl J Med 350:1655–1664

Schramek D et al (2010) Osteoclast differentiation factor RANKL controls development of progestin-driven mammary cancer. Nature 468:98–102

Simonet WS et al (1997) Osteoprotegerin: a novel secreted protein involved in the regulation of bone density. Cell 89:309–319

Smith MR et al (2012) Denosumab and bone-metastasis-free survival in men with castration-resistant prostate cancer: results of a phase 3, randomised, placebo-controlled trial. Lancet 379:39–46

Stopeck AT et al (2010) Denosumab compared with zoledronic acid for the treatment of bone metastases in patients with advanced breast cancer: a randomized, double-blind study. J Clin Oncol 28:5132–5139
Svedbom A et al (2013) Osteoporosis in the European Union: a compendium of country-specific reports. Arch Osteoporos 8:137
Thomas D et al (2010a) Denosumab in patients with giant-cell tumour of bone: an open-label, phase 2 study. Lancet Oncol 11:275–280
Thomas D, Carriere P, Jacobs I (2010b) Safety of denosumab in giant-cell tumour of bone. Lancet Oncol 11:815
Tsai JN et al (2013) Teriparatide and denosumab, alone or combined, in women with postmenopausal osteoporosis: the DATA study randomised trial. Lancet 382:50–56

Osteoporosis in Organ Transplant Patients

11

Jessica Furst and Elizabeth Shane

11.1 Skeletal Effects of Immunosuppressive Drugs

11.1.1 The Bone-Remodeling System

Bone remodeling is an orderly series of events in which old, damaged bone is replaced by new, mechanically stronger bone. Osteoclast first excavates small (0.5 mm^3) resorption pits or Howship's lacunae on cancellous and cortical bone surfaces, a process that takes 2–3 weeks (Ross 2006). After a brief rest period (the reversal phase), local mesenchymal marrow stem cells differentiate into osteoblasts and accumulate in the resorption pits (Aubin et al. 2006). Clusters of plump cuboidal osteoblasts first produce new bone matrix or osteoid and then mineralize it. Osteoclasts express receptors for NFkB ligand (RANKL), calcitonin, prostaglandins, calcium, and vitronectin (integrin $\alpha_1\beta_3$). Osteoblasts express receptors for several hormones (parathyroid hormone, estrogens, vitamin D_3), cell adhesion molecules (integrins), and cytokines. Bone remodeling is regulated by the tumor necrosis factor (TNF) ligand and receptor-signaling family: RANKL, RANK, and osteoprotegerin (OPG) (Gori et al. 2000; Hofbauer et al. 2000). RANKL is expressed by osteoblasts and by bone marrow stromal cells. In the presence of sufficient macrophage colony-stimulating factor (mCSF), RANKL binds to RANK receptors on surfaces of osteoclast lineage cells, resulting in rapid differentiation of osteoclast precursors to mature osteoclasts, increased osteoclast activity, and reduced apoptosis of mature

J. Furst, MD
Department of Medicine, Columbia University Medical Center, 630 west 168th street, PH8-864, New York, NY 10032, USA
e-mail: jrf2147@cumc.colmbia.edu

E. Shane, MD (✉)
Department of Medicine, Columbia University Medical Center, 630 West 168th Street, PH8-864, New York, NY 10032, USA
e-mail: es54@columbia.edu

P. Pietschmann (ed.), *Principles of Osteoimmunology*,
DOI 10.1007/978-3-319-34238-2_11

osteoclasts. RANKL also binds to another osteoblast product, OPG. Competitive binding of RANKL to either RANK or OPG regulates bone remodeling by increasing (RANK) or decreasing (OPG) osteoclastogenesis. Immunosuppressants exert their effects on remodeling by interacting with the RANK/RANKL/OPG system (Hofbauer et al. 2001). When bone remodeling becomes "uncoupled," such that the rate of resorption exceeds the rate of formation, bone loss occurs. Transplantation-related bone loss results from both an increase in the rate of bone resorption and a decrease in the rate of bone formation (Cohen et al. 2006a; Epstein 1996).

11.1.2 Glucocorticoids

Glucocorticoids (GCs), which are well recognized to cause osteoporosis, are included in the majority of posttransplant immunosuppression regimens. Prednisone or methylprednisolone is often prescribed in high doses (50–100 mg of prednisone or its equivalent daily) immediately after transplantation and during episodes of severe rejection, with gradual reduction over weeks to months. Total exposure varies with the organ transplanted, the number and management of rejection episodes, and the practice of individual transplantation programs.

The mechanisms by which GCs cause bone loss and fractures have been summarized in several reviews (Mazziotti et al. 2006; Sambrook 2006; van Staa 2006; Kulak et al. 2012; Amiche et al. 2016). Their predominant effect is an immediate and profound inhibition of bone formation: decreased osteoblast recruitment, differentiation and synthesis of type I collagen, induction of osteoblast and osteocyte apoptosis, and stimulation of osteocyte autophagy (Weinstein et al. 1998; Piemontese et al. 2015). Mechanistically, increased microRNA expression in osteoblasts exposed to GCs suppresses Wnt signaling, which is important for osteoblast expansion and function (Shi et al. 2015a; Westendorf et al. 2004). These effects are reflected biochemically by low serum levels of osteocalcin, a major non-collagenous bone matrix protein secreted by osteoblasts. GCs also inhibit growth hormone secretion and decrease production or bioactivity of certain skeletal growth factors (IGF-1, PGE_2, and TGF-β), actions that also reduce bone formation.

GCs also increase bone resorption by decreasing osteoblast expression of OPG and increasing osteoblast expression of RANKL. The effects of GCs on osteoclast formation and function are dose dependent and mediated via increases in reactive oxygen species, which facilitate osteoclastogenesis (Shi et al. 2015b). GC stimulatory effects on resorption are not as profound as their inhibitory effects on formation and are generally limited to the first 6–12 months. GCs may be associated with secondary hyperparathyroidism due to inhibition of intestinal calcium absorption and stimulation of urinary calcium excretion. Multiple studies have shown direct relationships between serum PTH levels and rates of bone loss after organ transplantation (Gupta et al. 2012; Torregrosa et al. 1995; Savaj and Ghods 2012; Obi et al. 2014; Mazzaferro and Pasquali 2016). GCs also cause hypogonadotrophic hypogonadism and reduced secretion of adrenal androgens and estrogens, which may also increase bone resorption. These contrasting effects of GCs upon bone formation (decreased) and resorption (increased) prevent osteoblasts from replacing the increased amount of bone resorbed at each remodeling site, and rapid bone loss ensues.

New data suggest that GCs decrease bone strength as well as bone mass (Mellibovsky et al. 2015). Reference point indentation of the anterior tibia allows tissue-level assessment of bone mechanical characteristics and detects subtle treatment-induced changes in bone material properties (Mellibovsky et al. 2015). In 52 patients tested within 4 weeks of GC initiation, strength declined in those on GCs for 7 and 20 weeks (Mellibovsky et al. 2015). In contrast, bone densitometry (BMD) by DXA did not decline, suggesting that subtle changes in bone material strength occur soon after GC initiation, before bone loss is detected by DXA (Mellibovsky et al. 2015).

GCs cause bone loss in individuals of all races, ages, and of both genders. However, postmenopausal Caucasian women, in whom GC effects are superimposed upon bone loss due to aging and estrogen deficiency, are at greatest risk for fracture (Sambrook 2006; van Staa 2006). In general, bone loss is most rapid during the first 12 months and is directly related to dose and duration of therapy. Areas of the skeleton rich in cancellous bone (ribs, vertebrae, and distal ends of long bones) and the cortical rim of the vertebral body are most severely affected and are the most common sites for fracture. This bone loss can be tempered with the use of anti-osteoporosis medications (Overman et al. 2015). One study found a 48 % decrease in fractures by 1 year and a 32 % decrease by 3 years if patients on GCs are treated with agents such as bisphosphonates, denosumab, or teriparatide within 90 days of GC initiation (Overman et al. 2015).

11.1.3 Calcineurin Inhibitors

Cyclosporine A (*CsA*) was introduced to posttransplant immunosuppression regimens in the early 1980s. CsA is a small fungal cyclic peptide that forms a heterodimer with its cytoplasmic receptor, cyclophilin, and inhibits phosphatase activity of calcineurin, which also regulates osteoblasts and osteoclasts (Kahan 1989; Sun et al. 2005; Sun et al. 2007). The calcineurin gene has been identified in osteoclasts and extracted whole rat bone, but does not appear to be affected by CsA administration (Awumey et al. 1999). When administered to rodents, albeit in doses higher than those currently used to prevent allograft rejection, CsA was associated with a marked increase in osteoclast-mediated bone resorption and rapid and severe cancellous bone loss (Epstein 1996; Tamler and Epstein 2006; Movsowitz et al. 1988, 1989), likely mediated by T lymphocytes (Buchinsky et al. 1996; Rucinski et al. 1994; Zahner et al. 1997). CsA also increases gene expression of the bone-resorbing cytokines, IL-1 and IL-6(Marshall et al. 1995). In contrast to GCs, bone formation is increased in CsA-treated animals, although insufficiently to compensate for the increased resorption. Rodent studies suggest that hepatic clearance of CsA was six times lower in male than female rats, leading to increased rates of osteopenia (Jager et al. 2012), although this has not been shown in humans. CsA-induced bone loss may be promoted by PTH (Epstein et al. 2001). Drugs that inhibit bone resorption (estrogen, raloxifene, calcitonin, alendronate) prevent or attenuate CsA-induced bone loss in the rat (Bowman et al. 1995; Joffe et al. 1992; Sass et al. 1997; Stein et al. 1991). Similarly, 1,25$(OH)_2$D and prostaglandin E2 prevent bone loss in CsA-treated rats (Epstein et al. 1990; Katz et al. 1992). Regimens that combine CsA with GCs found an additive rather than synergistic effect on bone loss (Shimizu et al. 2013).

Human studies have yielded conflicting results. Although kidney transplant patients receiving CsA in a GC-free regimen did not lose bone (Ponticelli and Aroldi 2001; Grotz et al. 1994; McIntyre et al. 1995; Cueto-Manzano et al. 2003), a prospective study found that cumulative CsA dose was associated with bone loss in the 2 years following transplant, independent of GCs (Josephson et al. 2004). In a recent study of patients on GC-sparing regimens, 27 % sustained a fracture within 6 months after transplant, similar to rates in patients on conventional GC-containing regimens (Edwards et al. 2011).

Tacrolimus (*FK506*), a macrolide that binds to an immunophilin FK-binding protein, blocks T-cell activation in a similar manner to CsA. FK506 causes bone loss in rodents similar in degree and mechanism to that observed with CsA (Cvetkovic et al. 1994). Rapid bone loss has been documented in patients after both cardiac (Stempfle et al. 1998) and liver (Park et al. 1996) transplantation. Rates of bone loss are directly associated with serum levels of FK506 (Luo et al. 2012). High serum FK506 levels were associated with high serum tartrate-resistant acid phosphatase (TRAP-5b) levels in male kidney transplant recipients, suggesting accelerated bone resorption as the underlying mechanism for FK506-mediated bone loss (Luo et al. 2012). FK506 may cause less bone loss than CsA (Goffin et al. 2002; Monegal et al. 2001a), likely because lower doses of GCs are required for immunosuppression. Whether FK506 confers any benefit over CsA with regard to fracture incidence is not known.

11.1.4 Other Immunosuppressive Agents

Sirolimus (rapamycin), a macrocyclic lactone, is structurally similar to FK506 and binds to the same binding protein, but induces immunosuppression via distinct mechanisms. In rat studies, low-dose CsA combined with rapamycin was bone sparing (Goodman et al. 2001). In an open-label study, markers of bone turnover (N-telopeptide and osteocalcin) were lower in kidney transplant recipients who received sirolimus than CsA; as BMD was not measured, no data on bone loss are available (Campistol et al. 2005). Mycophenolate mofetil does not have deleterious effects on bone in the rat (Dissanayake et al. 1998). The skeletal effects of other immunosuppressant agents such as mizoribine, deoxyspergualin, brequinar sodium, leflunomide, and azaspirane are unclear.

11.2 Vitamin D, Immunity, and Graft Rejection

Effects of vitamin D on immune function may be relevant after organ transplantation (Mazzaferro and Pasquali 2016; Lemire 1995; Stein and Shane 2011). Vitamin D potentiates the innate immune system and may protect against bacterial infections and tuberculosis (Bikle 2008). When 1,25(OH)$_2$D is produced by monocytes and macrophages in sufficient quantities, it has intracellular antimicrobial effects. 1,25(OH)$_2$D can also interact with and govern the cytokine profiles of activated T and B lymphocytes (Adams and Hewison 2010a). The ability of monocytes and macrophages to synthesize 1,25(OH)$_2$D is dependent on the availability of adequate serum concentrations of 25-OHD and increases in response to vitamin D supplementation (Adams and Hewison 2010a). When sufficient 25-OHD is not available, there are declines in both local production of 1,25(OH)$_2$D and activity of 1,25(OH)$_2$D against ingested microbes (Adams and Hewison 2010b). The antimicrobial actions of 1,25(OH)$_2$D also occur in barrier epithelial cells of the skin (Peric et al. 2008; Schauber et al. 2008), gut (Liu et al. 2008), and lungs (Hansdottir et al. 2008). Animals treated with the 1,25(OH)$_2$D analogue, calcitriol, were resistant to infection with *Candida albicans* and *Herpes simplex virus*-1 (Cantorna et al. 1998), two common opportunistic infections in transplant patients.

There is evidence supporting a role for vitamin D in the regulation of immune cell proliferation, differentiation, and responsiveness (Lemire 1995; Bikle 2011; Haroon and Fitzgerald 2012; Dimitrov et al. 2014; Adams et al. 2014; Wei and Christakos 2015). Animal studies suggest that 1,25(OH)$_2$D prevents acute allograft rejection following liver (Zhang et al. 2003; Redaelli et al. 2001), kidney (Becker et al. 2002), and heart (Hullett et al. 1998) transplantation. While there are limited data from human studies, kidney transplant recipients supplemented with calcitriol had fewer episodes of acute cellular rejection (Tanaci et al. 2003), required lower GC doses (Uyar et al. 2006), and had decreased expression of co-stimulatory and HLA-DR molecules (Ahmadpoor et al. 2009). In one study, patients treated with calcitriol after heart transplantation required less CsA (Briffa et al. 2003). In contrast, our study of heart transplant recipients found no differences between those randomized to calcitriol and alendronate with respect to CsA or prednisone dose (Shane et al. 2004). In a retrospective study of cardiac transplant patients, those with lower preoperative serum 25-OHD levels had more moderate-to-severe rejection episodes in the first 2 months after transplant (Bitetto et al. 2010). In 293 kidney transplant recipients, low serum 25-OHD levels predicted both decline in GFR and requirement for GCs to treat allograft rejection (Obi et al. 2014). In an observational study of liver transplant patients, those supplemented with cholecalciferol had fewer rejection episodes (Bitetto et al. 2010). Several studies have found increased BMD at the lumbar spine and femoral neck in renal transplant patients receiving activated vitamin D (calcitriol and alfacalcidol) (Sarno et al. 2015; Courbebaisse et al. 2014). The optimal amount of cholecalciferol that is both beneficial and safe in transplant recipients remains to be determined. The vitamin D supplementation in renAL transplant recipients (VITALE) trial is a prospective, multicenter, randomized trial of the efficacy and safety of cholecalciferol 100,000 IU monthly to maintain serum levels of 25-hydroxyvitamin D between 30 and 80 ng/mL (Courbebaisse et al. 2014). The results of this study will be interesting especially in light of studies that have found higher rates of falls and fractures in people treated with high doses of cholecalciferol (Sanders et al. 2010; Bischoff-Ferrari et al. 2016).

11.3 Effect of Transplantation on Bone and Mineral Metabolism

11.3.1 Bone Loss Before Transplantation

Many patients awaiting organ transplantation have multiple established risk factors for osteoporosis (older age, hypogonadism, physical inactivity, excessive tobacco and alcohol use, and exposure to drugs known to cause bone loss, such as GCs, heparin, loop diuretics). Most have already sustained considerable bone loss (Shane et al. 1997a; Ball et al. 2002; Hay 2003).

11.3.2 Kidney and Kidney-Pancreas Transplantation

Chronic kidney disease-mineral and bone disorder (CKD-MBD) refers to complex abnormalities in bone and mineral metabolism that affect most patients with chronic kidney disease (CKD) and end-stage renal disease (ESRD). Factors that contribute to development of CKD-MBD include disturbances in calcium and phosphate metabolism, low serum 25-OHD levels and decreased calcitriol synthesis, increased synthesis and secretion of PTH, metabolic acidosis, and defective bone mineralization (Martin et al. 2006; Beique et al. 2013). Patients with CKD-MBD may manifest extraskeletal calcifications, abnormal immune and cardiovascular function, and variable abnormalities in bone histology termed renal osteodystrophy (ROD). Almost 70–90 % of CKD stage 2–4 patients and all with stage 5 CKD have some form of ROD: high bone turnover with or without osteitis fibrosa (due to hyperparathyroidism), low bone turnover or adynamic bone disease, osteomalacia, or "mixed" renal osteodystrophy, a combination of one or more of the aforementioned lesions. The specific histological type depends on many factors that regulate bone remodeling and mineralization, such as PTH, acidosis, hypogonadism, and type I diabetes, as well as various therapeutic interventions used to manage patients with CKD and ESRD. Measurement of BMD by Dual-energy X-ray absorptiometry (DXA) does not distinguish among the various types of ROD. Fracture risk is estimated to be 4.4–14 times greater in patients with ESRD than the general population (Coco and Rush 2000; Mittalhenkle et al. 2004). In one study, 34 % of hemodialysis (HD) patients had a history of previous fracture (Kohlmeier et al. 1998). In another, fracture incidence was 0.1 per dialysis year in patients with osteitis fibrosa and 0.2 per dialysis year in patients with adynamic bone disease (Piraino et al. 1988). In people with moderate-to-severe CKD not on HD, hip fractures are twice as common as in those with normal kidney function (Nickolas et al. 2006).

As patients with ESRD are highly likely to have compromised bone health before kidney transplantation, it is not surprising that low spine and/or hip BMD has been reported in many cross-sectional studies of long-term kidney transplant recipients (Gupta et al. 2012; Grotz et al. 1994; Shane 2003; Cohen and Shane 2003; Epstein and Shane 2001; Shane and Epstein 2001; Braga Junior et al. 2006; Dolgos et al. 2008; Marcen et al. 2007; Dounousi et al. 2015; Boot et al. 1995; Cueto-Manzano et al. 1999), although the prognostic significance of low BMD in this group is unclear. Depending on the study and site of measurement, prevalence of low BMD long-term kidney transplant recipients ranges between 11 and 41 %. Factors associated with low BMD include increasing time since transplantation, high PTH concentrations, advanced age, female gender, and diabetic nephropathy.

A large number of longitudinal studies have documented high rates of bone loss at the spine and hip after kidney transplantation, ranging from 3 to 10 % during the first 6–12 months (Torregrosa et al. 1995; Savaj and Ghods 2012; Almond et al. 1994; Gallego et al. 2006; Horber et al. 1994; Kwan et al. 1992; Mikuls et al. 2003; Techawathanawanna et al. 2005; Yun et al. 1996; Bayat et al. 2007). Several studies have reported that GC dose is directly associated with bone loss (Rizzari et al. 2012; Woodle et al. 2008). However, by early 2000s, most kidney transplant programs were using lower GC doses and an increasing number of programs discontinue GCs altogether a few days after surgery. This shift in management seems to be associated with both reduced rates of bone loss and recovery of bone loss to pretransplant levels by 2 years after kidney transplantation in some (Aroldi et al. 1997; van den Ham et al. 2003; Nishioka et al. 2014), although not all studies (Meulen et al. 2004). For example, in 326 patients who had received a kidney transplant between 1996 and 2011, BMD Z scores were only slightly below average for age and sex at the baseline exam (approximately 6 months after transplant) and were not different from normal at 2.7 years (Naylor et al. 2014). In 47 patients managed without GCs and followed for 1 year after kidney transplant, DXA did not detect any bone loss at the spine or hip (Iyer et al. 2014). On a cautionary note, however, DXA did detect significant bone loss at the ultradistal and 1/3 radius sites, which are seldom measured (Iyer et al. 2014). Thus, despite the lower doses of GCs and alternative antirejection drugs used today (Gaston 2006; Paterson et al. 2005; Pescovitz 2006; Regazzi et al. 2005), significant bone loss occurs, although it may be less rapid than documented in the 1980s and 1990s (Van Staa et al. 2000). Moreover, high-resolution peripheral quantitative computed tomography (HRpQCT), which separately measures cortical and trabecular volumetric BMD and microarchitecture at the distal radius and tibia, documented cortical and trabecular bone loss and significant decreases in whole bone strength (Iyer et al. 2014). Cortical bone loss was directly related to an increase in cortical porosity at the endocortical surface and to higher PTH levels (Nishiyama et al. 2015). In contrast, seven patients, in whom GCs were stopped shortly after kidney transplantation and who underwent transiliac bone biopsy within 2 months and again 2–5 years after surgery, showed significant deterioration of trabecular bone microstructure, no significant changes in cortical bone, and no relationship between cortical bone parameters and PTH (Carvalho et al. 2015).

Fractures after renal transplantation typically affect appendicular sites (feet, ankles, long bones, hips) more than axial sites (spine, ribs) (Roe et al. 2005). In one study, non-vertebral fractures were far more prevalent in renal transplant recipients than in the normal population (Ramsey-Goldman et al. 1999), although another study found that renal transplant recipients had a lower risk of non-vertebral fracture compared to members of the general population who had previously sustained a non-vertebral fracture (Naylor et al. 2015). More recent data from a large cohort of first-time kidney transplant recipients suggests that hip fracture incidence may be lower in recent years with a hazard ratio for new hip fracture in the year 2010 of 0.56 (0.41–0.77) when compared to 1997 (Sukumaran Nair et al. 2014); this is noteworthy as patients transplanted in 2010 were older and had more medical comorbidities than the 1997 cohort (Sukumaran Nair et al. 2014). In a 3-year retrospective study of 35 kidney-pancreas recipients, approximately half had sustained from one to three symptomatic, non-vertebral fractures (Chiu et al. 1998). Men with type 1 diabetes who undergo kidney-pancreas transplants were recently reported to have approximately 30 % lower fracture rates than those who undergo kidney transplant alone; this relationship was not seen among women (Nikkel et al. 2011, 2013). Older age, white race, prior dialysis, and pretransplantation fracture were associated with increased fracture risk (Nikkel et al. 2013). Fracture incidence may be lower in patients managed with early GC withdrawal (Nikkel et al. 2012). A recent systematic review of fracture incidence in kidney transplant recipients analyzed 10 studies totaling 262,678 recipients and found a highly variable incidence rate ranging from 3.3 to 99.6 fractures per 1000 person-years (Naylor et al. 2013). Similarly, the 5-year cumulative incidence for fracture varied ranging from 0.85 to 27 %. Common factors associated with increased fracture risk were older age, female sex, the presence of diabetes, and receipt of dialysis before transplantation (Naylor et al. 2013). Other less common but statistically significant risk factors were a previous history of fracture and receipt of a kidney from a deceased versus living donor (Naylor et al. 2013). The authors concluded that there is poor consensus on the incidence and risk factors for fractures in kidney transplant recipients.

After kidney transplantation, there are several fairly consistent changes in biochemical indices of mineral metabolism and bone turnover (Julian et al. 1991; Kulak and Shane 2006; Sprague et al. 2008; Molnar et al. 2014; Alshayeb et al. 2013). PTH levels are often very high before transplantation, and while levels typically decline markedly immediately after transplantation (Akaberi et al. 2006; Rathi et al. 2015), they may remain above normal throughout the first year (Iyer et al. 2014) and may never completely normalize. Mild hypercalcemia and hypophosphatemia are common during the first few months after transplant (Rathi et al. 2015), but usually resolve within a few months. In long-term transplant recipients, persistent elevations in PTH may be associated with reduced hip BMD (Akaberi et al. 2006). Low calcitriol levels may persist after transplantation (Uyar et al. 2006; Sprague et al. 2008; Fleseriu and Licata 2007; Tripathi et al. 2008; Querings et al. 2006). Low serum 25-OHD levels were recently reported in nearly half of renal transplant recipients; this may represent improvement from older studies that reported vitamin D deficiency in over 75 % of renal transplant candidates, likely because of greater awareness of vitamin D (Beique et al. 2013). Sclerostin, secreted by osteocytes, is an inhibitor of the Wnt-signaling pathway which leads to decreased bone formation (Bonani et al. 2014). Sclerostin levels tend to be high in patients before renal transplantation. Although sclerostin levels decline within 15 days after transplant, it is unclear whether this is due to increased sclerostin catabolism or decreased sclerostin secretion, as phosphorus levels normalize (Bonani et al. 2014; Pelletier et al. 2013).

11.3.3 Cardiac Transplantation

The prevalence of osteoporosis in patients awaiting cardiac transplantation ranges from 4 to 23 %(Shane et al. 1997a, 2004; Dolgos et al. 2010; Anijar et al. 1999; Cohen and Shane 2005; Lee et al. 1994; Iqbal et al. 2008). Observational studies, primarily conducted during the 1990s, documented rates of bone loss ranging from 4 to 11 %, with the most rapid losses during the first 3–12 months, with greater losses at the hip than the spine (Stempfle et al. 1998; Shane et al. 2004; Berguer et al. 1994; Cremer et al. 1999; Henderson et al. 1995; Sambrook et al. 1994a; Shane et al. 1997b; Thiebaud et al. 1996; Valimaki et al. 1999a; Van Cleemput et al. 1995). The few longitudinal data after the first year suggest that the rate of bone loss slows or stops in the majority of patients, with partial recovery of bone mass at the spine though not at the hip (Shane et al. 2004; Shane et al. 1997b). More recent studies suggest lower rates of spine and hip bone loss (2–3 %) than previously reported, likely related to lower GC exposure than patients transplanted in the 1990s (Shane et al. 2004, 2012). However, while some GC-sparing protocols are associated with less bone loss, these results are not consistent across studies suggesting factors other than GCs are responsible for posttransplant bone loss (Baraldo et al. 2014).

Observational studies conducted during the 1990s found that approximately one-third of cardiac transplant recipients sustained a fragility fracture during the first 1–3 years after grafting (Shane et al. 1996a; Leidig-Bruckner et al. 2001). Fractures were associated with female gender, lower pretransplant BMD, and higher rates of bone loss after transplantation (Shane et al. 1996a; Leidig-Bruckner et al. 2001). However, many patients who fractured had normal BMD prior to transplant (Shane et al. 1996a; Leidig-Bruckner et al. 2001). In contrast, more recent studies report fracture incidence to be considerably lower than in the past, in the range of 12–14 %(Shane et al. 2004; Kerschan-Schindl et al. 2008). One study compared fracture risk between GC-free protocols, low-dose GC protocols (prednisone $\leq$5 mg daily by the end of the first posttransplant year), and high-dose GC protocols (prednisone>5 mg daily by the end of the first posttransplant year). At 2 years, there was no difference in fracture incidence among the three groups (1.1 %, 1.5 %, and 2.5 %, respectively) (Crespo Leiro et al. 2012). A recent retrospective study of 105 patients who received a heart transplant between 2005 and 2010 reported that 8.6 % sustained a fragility fracture, with a median time to first fracture of 12 months. A slightly larger proportion of women than men fractured (11.8 % versus 8.0 %). There was a high rate of bisphosphonate use in both fracture and non-fracture groups, 56 % versus 39 %, which may have reduced fracture rates (Hariman et al. 2014).

Early studies reported biochemical evidence of uncoupled bone remodeling (increased resorption and decreased formation) during the first 3–6 months after heart transplant, with restitution of coupling when glucocorticoid doses are lowered (Shane et al. 1997b; Valimaki et al. 1999a; Sambrook et al. 1994b). Later in the posttransplant period, many studies suggest that bone remodeling is increased (Lee et al. 1994; Thiebaud et al. 1996; Valimaki et al. 1999a; Sambrook et al. 1994b; Glendenning et al. 1999; Guo et al. 1998; Rich et al. 1992; Shane et al. 1993). Vitamin D deficiency is extremely common among heart and liver transplant recipients at the time of transplantation. In one study of patients assessed within the first 1–2 weeks after transplant, 55 % had deficiency with serum 25-OHD levels between 10 and 20 ng/ml, and 16 % had levels < 10 ng/ml (Stein et al. 2009).

11.3.4 Liver Transplantation

Osteoporosis is present in 20–43 % of patients with end-stage liver disease awaiting transplantation (Dolgos et al. 2010; Crosbie et al. 1999; Monegal et al. 1997; Bai et al. 2007; Guichelaar et al. 2006). A recent Spanish study that compared two groups of liver transplant candidates, one evaluated between 1992 and 1993 and the second between 2010 and 2011, found that the prevalence of osteoporosis (22 % versus 30 %, respectively) and vertebral fractures (36 % versus 33 %, respectively) was high and approximately the same in both (Monegal et al. 2013). Vitamin D deficiency is also very common, affecting >80 % of liver transplant candidates (Chaney et al. 2015; Venu et al. 2013). Chronic liver disease, of various etiologies, is associated with low levels of serum RANKL and high levels of OPG, possibly leading to an increased OPG/RANKL ratio, osteoclast inhibition, and a low bone-turnover state (Gatta et al. 2014). In fibrotic liver diseases, the synthesis of type I collagen is markedly increased and thus collagen-related bone-turnover markers do not reflect skeletal turnover in patients with liver disease (Guanabens et al. 1998). Serum osteocalcin and tartrate-resistant acid phosphatase (TRAP) may be more valid markers of bone remodeling activity.

In studies published before 2000 (Leidig-Bruckner et al. 2001; Eastell et al. 1991; Haagsma et al. 1988; Hawkins et al. 1994; Hussaini et al. 1999; Lopez et al. 1992; McDonald et al. 1991; Meys et al. 1994; Navasa et al. 1994; Porayko et al. 1991; Keogh et al. 1999), spine BMD fell by 2–24 %, primarily in the initial few months after liver transplantation, with recovery to levels equal to or even higher than before transplant by 1 year. More recent studies have reported lower rates of bone loss at the femoral neck (Keogh et al. 1999) and no significant bone loss at the spine (Ninkovic et al. 2002)·(Floreani et al. 2001). In a recent cohort study of 201 liver transplant recipients from the Netherlands, the majority treated with prednisone for at least 6 months and either tacrolimus or cyclosporine, lumbar spine BMD declined by 2.5 % but returned to pretransplant levels by 2 years. In contrast, femoral neck BMD had declined by 6.5 % at 6 months and remained stable thereafter, with no recovery over 5 years of follow-up (Krol et al. 2014a). Lower bone mass and/or higher rates of bone loss after liver transplant have been associated with older age, female gender, cholestatic liver disease, and higher prednisone dose (Smallwood et al. 2005).

In older studies, fracture incidence was extremely high, ranging from 24 to 65 %. Some recent studies suggest that fracture rates are lower (Ninkovic et al. 2002; Kaemmerer et al. 2012). However, in the Netherland cohort study, radiographic vertebral fractures were present in 56 % at screening, 68 % at 6 months, and 71 % at 12 months (Krol et al. 2014a), suggesting that vertebral fractures remain a significant problem (Krol et al. 2014a). Factors that have been associated with increased risk of fracture include male gender (Krol et al. 2014a), primary sclerosing cholangitis (Guichelaar et al. 2006) and alcoholic cirrhosis (Millonig et al. 2005), older age and lower pretransplant BMD (Ninkovic et al. 2002; Monegal et al. 2001b), prevalent pretransplant vertebral fractures, and longer survival (Leidig-Bruckner et al. 2001; Ninkovic et al. 2000).

PTH increases significantly during the first 3–6 months after liver transplant (Floreani et al. 2001; Monegal et al. 2001b; Compston et al. 1996). Serum 25-OHD concentrations are low in recent liver transplant recipients, likely because of disease-related factors such as malabsorption, impaired hepatic 25-hydroxylation of vitamin D, and reduced production of vitamin D-binding protein (Stein et al. 2009). Bone turnover, measured by markers of bone formation (osteocalcin and carboxyterminal peptide of type I collagen) and resorption are higher in liver transplant recipients than in normal controls in most (Haagsma et al. 1988; Abdelhadi et al. 1995; Watson et al. 1990; Valero et al. 1995), though not all studies (Rabinowitz et al. 1992). OPG and RANKL levels are significantly elevated in the first 2 weeks following liver transplant (Fabrega et al. 2006). In 21 patients who underwent tetracycline-labeled bone biopsies before and 3 months after transplantation, low turnover transitioned to a high turnover state (Compston et al. 1996; Vedi et al. 1999).

11.3.5 Lung Transplantation

Osteoporosis is extremely common among patients with end-stage lung disease. In cross-sectional studies, low bone mass and osteoporosis have been documented in 31–86 % of candidates for lung transplantation (Aris et al. 1996, 1998, 2000; Donovan et al. 1998; Ott and Aitken 1998; Shane et al. 1996b; Tschopp et al. 2006; Caplan-Shaw et al. 2006; Jastrzebski et al. 2010). All etiologies appear to be vulnerable. Markedly increased rates of all fractures and severe kyphosis due to vertebral fractures have been reported in adults with cystic fibrosis (Aris et al. 1998; Donovan et al. 1998; Ott and Aitken 1998). Low BMD and osteoporosis are also highly prevalent in patients with emphysema (Aris et al. 1996; Shane et al. 1996b), diffuse parenchymal lung disease (Caplan-Shaw et al. 2006), and primary pulmonary hypertension (Tschopp et al. 2006). The etiology of osteoporosis in patients with end-stage lung disease is multifactorial. In addition to prior GC exposure (Park et al. 1996; Aris et al. 1996; Shane et al. 1996b; Jastrzebski et al. 2010; Wang et al. 2013), low body weight, chronic hypoxemia, tobacco exposure, and chronic infection/inflammation have been implicated. In some studies, BMD correlated with pulmonary vascular resistance, functional measures, and walking distance (Tschopp et al. 2006; Caplan-Shaw et al. 2006). In patients with cystic fibrosis, additional etiologic factors include calcium malabsorption, hypogonadism, small body size, and vitamin D deficiency (Donovan et al. 1998).

Not surprisingly, therefore, osteoporosis affects a very high proportion of lung transplant recipients. In fact, a population-based cohort study from Taiwan found that those at highest risk for osteoporosis and various types of fracture after solid organ transplant were lung transplant recipients (Yu et al. 2014). Prospective studies have shown that spine and hip BMD declines by 4–5 % during the first year (Ferrari et al. 1996; Spira et al. 2000). An HRpQCT study comparing 58 lung transplant recipients and 60 controls found lung transplant recipients had greater cortical porosity, lower trabecular number and thickness, bone stiffness, and failure load (Fischer et al. 2015). In studies from the 1990s, fracture incidence varied from 11 to 42 %(Spira et al. 2000; Aringer et al. 1998; Rutherford et al. 2005; Shane et al. 1999). Risk factors for fracture and bone loss after lung transplant include female gender; low pretransplant BMD; pretransplant GC therapy; GC dose in some (Spira et al. 2000), but not all studies (Shane et al. 1999); pretransplant fractures; and high bone turnover after transplantation (Wang et al. 2013). In contrast, a retrospective study of patients transplanted between 2005 and 2010 reported a somewhat lower fracture rate of 8.0 %, perhaps because approximately 50 % were receiving bisphosphonates (Hariman et al. 2014).

11.3.6 Bone Marrow Transplantation (BMT)

As in the case of solid organ transplantation, low BMD is common in patients who require BMT. Bone loss may be related both to the underlying disease and to pretransplant myeloablative therapy with alkylating agents and/or total body irradiation, which may precipitate profound and frequently permanent hypogonadism, particularly in women, and may also have toxic effects on bone marrow osteoprogenitor cells (Banfi et al. 2001). Pretransplant exposure to GCs may also contribute. In patients studied after myeloablative chemotherapy but before BMT, osteopenia was present in 24 % and osteoporosis in 4 %(Schulte et al. 2000).

Adverse skeletal effects are more common after allogeneic BMT, in which donor and recipient are not genetically identical, than after autologous BMT, which involves removal and reinfusion of the patient's own stem cells after high-dose myeloablative therapy (Ebeling et al. 1999). After allogeneic BMT, there is a substantial risk of acute or chronic graft-versus-host disease (GVHD), in which donor immune cells react against the recipient. GVHD is typically treated with combinations of high-dose GCs, methotrexate, cyclosporine A, or tacrolimus. Several studies have documented low total body BMD and spine BMD measured by DXA and CT (Kelly et al. 1990; Bhatia et al. 1998; Yao et al. 2008; Nysom et al. 2000; Kauppila et al. 1999; Kerschan-Schindl et al. 2004). Those younger than 18 at transplantation may be more profoundly affected, perhaps because of failure to achieve optimal peak bone mass and smaller bone size (Bhatia et al. 1998, Frisk et al. 2012). Factors associated with low BMD 6 months after BMT include GVHD, GC use, low vitamin D levels, and family history of osteoporosis (Campos et al. 2014). In adult BMT recipients, hematopoietic stem cell transplantation before age 10 was associated with lower total body and femoral neck BMD than after age 18 (Petryk et al. 2014). Bone mass is low in hypogonadal women after BMT and hormone replacement therapy is associated with significant increases in BMD (Branco et al. 1996; Castaneda et al. 1997).

After BMT, spine bone loss ranges between 2 and 6 %, while FN BMD declines by 6–12 % (Serio et al. 2013; Ebeling 2005; Valimaki et al. 1999b; Kashyap et al. 2000). Duration of GC exposure may be more prognostic than cumulative dose (Serio et al. 2013). While there appears to be little bone loss after the first year (Kauppila et al. 1999), FN bone loss may not be regained. A recent study from the MD Anderson Cancer Center found that 8 % of patients fractured after BMT (Pundole et al. 2015). Risk factors included age over 50, multiple myeloma, solid organ tumors, and autologous BMT (Pundole et al. 2015). Vertebral fractures were slightly more common (53 %) than non-vertebral fractures (47 %) (Pundole et al. 2015), which mostly affected the ribs, upper limbs, and femur (Pundole et al. 2015).

The pathogenesis of post-BMT bone loss is complex. Cellular or cytokine-mediated bone marrow dysfunction may affect bone remodeling and loss after BMT (Lee et al. 2002a). High-dose chemotherapy, total body irradiation, and treatment with GCs and/or CsA may reduce osteoblast differentiation. Colony-forming unit fibroblasts (CFU-f) are reduced for up to 12 years following BMT (Banfi et al. 2001; Tauchmanova et al. 2002). During the first 3 months after BMT, bone formation markers decrease and resorption markers increase (Valimaki et al. 1999b), a pattern consistent with uncoupling of formation from resorption. After 3 months, bone formation markers rebound and elevated turnover persists during the latter half of the year (Kauppila et al. 1999; Valimaki et al. 1999b; Carlson et al. 1994; Withold et al. 1996; Lee et al. 2002b; Kang et al. 2000). Long-term survivors after BMT may have persistent abnormalities in bone turnover and vitamin D (Kananen et al. 2002).

11.4 Evaluation and Management of Osteoporosis in Patients Awaiting Transplantation

Candidates for all types of transplantation should be evaluated so that potentially treatable abnormalities of bone mineral metabolism may be addressed and bone health optimized preoperatively (Table 11.1). BMD of the spine and hip should be measured by DXA and spine x-rays or vertebral fracture analysis (VFA) performed to detect prevalent fractures, which are associated with an increased risk of incident fractures. If pretransplant BMD is low, a thorough evaluation should be performed to detect potentially treatable causes of low bone mass and guide therapy. On a cautionary note, however, one study found no association between BMD and vertebral fractures in OLT recipients suggesting that BMD may not provide a complete radiologic assessment of pretransplant bone health (Krol et al. 2014b). Patients found to have osteoporosis before transplantation should receive treatment, typically antiresorptive therapy with a bisphosphonate. Teriparatide can be considered for patients with very low BMD or fractures if they do not have

Table 11.1 Skeletal evaluation of the candidate for organ transplantation

In all candidates:
Assess risk factors for osteoporosis, including menstrual/menopausal history in women, history of low-trauma fractures
Measure bone densitometry (BMD) of the spine and hip by DXA
Obtain thoracic and lumbar spine radiographs or vertebral fracture analysis by DXA
If BMD testing reveals osteoporosis or if there are prevalent vertebral or non-vertebral fractures:
Serum electrolytes, BUN creatinine, calcium, parathyroid hormone, 25-hydyroxyvitamin D, thyroid function tests (see text)
In men, serum total and/or testosterone, FSH, and LH
Urine for calcium and creatinine

hyperparathyroidism. The pretransplant waiting period is often long enough (1–2 years) for significant improvement in BMD before transplantation. Patients with renal osteodystrophy should be managed in accordance with accepted clinical guidelines (Kidney Disease: Improving Global Outcomes 2009).

11.5 Prevention and Treatment of Transplantation Osteoporosis

As bone loss is most rapid during the first 6–12 months after transplantation and fractures may also occur early, strategies to prevent bone loss should be instituted immediately after transplantation, particularly in patients with low BMD and those who will be on GC therapy (Table 11.2).

The most commonly tested therapies for transplantation osteoporosis include bisphosphonates and active vitamin D analogues. While teriparatide is approved for the treatment of GC-induced osteoporosis with increases in BMD and decreases in non-vertebral fractures (Saag et al. 2009), the only study of teriparatide after organ transplantation found that FN BMD did not decline in kidney transplant patients randomized to teriparatide but decreased significantly in those on placebo (Cejka et al. 2008).

11.5.1 Vitamin D Analogues

Active vitamin D analogues prevent GC-induced decreases in intestinal calcium absorption, reduce secondary hyperparathyroidism, promote differentiation of osteoblast precursors into mature cells, and may potentiate immunosuppressive effects of CsA (Briffa et al. 2003; Lemire 1992; Lemire et al. 1994). Alfacalcidol (1-α-OHD) prevented or attenuated bone loss at the LS and FN (El-Agroudy et al. 2003; De Sevaux et al. 2002; El-Agroudy et al. 2005). Calcitriol (0.5–0.75 ug/day) prevented bone loss at the spine and hip during the first 6 months after heart or lung

Table 11.2 Primary prevention of bone loss in transplant recipients

Measure BMD before or immediately after transplantation and then annually
Utilize the lowest dose of glucocorticoids possible
Consider alternative therapies for rejection
Calcium intake (diet + supplements) of 1000 mg/day both before and after transplantation
Vitamin D intake of 400–1000 IU or as needed to maintain serum 25-OHD concentrations above 30 ng/ml (80 nmol/ml)
Physical rehabilitation program both before and after transplantation
Consider gonadal steroid replacement in hypogonadal men and premenopausal women
Consider pharmacologic therapy in all patients with low bone mass (*T* score between −1.0 and −2.5) or osteoporosis (*T* score < −2.5)
Begin antiresorptive therapy, preferably a bisphosphonate, before transplantation in patients with osteoporosis
Begin antiresorptive therapy, preferably a bisphosphonate, immediately after transplantation in patients at high risk for fracture
Osteoporosis (*T* score < −2.5)
History of fragility fracture
FRAX score >20 %
Postmenopausal women or men >50 years with osteopenia and no history of fragility fracture with osteopenia who are expected to continue on prednisone for greater than 1 month
Premenopausal women and men <50 years with osteopenia and no history of fragility fracture who are expected to continue on prednisone for greater than 1 month

transplantation (Henderson et al. 2001); those who received it for 24 months sustained significantly less FN bone loss, while those who stopped at 12 months had no evidence of benefit at the FN at 24 months (Sambrook et al. 2000). Calcitriol was as effective as daily alendronate in preventing bone loss after heart transplantation (Cohen et al. 2006b). In one study, daily calcitriol was associated with an increase in BMD at the LS, FN, and forearm during the first year after kidney transplantation (Josephson et al. 2004), while in another, intermittent calcitriol prevented bone loss at the hip but not the spine after kidney transplant (Torres et al. 2004). As hypercalcemia and hypercalciuria may develop rapidly and at any time during therapy with active vitamin D analogues, frequent monitoring of serum and urine calcium is necessary.

11.5.2 Bisphosphonates

Two meta-analyses of bisphosphonate trials in kidney transplant recipients found that bisphosphonates prevented bone loss at the spine and hip (Mitterbauer et al. 2006; Palmer et al. 2005). Similarly, a meta-analysis of randomized controlled trials in diverse transplant types showed that treatment with bisphosphonates improved spine and hip BMD by approximately 3 % on average (Cohen et al. 2011).

Intravenous bisphosphonates consistently prevent bone loss after transplantation (Valero et al. 1995; Aris et al. 2000; Fan et al. 2000, 2003; Garcia-Delgado et al.

1997; Shane et al. 1998; Bianda et al. 2000; Hommann et al. 2002; Grotz et al. 2001; Arlen et al. 2001; Trombetti et al. 2000; Krieg et al. 2001; Haas et al. 2003; Coco et al. 2003; Walsh et al. 2009; Jeon et al. 2013; Okamoto et al. 2014). Repeated doses of intravenous pamidronate prevented spine and hip bone loss in the kidney (Fan et al. 2000, 2003; Coco et al. 2003; Torregrosa et al. 2011), heart (Bianda et al. 2000; Krieg et al. 2001), liver (Pennisi et al. 2007; Monegal et al. 2009), lung (Aris et al. 2000; Trombetti et al. 2000), and bone marrow transplant recipients (Kananen et al. 2005; Grigg et al. 2004). A meta-analysis of six studies (144 kidney transplant patients) found that pamidronate, given 3–7 times at 30–90 mg per dose, was associated with significant reduction of bone loss in the spine without adverse effects on graft function (Wang et al. 2014). Similarly, both zoledronic acid and ibandronate prevent bone loss at 6 and 12 months in recipients of heart (Fahrleitner-Pammer et al. 2009), liver (Hommann et al. 2002; Crawford et al. 2006; Bodingbauer et al. 2007), and kidney (Grotz et al. 2001; Haas et al. 2003; Walsh et al. 2009; Smerud et al. 2012) transplants. Zoledronic acid administered monthly for the first 6 months and at 9 and 12 months after liver transplantation was associated with stable BMD at the spine and reduced losses at the FN compared to controls and was associated with a reduction in vertebral fractures (Bodingbauer et al. 2007). Ibandronate, 2 mg IV every 3 months for 1 year, maintained spine BMD and attenuated hip bone loss after liver transplantation, with a significant reduction in number of fractures (Kaemmerer et al. 2010). In long-term liver transplant recipients, quarterly IV ibandronate and calcitriol were associated with improvements in BMD at the hip and a reduction in vertebral fractures over 3 years compared to an untreated reference group (Wagner et al. 2012).

In two large prospective studies of patients after allogeneic BMT, IV pamidronate prevented bone loss at the spine and ameliorated losses at the hip (Kananen et al. 2005; Grigg et al. 2004). Similarly, IV zoledronic acid (4 mg) given 12 months after grafting and in another study, given every 3 months for 2 years, prevented spine and hip bone loss (Tauchmanova et al. 2005; Hausmann et al. 2012). In a small uncontrolled study, BMD was stable over 3 years after BMT in patients who received zoledronic acid once at transplant and 6 months later (Ganguly et al. 2012). When BMT patients were treated with zoledronic acid, FN BMD increased at 12 months (Lee et al. 2002c; Hari et al. 2013). Similarly, monthly zoledronic acid infusions for the first 3 months after BMT restored BMD at the LS and FN (Serio et al. 2013).

Results with oral bisphosphonates are less compelling. There was no change in BMD in 42 renal transplant patients randomized to risedronate or placebo, although risedronate appeared to preserve BMD in women (Coco et al. 2012). Monthly oral ibandronate (150 mg) or weekly oral risedronate (35 mg) initiated 12 months after kidney transplant provided comparable increases in spine and hip BMD after 1 year (Sanchez-Escuredo et al. 2015). A study of 76 kidney transplant recipients treated with weekly oral alendronate (70 mg) found improvements in lumbar spine BMD both in patients with and without preexisting osteoporosis after 14 months of therapy (Huang et al. 2012). Alendronate reduced bone-turnover markers in 24 kidney transplant recipients after 36 months of treatment (Yamamoto et al. 2013). Kidney transplant patients treated with alendronate (10 mg daily), calcitriol (0.25 μg daily), and calcium carbonate (2 g daily) had marked increases in spine BMD compared to decreases in those who received only calcium and calcitriol (Kovac et al. 2001). There were similar increases in spine BMD in patients treated with alendronate or risedronate after kidney transplant (Nowacka-Cieciura et al. 2006; Torregrosa et al. 2007; Giannini et al. 2001). Weekly alendronate (70 mg) improved BMD in kidney (Toro et al. 2005) and liver transplant recipients (Atamaz et al. 2006). Risedronate prevented bone loss at the spine but not the hip 12 months after liver transplant (Guadalix et al. 2011). In contrast, monthly oral ibandronate increased spine and hip BMD in 74 liver transplant patients, and fracture rates were relatively low (5.4 %) (Kaemmerer et al. 2012). Both alendronate (10 mg daily) and calcitriol (0.25 μg twice daily) initiated immediately after heart transplant prevented bone loss at the spine and hip compared with a reference group receiving only calcium and vitamin D (Shane et al. 2004); BMD remained stable during the second year after alendronate and calcitriol were discontinued (Cohen et al. 2006a). A single infusion of zoledronic acid (5 mg) or weekly alendronate (70 mg) initiated immediately after heart and liver transplantation prevented bone loss at the hip and, in liver transplant patients, at the spine compared to a reference group that received only calcium and vitamin D (Cohen et al. 2011). In heart transplant patients, spine bone mineral density increased with IV zoledronic acid and decreased on alendronate (Cohen et al. 2011). In BMT patients, risedronate started 12 months after BMT improved LS BMD and prevented loss at the FN (Tauchmanova et al. 2003).

Although fracture is a very important clinical outcome, few studies have had adequate power to detect differences in fracture. A meta-analysis of 11 randomized controlled clinical trials showed that treatment with either bisphosphonates or vitamin D analogues significantly reduced the number of subjects with fracture by 50 % and number of vertebral fractures by 76 %. When vitamin D analogue trials were excluded and bisphosphonate trials examined separately, there was a significant 47 % reduction in number of subjects with fractures but the 66 % reduction in vertebral fractures was not significant (Cohen et al. 2011). In our opinion, bisphosphonates represent the most promising approach to the prevention of transplantation osteoporosis, but data on fracture prevention is needed.

11.6 Summary and Conclusions

The prevalence of osteoporosis among candidates for solid organ and bone marrow transplantation is high. During the 1990s, prospective longitudinal studies demonstrated rapid bone loss and high fracture rates, particularly during the first posttransplant year. More recent studies suggest that rates of bone loss and fracture are considerably lower than in the past, although rates remain unacceptably high. All patients should be evaluated before transplantation and receive treatment for prevalent osteoporosis, if present. Primary prevention therapy should be initiated immediately after transplantation in patients with low BMD, advanced age, or other significant risk factors, as the most rapid bone loss occurs in the first few months after grafting.

Table 11.3 Management of the long-term organ transplant recipient

In all patients:
Assess risk factors for osteoporosis
BMD of the spine and hip by DXA
Thoracic and lumbar spine radiographs
Calcium intake of 1000 mg/day both before and after transplantation
Vitamin D intake of 400–1000 IU or as needed to maintain serum 25-OHD concentrations above 30 ng/ml (80 nmol/ml)
Physical rehabilitation program
If BMD testing reveals osteoporosis or there are prevalent vertebral fractures:
Serum electrolytes, BUN creatinine, calcium, parathyroid hormone, 25-hydyroxyvitamin D, thyroid function tests
In men, serum total and/or testosterone, FSH, and LH
Urine for calcium and creatinine
Replace gonadal steroids (in men and hypogonadal women, if appropriate)
Begin antiresorptive therapy, preferably a bisphosphonate
These recommendations should not be applied to kidney transplant recipients, in whom the risk of the adynamic bone lesion is high and benefits of bisphosphonates are controversial

Long-term transplant recipients should be monitored and treated for bone disease as well (Table 11.3). At present, bisphosphonates are the most consistently effective agents for prevention and treatment of bone loss in organ transplant recipients. In the future, the use of denosumab or the introduction of newer, bone-sparing immunosuppressive agents may further reduce rates of bone loss after transplant. It is hoped that these advances will further reduce morbidity from transplantation osteoporosis.

References

Abdelhadi M, Eriksson SA, Ljusk Eriksson S, Ericzon BG, Nordenstrom J (1995) Bone mineral status in end-stage liver disease and the effect of liver transplantation. Scand J Gastroenterol 30:1210–1215

Adams JS, Hewison M (2010) Update in vitamin D. J Clin Endocrinol Metab 95:471–478

Adams JS, Rafison B, Witzel S et al (2014) Regulation of the extrarenal CYP27B1-hydroxylase. J Steroid Biochem Mol Biol 144(Pt A):22–27

Ahmadpoor P, Ilkhanizadeh B, Ghasemmahdi L, Makhdoomi K, Ghafari A (2009) Effect of active vitamin D on expression of co-stimulatory molecules and HLA-DR in renal transplant recipients. Exp Clin Transplant 7:99–103

Akaberi S, Lindergard B, Simonsen O, Nyberg G (2006) Impact of parathyroid hormone on bone density in long-term renal transplant patients with good graft function. Transplantation 82:749–752

Almond MK, Kwan JTC, Evans K, Cunningham J (1994) Loss of regional bone mineral density in the first 12 months following renal transplantation. Nephron 66:52–57

Alshayeb HM, Josephson MA, Sprague SM (2013) CKD-mineral and bone disorder management in kidney transplant recipients. Am J Kidney Dis 61:310–325

Amiche MA, Albaum JM, Tadrous M et al (2016) Efficacy of osteoporosis pharmacotherapies in preventing fracture among oral glucocorticoid users: a network meta-analysis. Osteoporos Int 6:1989–98

Anijar JR, Szejnfeld VL, Almeida DR, Fernandes AR, Ferraz MB (1999) Reduced bone mineral density in men after heart transplantation. Braz J Med Biol Res 32:413–420
Aringer M, Kiener H, Koeller M et al (1998) High turnover bone disease following lung transplantation. Bone 23:485–488
Aris R, Neuringer I, Weiner M, Egan T, Ontjes D (1996) Severe osteoporosis before and after lung transplantation. Chest 109:1176–1183
Aris RM, Renner JB, Winders AD et al (1998) Increased rate of fractures and severe kyphosis: sequelae of living into adulthood with cystic fibrosis. Ann Intern Med 128:186–193
Aris RM, Lester GE, Renner JB et al (2000) Efficacy of pamidronate for osteoporosis in patients with cystic fibrosis following lung transplantation. Am J Respir Crit Care Med 162:941–946
Arlen DJ, Lambert K, Ioannidis G, Adachi JD (2001) Treatment of established bone loss after renal transplantation with etidronate. Transplantation 71:669–673
Aroldi A, Tarantino A, Montagnino G, Cesana B, Cocucci C, Ponticelli C (1997) Effects of three immunosuppressive regimens on vertebral bone density in renal transplant recipients: a prospective study. Transplantation 63:380–386
Atamaz F, Hepguler S, Karasu Z, Kilic M, Tokat Y (2006) The prevention of bone fractures after liver transplantation: experience with alendronate treatment. Transplant Proc 38:1448–1452
Aubin J, Lian J, Stein G (2006) Bone formation: maturation and functional activities of osteoblast lineage cells. In: Favus M (ed) Primer on the metabolic bone diseases and other disorders of bone and mineral metabolism. Wiley-Blackwell. Iowa,USA pp 20–29
Awumey E, Moonga B, Sodam B et al (1999) Molecular and functional evidence for calcineurin alpha and beta isoforms in the osteoclasts. Novel insights into the mode of action of cyclosporine A. Biochem Biophys Res Commun 254:148–252
Bai XL, Liang TB, Wu LH et al (2007) Elevation of intact parathyroid hormone level is a risk factor for low bone mineral density in pretransplant patients with liver diseases. Transplant Proc 39:3182–3185
Ball AM, Gillen DL, Sherrard D et al (2002) Risk of hip fracture among dialysis and renal transplant recipients. JAMA 288:3014–3018
Banfi A, Podesta M, Fazzuoli L et al (2001) High-dose chemotherapy shows a dose dependent toxicity to bone marrow osteoprogenitors: a mechanism for post-bone marrow transplantation osteopenia. Cancer 92:2419–2428
Baraldo M, Gregoraci G, Livi U (2014) Steroid-free and steroid withdrawal protocols in heart transplantation: the review of literature. Transpl Int 27:515–529
Bayat N, Einollahi B, Pourfarzian V et al (2007) Bone mineral density changes within 11 months of renal transplantation in Iranian patients. Transplant Proc 39:1039–1043
Becker BN, Hullett DA, O'Herrin JK, Malin G, Sollinger HW, DeLuca H (2002) Vitamin D as immunomodulatory therapy for kidney transplantation. Transplantation 74:1204–1206
Belque LC, Kline GA, Dalton B, Duggan K, Yilmaz S (2013) Predicting deficiency of vitamin D in renal transplant recipients in northern climates. Transplantation 95:1479–1484
Berguer DG, Krieg MA, Thiebaud D et al (1994) Osteoporosis in heart transplant recipients: a longitudinal study. Transplant Proc 26:2649–2651
Bhatia S, Ramsay NK, Weisdorf D, Griffiths H, Robison LL (1998) Bone mineral density in patients undergoing bone marrow transplantation for myeloid malignancies. Bone Marrow Transplant 22:87–90
Bianda T, Linka A, Junga G et al (2000) Prevention of osteoporosis in heart transplant recipients: a comparison of calcitriol with calcitonin and pamidronate. Calcif Tissue Int 67:116–121
Bikle DD (2008) Vitamin D, and the immune system: role in protection against bacterial infection. Curr Opin Nephrol Hypertens 17:348–352
Bikle DD (2011) Vitamin D, regulation of immune function. Vitam Horm 86:1–21
Bischoff-Ferrari HA, Dawson-Hughes B, Orav EJ et al (2016) Monthly high-dose vitamin D treatment for the prevention of functional decline: a randomized clinical trial. JAMA Intern Med 176:175–183
Bitetto D, Fabris C, Falleti E et al (2010) Vitamin D and the risk of acute allograft rejection following human liver transplantation. Liver Int 30:417–444

Bodingbauer M, Wekerle T, Pakrah B et al (2007) Prophylactic bisphosphonate treatment prevents bone fractures after liver transplantation. Am J Transplant 7:1763–1769

Bonani M, Rodriguez D, Fehr T et al (2014) Sclerostin blood levels before and after kidney transplantation. Kidney Blood Press Res 39:230–239

Boot AM, Nauta J, Hokken-Koelega ACS, Pols HAP, Ridder MAJ, Keizer-Schrama SMPF (1995) Renal transplantation and osteoporosis. Arch Dis Child 72:502–506

Bowman A, Sass D, Marshall I et al (1995) Raloxifene analog (Ly 117018-HCL) ameliorates cyclosporin A induced osteopenia. J Bone Miner Res 10(Suppl 1):350

Braga Junior JW, Neves RM, Pinheiro MM et al (2006) Prevalence of low trauma fractures in long-term kidney transplant patients with preserved renal function. Braz J Med Biol Res 39:137–147

Branco CC, Rovira M, Pons F et al (1996) The effect of hormone replacement therapy on bone mass in patients with ovarian failure due to bone marrow transplantation. Maturitas 23:307–312

Briffa NK, Keogh AM, Sambrook PN, Eisman JA (2003) Reduction of immunosuppressant therapy requirement in heart transplantation by calcitriol. Transplantation 75:2133–2134

Buchinsky FJ, Ma Y, Mann GN et al (1996) T lymphocytes play a critical role in the development of cyclosporin A-induced osteopenia. Endocrinology 137:2278–2285

Campistol JM, Holt DW, Epstein S, Gioud-Paquet M, Rutault K, Burke JT (2005) Bone metabolism in renal transplant patients treated with cyclosporine or sirolimus. Transpl Int 18:1028–1035

Campos DJ, Boguszewski CL, Funke VA et al (2014) Bone mineral density, vitamin D, and nutritional status of children submitted to hematopoietic stem cell transplantation. Nutrition 30:654–659

Cantorna MT, Hullett DA, Redaelli C et al (1998) 1,25-Dihydroxyvitamin D3 prolongs graft survival without compromising host resistance to infection or bone mineral density. Transplantation 66:828–831

Caplan-Shaw CE, Arcasoy SM, Shane E et al (2006) Osteoporosis in diffuse parenchymal lung disease. Chest 129:140–146

Carlson K, Simonsson B, Ljunghall S (1994) Acute effects of high dose chemotherapy followed by bone marrow transplantation on serum markers of bone metabolism. Calcif Tissue Int 55:408–411

Carvalho C, Magalhaes J, Pereira L, Simoes-Silva L, Castro-Ferreira I, Frazao JM (2015) Evolution of bone disease after kidney transplantation: a prospective histomorphometric analysis of trabecular and cortical bone. Nephrology (Carlton) 21:55–61

Castaneda S, Carmona L, Carjaval I, Arranz B, Diaz A, Garcia-Vadillo A (1997) Reduction of bone mass in women after bone marrow transplantation. Calcif Tissue Int 60:343–347

Cejka D, Benesch T, Krestan C et al (2008) Effect of teriparatide on early bone loss after kidney transplantation. Am J Transplant 8:1864–1870

Chaney A, Heckman MG, Diehl NN, Meek S, Keaveny AP (2015) Effectiveness and outcomes of current practice in treating vitamin D deficiency in patients listed for liver transplantation. Endocr Pract 21:761–769

Chiu MY, Sprague SM, Bruce DS, Woodle ES, Thistlethwaite JR Jr, Josephson MA (1998) Analysis of fracture prevalence in kidney-pancreas allograft recipients. J Am Soc Nephrol 9:677–683

Coco M, Rush H (2000) Increased incidence of hip fractures in dialysis patients with low serum parathyroid hormone. Am J Kidney Dis 36:1115–1121

Coco M, Glicklich D, Faugere MC et al (2003) Prevention of bone loss in renal transplant recipients: a prospective, randomized trial of intravenous pamidronate. J Am Soc Nephrol 14:2669–2676

Coco M, Pullman J, Cohen HW et al (2012) Effect of risedronate on bone in renal transplant recipients. J Am Soc Nephrol 23:1426–1437

Cohen A, Shane E (2003) Osteoporosis after solid organ and bone marrow transplantation. Osteoporos Int 14:617–630

Cohen A, Shane E (2005) Bone disease in patients before and after cardiac transplantation. In: Compston JE, Shane E (eds) Bone disease of organ transplantation. Elsevier Academic Press, Burlington, pp 287–301

Cohen A, Ebeling P, Sprague S, Shane E (2006a) Transplantation osteoporosis. In: Favus M (ed) Primer on the metabolic bone diseases and disorders of bone and mineral metabolism. Wiley-Blackwell. Iowa, USA pp 302–309

Cohen A, Addesso V, McMahon DJ et al (2006b) Discontinuing antiresorptive therapy one year after cardiac transplantation: effect on bone density and bone turnover. Transplantation. Wiley-Blackwell. Iowa, USA 81:686–691

Cohen A, Stein EM, Boutroy S et al (2011) Weekly alendronate versus a single infusion of zoledronic acid: effects on bone turnover markers, areal and volumetric BMD and bone microarchitecture during the first year after heart or liver transplantation. J Bone Miner Res 26

Compston J, Greer S, Skingle S et al (1996) Early increase in plasma parathyroid hormone level following liver transplantation. J Hepatol 25:715–718

Courbebaisse M, Alberti C, Colas S et al (2014) VITamin D supplementation in renAL transplant recipients (VITALE): a prospective, multicentre, double-blind, randomized trial of vitamin D estimating the benefit and safety of vitamin D3 treatment at a dose of 100,000 UI compared with a dose of 12,000 UI in renal transplant recipients: study protocol for a double-blind, randomized, controlled trial. Trials 15:430

Crawford BA, Kam C, Pavlovic J et al (2006) Zoledronic acid prevents bone loss after liver transplantation: a randomized, double-blind, placebo-controlled trial. Ann Intern Med 144:239–248

Cremer J, Struber M, Wagenbreth I et al (1999) Progression of steroid-associated osteoporosis after heart transplantation. Ann Thorac Surg 67:130–133

Crespo Leiro MG, Bonet LA, Paniagua Martin MJ et al (2012) Steroid withdrawal during 5 years following heart transplantation, and the relationship between steroid dosage at 1-year follow-up and complications during the next 2 years: results from the RESTCO study. Transplant Proc 44:2631–2634

Crosbie OM, Freaney R, McKenna MJ, Hegarty JE (1999) Bone density, vitamin D status, and disordered bone remodeling in end- stage chronic liver disease. Calcif Tissue Int 64:295–300

Cueto-Manzano A, Konel S, Hutchinson AJ et al (1999) Bone loss in long term renal transplantation. Histopathology and densitometry analysis. Kidney Int 55:2021–2029

Cueto-Manzano AM, Konel S, Crowley V et al (2003) Bone histopathology and densitometry comparison between cyclosporine a monotherapy and prednisolone plus azathioprine dual immunosuppression in renal transplant patients. Transplantation 75:2053–2058

Cvetkovic M, Mann GN, Romero DF et al (1994) The deleterious effects of long term cyclosporin A, cyclosporin G and FK506 on bone mineral metabolism in vivo. Transplantation 57:1231–1237

De Sevaux RG, Hoitsma AJ, Corstens FH, Wetzels JF (2002) Treatment with vitamin D and calcium reduces bone loss after renal transplantation: a randomized study. J Am Soc Nephrol 13:1608–1614

Dimitrov V, Salehi-Tabar R, An BS, White JH (2014) Non-classical mechanisms of transcriptional regulation by the vitamin D receptor: insights into calcium homeostasis, immune system regulation and cancer chemoprevention. J Steroid Biochem Mol Biol 144(Pt A):74–80

Dissanayake IR, Goodman GR, Bowman AR et al (1998) Mycophenolate mofetil; a promising new immunosuppressant that does not cause bone loss in the rat. Transplantation 65:275–278

Dolgos S, Hartmann A, Bonsnes S et al (2008) Determinants of bone mass in end-stage renal failure patients at the time of kidney transplantation. Clin Transplant 22:462–468

Dolgos S, Hartmann A, Isaksen GA et al (2010) Osteoporosis is a prevalent finding in patients with solid organ failure awaiting transplantation – a population based study. Clin Transplant 24:E145–E152

Donovan DS Jr, Papadopoulos A, Staron RB et al (1998) Bone mass and vitamin D deficiency in adults with advanced cystic fibrosis lung disease. Am J Respir Crit Care Med 157:1892–1899

Dounousi E, Leivaditis K, Eleftheriadis T, Liakopoulos V (2015) Osteoporosis after renal transplantation. Int Urol Nephrol 47:503–511

Eastell R, Dickson RE, Hodgson SF et al (1991) Rates of vertebral bone loss before and after liver transplantation in women with primary biliary cirrhosis. Hepatology 14:296–300

Ebeling PR (2005) Defective osteoblast function may be responsible for bone loss from the proximal femur despite pamidronate therapy. J Clin Endocrinol Metab 90:4414–4416

Ebeling P, Thomas D, Erbas B, Hopper L, Szer J, Grigg A (1999) Mechanism of bone loss following allogeneic and autologous hematopoeitic stem cell transplantation. J Bone Miner Res 14:342–350

Edwards BJ, Desai A, Tsai J et al (2011) Elevated incidence of fractures in solid-organ transplant recipients on glucocorticoid-sparing immunosuppressive regimens. J Osteoporos 2011:591793

El-Agroudy AE, El-Husseini AA, El-Sayed M, Ghoneim MA (2003) Preventing bone loss in renal transplant recipients with vitamin D. J Am Soc Nephrol 14:2975–2979

El-Agroudy AE, El-Husseini AA, El-Sayed M, Mohsen T, Ghoneim MA (2005) A prospective randomized study for prevention of postrenal transplantation bone loss. Kidney Int 67:2039–2045

Epstein S (1996) Post-transplantation bone disease: the role of immunosuppressive agents on the skeleton. J Bone Miner Res 11:1–7

Epstein S, Shane E (2001) Transplantation osteoporosis. In: Marcus R, Feldman D, Kelsey J (eds) Osteoporosis. Academic Press, San Diego, pp 327–340

Epstein S, Schlosberg M, Fallon M, Thomas S, Movsowitz C, Ismail F (1990) 1,25 dihydroxyvitamin D3 modifies cyclosporine induced bone loss. Calcif Tissue Int 47:152–157

Epstein S, Dissanayake A, Goodman GR et al (2001) Effect of the interaction of parathyroid hormone and cyclosporine A on bone mineral metabolism in the rat. Calcif Tissue Int 68:240–247

Fabrega E, Orive A, Garcia-Unzueta M, Amado JA, Casafont F, Pons-Romero F (2006) Osteoprotegerin and receptor activator of nuclear factor-kappaB ligand system in the early post-operative period of liver transplantation. Clin Transplant 20:383–388

Fahrleitner-Pammer A, Piswanger-Soelkner JC, Pieber TR et al (2009) Ibandronate prevents bone loss and reduces vertebral fracture risk in male cardiac transplant patients: a randomized double-blind, placebo-controlled trial. J Bone Miner Res 24:1335–1344

Fan S, Almond MK, Ball E, Evans K, Cunningham J (2000) Pamidronate therapy as prevention of bone loss following renal transplantation. Kidney Int 57:684–690

Fan SL, Kumar S, Cunningham J (2003) Long-term effects on bone mineral density of pamidronate given at the time of renal transplantation. Kidney Int 63:2275–2279

Ferrari SL, Nicod LP, Hamacher J et al (1996) Osteoporosis in patients undergoing lung transplantation. Eur Respir J 9:2378–2382

Fischer L, Valentinitsch A, DiFranco MD et al (2015) High-resolution peripheral quantitative CT imaging: cortical porosity, poor trabecular bone microarchitecture, and low bone strength in lung transplant recipients. Radiology 274:473–481

Fleseriu M, Licata AA (2007) Failure of successful renal transplant to produce appropriate levels of 1,25-dihydroxyvitamin D. Osteoporos Int 18:363–368

Floreani A, Mega A, Tizian L et al (2001) Bone metabolism and gonad function in male patients undergoing liver transplantation: a two-year longitudinal study. Osteoporos Int 12:749–754

Frisk P, Arvidson J, Ljunggren O, Gustafsson J (2012) Decreased bone mineral density in young adults treated with SCT in childhood: the role of 25-hydroxyvitamin D. Bone Marrow Transplant 47:657–662

Gallego R, Oliva E, Vega N et al (2006) Steroids and bone density in patients with functioning kidney allografts. Transplant Proc 38:2434–2437

Ganguly S, Divine CL, Aljitawi OS, Abhyankar S, McGuirk JP, Graves L (2012) Prophylactic use of zoledronic acid to prevent early bone loss is safe and feasible in patients with acute myeloid leukemia undergoing allogeneic stem cell transplantation. Clin Transplant 26:447–453

Garcia-Delgado I, Prieto S, Fragnas LG, Robles E, Rufilanchas T, Hawkins F (1997) Calcitonin, etidronate and calcidiol treatment in bone loss after cardiac transplantation. Calcif Tissue Int 60:155–159

Gaston RS (2006) Current and evolving immunosuppressive regimens in kidney transplantation. Am J Kidney Dis 47:S3–S21

Gatta A, Verardo A, Di Pascoli M, Giannini S, Bolognesi M (2014) Hepatic osteodystrophy. Clin Cases Miner Bone Metab 11:185–191

Giannini S, Dangel A, Carraro G et al (2001) Alendronate prevents further bone loss in renal transplant recipients. J Bone Miner Res 16:2111–2117

Glendenning P, Kent GN, Adler BD et al (1999) High prevalence of osteoporosis in cardiac transplant recipients and discordance between biochemical turnover markers and bone histomorphometry. Clin Endocrinol (Oxf) 50:347–355

Goffin E, Devogelaer JP, Lalaoui A et al (2002) Tacrolimus and low-dose steroid immunosuppression preserves bone mass after renal transplantation. Transpl Int 15:73–80

Goodman GR et al (2001) Immunosuppressant use without bone loss implications for bone loss after transplantation. J Bone Miner Res. 16:72–78

Gori F, Hofbauer LC, Dunstan CR, Spelsberg TC, Khosla S, Riggs BL (2000) The expression of osteoprotegerin and RANK ligand and the support of osteoclast formation by stromal-osteoblast lineage cells is developmentally regulated. Endocrinology 141:4768–4776

Grigg AC, Shuttleworth P, Reynolds J et al (2004) Pamidronate therapy for one year after allogeneic bone marrow transplantation (AlloBMT) reduces bone loss from the lumbar spine, femoral neck and total hip. Blood 104:A2253

Grotz WH, Mundinger A, Gugel B, Exner V, Kirste G, Schollmeyer PJ (1994) Bone fracture and osteodensitometry with dual energy x-ray absorptiometry in kidney transplant recipients. Transplantation 58:912–915

Grotz W, Nagel C, Poeschel D et al (2001) Effect of ibandronate on bone loss and renal function after kidney transplantation. J Am Soc Nephrol 12:1530–1537

Guadalix S, Martinez-Diaz-Guerra G, Lora D et al (2011) Effect of early risedronate treatment on bone mineral density and bone turnover markers after liver transplantation: a prospective single-center study. Transpl Int 24:657–665

Guanabens N, Pares A, Alvarez L et al (1998) Collagen-related markers of bone turnover reflect the severity of liver fibrosis in patients with primary biliary cirrhosis. J Bone Miner Res 13:731–738

Guichelaar MM, Kendall R, Malinchoc M, Hay JE (2006) Bone mineral density before and after OLT: long-term follow-up and predictive factors. Liver Transpl 12:1390–1402

Guo C, Johnson A, Locke T, Eastell R (1998) Mechanism of bone loss after cardiac transplantation. Bone 22:267–271

Gupta AK, Huang M, Prasad GV (2012) Determinants of bone mineral density in stable kidney transplant recipients. J Nephrol 25:373–383

Haagsma EB, Thijn CJP, Post JG, Slooff MJH, Gisp CH (1988) Bone disease after liver transplantation. J Hepatol 6:94–100

Haas M, Leko-Mohr Z, Roschger P et al (2003) Zoledronic acid to prevent bone loss in the first 6 months after renal transplantation. Kidney Int 63:1130–1136

Hansdottir S, Monick MM, Hinde SL, Lovan N, Look DC, Hunninghake GW (2008) Respiratory epithelial cells convert inactive vitamin D to its active form: potential effects on host defense. J Immunol 181:7090–7099

Hari P, DeFor TE, Vesole DH, Bredeson CN, Burns LJ (2013) Intermittent zoledronic acid prevents bone loss in adults after allogeneic hematopoietic cell transplantation. Biol Blood Marrow Transplant 19:1361–1367

Hariman A, Alex C, Heroux A, Camacho P (2014) Incidence of fractures after cardiac and lung transplantation: a single center experience. J Osteoporos 2014:573041

Haroon M, Fitzgerald O (2012) Vitamin D and its emerging role in immunopathology. Clin Rheumatol 31:199–202

Hausmann A, Hill W, Stemmler HJ et al (2012) Bone loss after allogeneic haematopoietic stem cell transplantation: a pilot study on the use of zoledronic acid. Chemother Res Pract 2012:858590

Hawkins FG, Leon M, Lopez MB et al (1994) Bone loss and turnover in patients with liver transplantation. Hepatogastroenterology 41:158–161

Hay JE (2003) Osteoporosis in liver diseases and after liver transplantation. J Hepatol 38:856–865

Henderson NK, Sambrook PN, Kelly PJ et al (1995) Bone mineral loss and recovery after cardiac transplantation [letter]. Lancet 346:905

Henderson K, Eisman J, Keogh A et al (2001) Protective effect of short-tem calcitriol or cyclical etidronate on bone loss after cardiac or lung transplantation. J Bone Miner Res 16:565–571

Hofbauer LC, Khosla S, Dunstan CR, Lacey DL, Boyle WJ, Riggs BL (2000) The roles of osteoprotegerin and osteoprotegerin ligand in the paracrine regulation of bone resorption. J Bone Miner Res 15:2–12

Hofbauer LC, Shui C, Riggs BL et al (2001) Effects of immunosuppressants on receptor activator of NF-kappaB ligand and osteoprotegerin production by human osteoblastic and coronary artery smooth muscle cells. Biochem Biophys Res Commun 280:334–339

Hommann M, Abendroth K, Lehmann G et al (2002) Effect of transplantation on bone: osteoporosis after liver and multivisceral transplantation. Transplant Proc 34:2296–2298

Horber FF, Casez JP, Steiger U, Czerniack A, Montandon A, Jaeger PH (1994) Changes in bone mass early after kidney transplantation. J Bone Miner Res 9:1–9

Huang WH, Lee SY, Weng CH, Lai PC (2012) Use of alendronate sodium (Fosamax) to ameliorate osteoporosis in renal transplant patients: a case–control study. PLos One 7:e48481

Hullett DA, Cantorna MT, Redaelli C et al (1998) Prolongation of allograft survival by 1,25-dihydroxyvitamin D3. Transplantation 66:824–828

Hussaini SH, Oldroyd B, Stewart SP et al (1999) Regional bone mineral density after orthotopic liver transplantation. Eur J Gastroenterol Hepatol 11:157–163

Iqbal N, Ducharme J, Desai S et al (2008) Status of bone mineral density in patients selected for cardiac transplantation. Endocr Pract 14:704–712

Iyer SP, Nikkel LE, Nishiyama KK et al (2014) Kidney transplantation with early corticosteroid withdrawal: paradoxical effects at the central and peripheral skeleton. J Am Soc Nephrol 25:1331–1341

Jager W, Xu H, Wlcek K, Schuler C, Rubel F, Erben RG (2012) Gender- and dose-related effects of cyclosporin A on hepatic and bone metabolism. Bone 50:140–148

Jastrzebski D, Lutogniewska W, Ochman M et al (2010) Osteoporosis in patients referred for lung transplantation. Eur J Med Res 15(Suppl 2):68–71

Jeon HJ, Han M, Jeong JC et al (2013) Impact of vitamin D, bisphosphonate, and combination therapy on bone mineral density in kidney transplant patients. Transplant Proc 45:2963–2967

Joffe I, Katz I, Jacobs T et al (1992) 17 beta estradiol prevents osteopenia in the oophorectomized rat treated with cyclosporin A. Endocrinology 130:1578–1586

Josephson MA, Schumm LP, Chiu MY, Marshall C, Thistlethwaite JR, Sprague SM (2004) Calcium and calcitriol prophylaxis attenuates posttransplant bone loss. Transplantation 78:1233–1236

Julian BA, Laskow DA, Dubovsky J, Dubovsky EV, Curtis JJ, Quarrles LD (1991) Rapid loss of vertebral bone density after renal transplantation. N Engl J Med 325:544–550

Kaemmerer D, Lehmann G, Wolf G, Settmacher U, Hommann M (2010) Treatment of osteoporosis after liver transplantation with ibandronate. Transpl Int 23:753–759

Kaemmerer D, Schmidt B, Lehmann G, Wolf G, Hommann M, Settmacher U (2012) Monthly ibandronate for the prevention of bone loss in patients after liver transplantation. Transplant Proc 44:1362–1367

Kahan BD (1989) Cyclosporine. N Engl J Med 321:1725–1738

Kananen K, Volin L, Tahtela R, Laitinen K, Ruutu T, Valimaki MJ (2002) Recovery of bone mass and normalization of bone turnover in long-term survivors of allogeneic bone marrow transplantation. Bone Marrow Transplant 29:33–39

Kananen K, Volin L, Laitinen K, Alfthan H, Ruutu T, Valimaki MJ (2005) Prevention of bone loss after allogeneic stem cell transplantation by calcium, vitamin D, and sex hormone replacement with or without pamidronate. J Clin Endocrinol Metab 90:3877–3885

Kang MI, Lee WY, Oh KW et al (2000) The short-term changes of bone mineral metabolism following bone marrow transplantation. Bone 26:275–279

Kashyap A, Kandeel F, Yamauchi D et al (2000) Effects of allogeneic bone marrow transplantation on recipient bone mineral density: a prospective study. Biol Blood Marrow Transplant 6:344–351

Katz IA, Jee WSS, Joffe I et al (1992) Prostaglandin E2 alleviates cyclosporin A-induced bone loss in the rat. J Bone Miner Res 4:1191–1200

Kauppila M, Irjala K, Koskinen P et al (1999) Bone mineral density after allogeneic bone marrow transplantation. Bone Marrow Transplant 24:885–889

Kelly PJ, Atkinson K, Ward RL, Sambrook PN, Biggs JC, Eisman JA (1990) Reduced bone mineral density in men and women with allogeneic bone marrow. Transplantation 50:881–883

Keogh JB, Tsalamandris C, Sewell RB et al (1999) Bone loss at the proximal femur and reduced lean mass following liver transplantation: a longitudinal study. Nutrition 15:661–664

Kerschan-Schindl K, Mitterbauer M, Fureder W et al (2004) Bone metabolism in patients more than five years after bone marrow transplantation. Bone Marrow Transplant 34:491–496

Kerschan-Schindl K, Ruzicka M, Mahr S et al (2008) Unexpected low incidence of vertebral fractures in heart transplant recipients: analysis of bone turnover. Transpl Int 21:255–262

Kidney Disease: Improving Global Outcomes CKDMBDWG (2009) KDIGO clinical practice guideline for the diagnosis, evaluation, prevention, and treatment of Chronic Kidney Disease-Mineral and Bone Disorder (CKD-MBD). Kidney Int Suppl 113:S1–S130

Kohlmeier M, Saupe J, Schaefer K, Asmus G (1998) Bone fracture history and prospective bone fracture risk of hemodialysis patients are related to apolipoprotein E genotype. Calcif Tissue Int 62:278–281

Kovac D, Lindic J, Kandus A, Bren AF (2001) Prevention of bone loss in kidney graft recipients. Transplant Proc 33:1144–1145

Krieg M, Seydoux C, Sandini L et al (2001) Intravenous pamidronate as a treatment for osteoporosis after heart transplantation: a prospective study. Osteoporos Int 12:112–116

Krol CG, Dekkers OM, Kroon HM, Rabelink TJ, van Hoek B, Hamdy NA (2014a) Longitudinal changes in BMD and fracture risk in orthotopic liver transplant recipients not using bone-modifying treatment. J Bone Miner Res 29:1763–1769

Krol CG, Dekkers OM, Kroon HM, Rabelink TJ, van Hoek B, Hamdy NAT (2014b) No association between BMD and prevalent vertebral fractures in liver transplant recipients at time of screening before transplantation. J Clin Endocrinol Metab 99:3677–3685

Kulak C, Shane E (2006) Transplantation osteoporosis: biochemical correlates of pathogenesis and treatment. In: Seibel M, Robins S, Bilezikian J (eds) Dynamics of bone and cartilage metabolism: principles and clinical applications, 2nd edn. Elsevier, San Diego, pp 701–716

Kulak CA, Borba VZ, Kulak J Jr, Custodio MR (2012) Osteoporosis after transplantation. Curr Osteoporos Rep 10:48–55

Kwan JTC, Almond MK, Evans K, Cunningham J (1992) Changes in total body bone mineral content and regional bone mineral density in renal patients following renal transplantation. Miner Electrolyte Metab 18:166–168

Lee AH, Mull RL, Keenan GF et al (1994) Osteoporosis and bone morbidity in cardiac transplant recipients. Am J Med 96:35–41

Lee WY, Kang MI, Oh ES et al (2002a) The role of cytokines in the changes in bone turnover following bone marrow transplantation. Osteoporos Int 13:62–68

Lee WY, Cho SW, Oh ES et al (2002b) The effect of bone marrow transplantation on the osteoblastic differentiation of human bone marrow stromal cells. J Clin Endocrinol Metab 87:329–335

Lee WY, Kang MI, Baek KH et al (2002c) The skeletal site-differential changes in bone mineral density following bone marrow transplantation: 3-year prospective study. J Korean Med Sci 17:749–754

Leidig-Bruckner G, Hosch S, Dodidou P et al (2001) Frequency and predictors of osteoporotic fractures after cardiac or liver transplantation: a follow-up study. Lancet 357:342–347

Lemire JM (1992) Immunomodulatory role of 1,25 Dihydroxyvitamin D3. J Cell Biochem 49:26–31

Lemire JM (1995) Immunomodulatory actions of 1,25-dihydroxyvitamin D3. J Steroid Biochem Mol Biol 53:599–602

Lemire JM, Archer DC, Reddy GS (1994) Dihydroxy-24-oxo-16-ene-vitamin D3, a renal metabolite of the vitamin D analog 1,25-dihydroxy-16ene-vitamin D3, exerts immunosuppressive activity equal to its parent without causing hypercalcemia in vivo. Endocrinology 135:2818–2821

Liu N, Nguyen L, Chun RF et al (2008) Altered endocrine and autocrine metabolism of vitamin D in a mouse model of gastrointestinal inflammation. Endocrinology 149:4799–4808

Lopez MB, Pinto IG, Hawkins F et al (1992) Effect of liver transplantation and immunosuppressive treatment on bone mineral density. Transplant Proc 24:3044–3046

Luo L, Shi Y, Bai Y et al (2012) Impact of tacrolimus on bone metabolism after kidney transplantation. Int Immunopharmacol 13:69–72

Marcen R, Caballero C, Uriol O et al (2007) Prevalence of osteoporosis, osteopenia, and vertebral fractures in long-term renal transplant recipients. Transplant Proc 39:2256–2258

Marshall I, Isserow JA, Buchinsky FJ, Paynton BV, Epstein S (1995) Expression of interleukin 1 and interleukin 6 in bone from normal and cyclosporine treated rats. XIIth International Conference on calcium regulating hormones, Melbourne

Martin K, Al-Aly Z, Gonzalez E (2006) Renal osteodystrophy. In: Favus M (ed) Primer on the metabolic bone diseases and other disorders of bone and mineral metabolism, Wiley-Blackwell. Iowa, USA pp 359–366

Mazzaferro S, Pasquali M (2016) Vitamin D: a dynamic molecule. How relevant might the dynamism for a vitamin be? Nephrol Dial Transplant 31:23–30

Mazziotti G, Angeli A, Bilezikian JP, Canalis E, Giustina A (2006) Glucocorticoid-induced osteoporosis: an update. Trends Endocrinol Metab 17:144–149

McDonald JA, Dunstan CR, Dilworth P et al (1991) Bone loss after liver transplantation. Hepatology 14:613–619

McIntyre HD, Menzies B, Rigby R, Perry-Keene DA, Hawley CM, Hardie IR (1995) Long-term bone loss after renal transplantation: comparison of immunosuppressive regimens. Clin Transplant 9:20–24

Mellibovsky L, Prieto-Alhambra D, Mellibovsky F et al (2015) Bone tissue properties measurement by reference point indentation in glucocorticoid-induced osteoporosis. J Bone Miner Res 30:1651–1656

Meys E, Fontanges E, Fourcade N, Thomasson A, Pouyet M, Delmas P (1994) Bone loss after orthotopic liver transplantation. Am J Med 97:445–450

Mikuls TR, Julian BA, Bartolucci A, Saag KG (2003) Bone mineral density changes within six months of renal transplantation. Transplantation 75:49–54

Millonig G, Graziadei IW, Eichler D et al (2005) Alendronate in combination with calcium and vitamin D prevents bone loss after orthotopic liver transplantation: a prospective single-center study. Liver Transpl 11:960–966

Mittalhenkle A, Gillen DL, Stehman-Breen CO (2004) Increased risk of mortality associated with hip fracture in the dialysis population. Am J Kidney Dis 44:672–679

Mitterbauer C, Schwarz C, Haas M, Oberbauer R (2006) Effects of bisphosphonates on bone loss in the first year after renal transplantation – a meta-analysis of randomized controlled trials. Nephrol Dial Transplant 21:2275–2281

Molnar MZ, Naser MS, Rhee CM, Kalantar-Zadeh K, Bunnapradist S (2014) Bone and mineral disorders after kidney transplantation: therapeutic strategies. Transplant Rev (Orlando) 28:56–62

Monegal A, Navasa M, Guanabens N et al (1997) Osteoporosis and bone mineral metabolism in cirrhotic patients referred for liver transplantation. Calcif Tissue Int 60:148–154

Monegal A, Navasa M, Guanabens N et al (2001a) Bone mass and mineral metabolism in liver transplant patients treated with FK506 or cyclosporine A. Calcif Tissue Int 68:83–86

Monegal A, Navasa M, Guanabens N et al (2001b) Bone disease after liver transplantation: a long-term prospective study of bone mass changes, hormonal status and histomorphometric characteristics. Osteoporos Int 12:484–492

Monegal A, Guanabens N, Suarez MJ et al (2009) Pamidronate in the prevention of bone loss after liver transplantation: a randomized controlled trial. Transpl Int 22:198–206

Monegal A, Navasa M, Peris P et al (2013) Bone disease in patients awaiting liver transplantation. Has the situation improved in the last two decades? Calcif Tissue Int 93:571–576

Movsowitz C, Epstein S, Fallon M, Ismail F, Thomas S (1988) Cyclosporin a in vivo produces severe osteopenia in the rat: effect of dose and duration of administration. Endocrinology 123:2571–2577

Movsowitz C, Epstein S, Ismail F, Fallon M, Thomas S (1989) Cyclosporin A in the oophorectomized rat: unexpected severe bone resorption. J Bone Miner Res 4:393–398

Navasa M, Monegal A, Guanabens N et al (1994) Bone fractures in liver transplant patients. Br J Rheumatol 33:52–55

Naylor KL, Li AH, Lam NN, Hodsman AB, Jamal SA, Garg AX (2013) Fracture risk in kidney transplant recipients: a systematic review. Transplantation 95:1461–1470

Naylor KL, Garg AX, Hodsman AB, Rush DN, Leslie WD (2014) Long-term changes in bone mineral density in kidney transplant recipients. Transplantation 98:1279–1285

Naylor KL, Jamal SA, Zou G et al (2015) Fracture incidence in adult kidney transplant recipients. Transplantation 100:167–175

Nickolas TL, McMahon DJ, Shane E (2006) Relationship between moderate to severe kidney disease and hip fracture in the United States. J Am Soc Nephrol 17:3223–3232

Nikkel L, Mohan S, Tanriover B et al (2011) Does simultaneous pancreas-kidney transplantation reduce fracture risk? An analysis of the USRDS. Am Soc Nephrol

Nikkel LE, Mohan S, Zhang A et al (2012) Reduced fracture risk with early corticosteroid withdrawal after kidney transplant. Am J Transplant 12:649–659

Nikkel LE, Iyer SP, Mohan S et al (2013) Pancreas-kidney transplantation is associated with reduced fracture risk compared with kidney-alone transplantation in men with type 1 diabetes. Kidney Int 83:471–478

Ninkovic M, Skingle SJ, Bearcroft PW, Bishop N, Alexander GJ, Compston JE (2000) Incidence of vertebral fractures in the first three months after orthotopic liver transplantation. Eur J Gastroenterol Hepatol 12:931–935

Ninkovic M, Love S, Tom BD, Bearcroft PW, Alexander GJ, Compston JE (2002) Lack of effect of intravenous pamidronate on fracture incidence and bone mineral density after orthotopic liver transplantation. J Hepatol 37:93–100

Nishioka S, Sofue T, Inui M et al (2014) Mineral and bone disorder is temporary in patients treated with early rapid corticosteroid reduction after kidney transplantation: a single-center experience. Transplant Proc 46:514–520

Nishiyama KK, Pauchard Y, Nikkel LE et al (2015) Longitudinal HR-pQCT and image registration detects endocortical bone loss in kidney transplantation patients. J Bone Miner Res 30:554–561

Nowacka-Cieciura E, Cieciura T, Baczkowska T et al (2006) Bisphosphonates are effective prophylactic of early bone loss after renal transplantation. Transplant Proc 38:165–167

Nysom K, Holm K, Michaelsen KF et al (2000) Bone mass after allogeneic BMT for childhood leukaemia or lymphoma. Bone Marrow Transplant 25:191–196

Obi Y, Hamano T, Ichimaru N et al (2014) Vitamin D deficiency predicts decline in kidney allograft function: a prospective cohort study. J Clin Endocrinol Metab 99:527–535

Okamoto M, Yamanaka S, Yoshimoto W, Shigematsu T (2014) Alendronate as an effective treatment for bone loss and vascular calcification in kidney transplant recipients. J Transplant 2014:269613

Ott SM, Aitken ML (1998) Osteoporosis in patients with cystic fibrosis. Clin Chest Med 19:555–567

Overman RA, Gourlay ML, Deal CL, Farley JF, Brookhart MA, Layton JB (2015) Fracture rate associated with quality metric-based anti-osteoporosis treatment in glucocorticoid-induced osteoporosis. Osteoporos Int 26:1515–1524

Palmer SC, Strippoli GF, McGregor DO (2005) Interventions for preventing bone disease in kidney transplant recipients: a systematic review of randomized controlled trials. Am J Kidney Dis 45:638–649

Park KM, Hay JE, Lee SG et al (1996) Bone loss after orthotopic liver transplantation: FK 506 versus cyclosporine. Transplant Proc 28:1738–1740

Paterson AM, Yates SF, Nolan KF, Waldmann H (2005) The new immunosuppression: intervention at the dendritic cell-T-cell interface. Curr Drug Targets Immune Endocr Metabol Disord 5:397–411

Pelletier S, Dubourg L, Carlier MC, Hadj-Aissa A, Fouque D (2013) The relation between renal function and serum sclerostin in adult patients with CKD. Clin J Am Soc Nephrol 8:819–823

Pennisi P, Trombetti A, Giostra E, Mentha G, Rizzoli R, Fiore CE (2007) Pamidronate and osteoporosis prevention in liver transplant recipients. Rheumatol Int 27:251–256

Peric M, Koglin S, Kim SM et al (2008) IL-17A enhances vitamin D3-induced expression of cathelicidin antimicrobial peptide in human keratinocytes. J Immunol 181:8504–8512

Pescovitz MD (2006) B cells: a rational target in alloantibody-mediated solid organ transplantation rejection. Clin Transplant 20:48–54

Petryk A, Polgreen LE, Zhang L et al (2014) Bone mineral deficits in recipients of hematopoietic cell transplantation: the impact of young age at transplant. Bone Marrow Transplant 49:258–263

Piemontese M, Onal M, Xiong J et al (2015) Suppression of autophagy in osteocytes does not modify the adverse effects of glucocorticoids on cortical bone. Bone 75:18–26

Piraino B, Chen T, Cooperstein L, Segre G, Puschett J (1988) Fractures and vertebral bone mineral density in patients with renal osteodystrophy. Clin Nephrol 30:57–62

Ponticelli C, Aroldi A (2001) Osteoporosis after organ transplantation. Lancet 357:1623

Porayko MK, Wiesner RH, Hay JE et al (1991) Bone disease in liver transplant recipients: incidence, timing, and risk factors. Transplant Proc 23:1462–1465

Pundole XN, Barbo AG, Lin H, Champlin RE, Lu H (2015) Increased incidence of fractures in recipients of hematopoietic stem-cell transplantation. J Clin Oncol 33:1364–1370

Querings K, Girndt M, Geisel J, Georg T, Tilgen W, Reichrath J (2006) 25-hydroxyvitamin D deficiency in renal transplant recipients. J Clin Endocrinol Metab 91:526–529

Rabinowitz M, Shapiro J, Lian J, Bloch GD, Merkel IS, Thiel DHV (1992) Vitamin D and osteocalcin levels in liver transplant recipients. Is osteocalcin a reliable marker of bone turnover in such cases? J Hepatol 16:50–55

Ramsey-Goldman R, Dunn JE, Dunlop DD et al (1999) Increased risk of fracture in patients receiving solid organ transplants. J Bone Miner Res 14:456–463

Rathi M, Kumar D, Bhadada SK et al (2015) Sequential changes in bone biochemical parameters and bone mineral density after renal transplant. Saudi J Kidney Dis Transpl 26:671–677

Redaelli CA, Wagner M, Tien YH et al (2001) 1 alpha,25-Dihydroxycholecalciferol reduces rejection and improves survival in rat liver allografts. Hepatology 34:926–934

Regazzi MB, Alessiani M, Rinaldi M (2005) New strategies in immunosuppression. Transplant Proc 37:2675–2678

Rich GM, Mudge GH, Laffel GL, LeBoff MS (1992) Cyclosporine A and prednisone-associated osteoporosis in heart transplant recipients. J Heart Lung Transplant 11:950–958

Rizzari MD, Suszynski TM, Gillingham KJ et al (2012) Ten-year outcome after rapid discontinuation of prednisone in adult primary kidney transplantation. Clin J Am Soc Nephrol 7:494–503

Roe SD, Porter CJ, Godber IM, Hosking DJ, Cassidy MJ (2005) Reduced bone mineral density in male renal transplant recipients: evidence for persisting hyperparathyroidism. Osteoporos Int 16:142–148

Ross F (2006) Osteoclast biology and bone resorption. In: Favus M (ed) Primer on the metabolic bone diseases and other disorders of bone and mineral metabolism, Wiley-Blackwell. Iowa, USA pp 30–35

Rucinski B, Liu CC, Epstein S (1994) Utilization of cyclosporine H to elucidate the possible mechanisms of cyclosporine A induced osteopenia in the rat. Metabolism 43:1114–1118

Rutherford RM, Fisher AJ, Hilton C et al (2005) Functional status and quality of life in patients surviving 10 years after lung transplantation. Am J Transplant 5:1099–1104

Saag KG, Zanchetta JR, Devogelaer JP et al (2009) Effects of teriparatide versus alendronate for treating glucocorticoid-induced osteoporosis: thirty-six-month results of a randomized, double-blind, controlled trial. Arthritis Rheum 60:3346–3355

Sambrook P (2006) Glucocorticoid-induced osteoporosis. In: Favus M (ed) Primer on the metabolic bone diseases and other disorders of bone and mineral metabolism, Wiley-Blackwell. Iowa, USA pp 296–302

Sambrook PN, Kelly PJ, Keogh A et al (1994a) Bone loss after cardiac transplantation: a prospective study. J Heart Lung Transplant 13:116–121

Sambrook PN, Kelly PJ, Fontana D et al (1994b) Mechanisms of rapid bone loss following cardiac transplantation. Osteoporos Int 4:273–276

Sambrook P, Henderson NK, Keogh A et al (2000) Effect of calcitriol on bone loss after cardiac or lung transplantation. J Bone Miner Res 15:1818–1824

Sanchez-Escuredo A, Fuster D, Rubello D et al (2015) Monthly ibandronate versus weekly risedronate treatment for low bone mineral density in stable renal transplant patients. Nucl Med Commun 36:815–818

Sanders KM, Stuart AL, Williamson EJ et al (2010) Annual high-dose oral vitamin D and falls and fractures in older women: a randomized controlled trial. JAMA 303:1815–1822

Sarno G, Daniele G, Tirabassi G et al (2015) The impact of vitamin D deficiency on patients undergoing kidney transplantation: focus on cardiovascular, metabolic, and endocrine outcomes. Endocrine 50:568–574

Sass DA, Bowman AR, Marshall I et al (1997) Alendronate prevents cyclosporin-induced osteopenia in the rat. Bone 21:65–70

Savaj S, Ghods FJ (2012) Vitamin D, parathyroid hormone, and bone mineral density status in kidney transplant recipients. Iran J Kidney Dis 6:295–299

Schauber J, Oda Y, Buchau AS et al (2008) Histone acetylation in keratinocytes enables control of the expression of cathelicidin and CD14 by 1,25-dihydroxyvitamin D3. J Invest Dermatol 128:816–824

Schulte C, Beelen DW, Schaefer UW, Mann K (2000) Bone loss in long-term survivors after transplantation of hematopoietic stem cells: a prospective study. Osteoporos Int 11:344–353

Serio B, Pezzullo L, Fontana R et al (2013) Accelerated bone mass senescence after hematopoietic stem cell transplantation. Transl Med UniSa 5:7–13

Shane E (2003) Transplantation osteoporosis. In: Orwoll E, Bliziotes M (eds) Osteoporosis: pathophysiology and clinical management. Humana Press, Totowa, pp 537–567

Shane E, Epstein S (2001) Transplantation osteoporosis. Transplant Rev 15:11–32

Shane E, Rivas MDC, Silverberg SJ, Kim TS, Staron RB, Bilezikian JP (1993) Osteoporosis after cardiac transplantation. Am J Med 94:257–264

Shane E, Rivas M, Staron RB et al (1996a) Fracture after cardiac transplantation: a prospective longitudinal study. J Clin Endocrinol Metab 81:1740–1746

Shane E, Silverberg SJ, Donovan D et al (1996b) Osteoporosis in lung transplantation candidates with end stage pulmonary disease. Am J Med 101:262–269

Shane E, Mancini D, Aaronson K et al (1997a) Bone mass, vitamin D deficiency and hyperparathyroidism in congestive heart failure. Am J Med 103:197–207

Shane E, Rivas M, McMahon DJ et al (1997b) Bone loss and turnover after cardiac transplantation. J Clin Endocrinol Metab 82:1497–1506

Shane E, Rodino MA, McMahon DJ et al (1998) Prevention of bone loss after heart transplantation with antiresorptive therapy: a pilot study. J Heart Lung Transplant 17:1089–1096

Shane E, Papadopoulos A, Staron RB et al (1999) Bone loss and fracture after lung transplantation. Transplantation 68:220–227

Shane E, Addesso V, Namerow PB et al (2004) Alendronate versus calcitriol for the prevention of bone loss after cardiac transplantation. N Engl J Med 350:767–776
Shane E, Cohen A, Stein EM et al (2012) Zoledronic acid versus alendronate for the prevention of bone loss after heart or liver transplantation. J Clin Endocrinol Metab 97:4481–4490
Shi C, Huang P, Kang H et al (2015a) Glucocorticoid inhibits cell proliferation in differentiating osteoblasts by microRNA-199a targeting of WNT signaling. J Mol Endocrinol 54:325–337
Shi J, Wang L, Zhang H et al (2015b) Glucocorticoids: dose-related effects on osteoclast formation and function via reactive oxygen species and autophagy. Bone 79:222–232
Shimizu C, Fujita T, Fuke Y et al (2013) Effects of cyclosporine on bone mineral density in patients with glucocorticoid-dependent nephrotic syndrome in remission. Int Urol Nephrol 45:803–808
Smallwood GA, Burns D, Fasola CG, Steiber AC, Heffron TG (2005) Relationship between immunosuppression and osteoporosis in an outpatient liver transplant clinic. Transplant Proc 37:1910–1911
Smerud KT, Dolgos S, Olsen IC et al (2012) A 1-year randomized, double-blind, placebo-controlled study of intravenous ibandronate on bone loss following renal transplantation. Am J Transplant 12:3316–3325
Spira A, Gutierrez C, Chaparro C, Hutcheon MA, Chan CK (2000) Osteoporosis and lung transplantation: a prospective study. Chest 117:476–481
Sprague SM, Belozeroff V, Danese MD, Martin LP, Olgaard K (2008) Abnormal bone and mineral metabolism in kidney transplant patients--a review. Am J Nephrol 28:246–253
Stein EM, Shane E (2011) Vitamin D in organ transplantation. Osteoporos Int 22:2107–2118
Stein B, Halloran P, Reinhardt T et al (1991) Cyclosporin A increases synthesis of 1,25 dihydroxyvitamin D3 in the rat and mouse. Endocrinology 128:1369–1373
Stein EM, Cohen A, Freeby M et al (2009) Severe vitamin D deficiency among heart and liver transplant recipients. Clin Transplant 23:861–865
Stempfle HU, Werner C, Echtler S et al (1998) Rapid trabecular bone loss after cardiac transplantation using FK506 (tacrolimus)-based immunosuppression. Transplant Proc 30:1132–1133
Sukumaran Nair S, Lenihan CR, Montez-Rath ME, Lowenberg DW, Chertow GM, Winkelmayer WC (2014) Temporal trends in the incidence, treatment and outcomes of hip fracture after first kidney transplantation in the United States. Am J Transplant 14:943–951
Sun L, Blair HC, Peng Y et al (2005) Calcineurin regulates bone formation by the osteoblast. Proc Natl Acad Sci U S A 102:17130–17135
Sun L, Peng Y, Zaidi N et al (2007) Evidence that calcineurin is required for the genesis of bone-resorbing osteoclasts. Am J Physiol Renal Physiol 292:F285–F291
Tamler R, Epstein S (2006) Nonsteroid immune modulators and bone disease. Ann N Y Acad Sci 1068:284–296
Tanaci N, Karakose H, Guvener N, Tutuncu NB, Colak T, Haberal M (2003) Influence of 1,25-dihydroxyvitamin D3 as an immunomodulator in renal transplant recipients: a retrospective cohort study. Transplant Proc 35:2885–2887
Tauchmanova L, Serio B, Del Puente A et al (2002) Long-lasting bone damage detected by dual-energy x-ray absorptiometry, phalangeal osteosonogrammetry, and in vitro growth of marrow stromal cells after allogeneic stem cell transplantation. J Clin Endocrinol Metab 87:5058–5065
Tauchmanova L, Selleri C, Esposito M et al (2003) Beneficial treatment with risedronate in long-term survivors after allogeneic stem cell transplantation for hematological malignancies. Osteoporos Int 14:1013–1019
Tauchmanova L, Ricci P, Serio B et al (2005) Short-term zoledronic acid treatment increases bone mineral density and marrow clonogenic fibroblast progenitors after allogeneic stem cell transplantation. J Clin Endocrinol Metab 90:627–634
Techawathanawanna N, Avihingsanon Y, Praditpornsilpa K et al (2005) The prevalence and risk factors of osteoporosis in Thai renal-transplant patients. J Med Assoc Thai 88(Suppl 4):S103–S109

ter Meulen CG, van Riemsdijk I, Hene RJ et al (2004) No important influence of limited steroid exposure on bone mass during the first year after renal transplantation: a prospective, randomized, multicenter study. Transplantation 78:101–106

Thiebaud D, Krieg M, Gillard-Berguer D, Jaquet A, Goy JJ, Burckhardt P (1996) Cyclosporine induces high turnover and may contribute to bone loss after heart transplantation. Eur J Clin Invest 26:549–555

Toro J, Gentil MA, Garcia R et al (2005) Alendronate in kidney transplant patients: a single-center experience. Transplant Proc 37:1471–1472

Torregrosa JV, Campistol JM, Montesinos M et al (1995) Factors involved in the loss of bone mineral density after renal transplantation. Transplant Proc 27:2224–2225

Torregrosa JV, Fuster D, Pedroso S et al (2007) Weekly risedronate in kidney transplant patients with osteopenia. Transpl Int 20:708–711

Torregrosa JV, Fuster D, Monegal A et al (2011) Efficacy of low doses of pamidronate in osteopenic patients administered in the early post-renal transplant. Osteoporos Int 22:281–287

Torres A, Garcia S, Gomez A et al (2004) Treatment with intermittent calcitriol and calcium reduces bone loss after renal transplantation. Kidney Int 65:705–712

Tripathi SS, Gibney EM, Gehr TW, King AL, Beckman MJ (2008) High prevalence of vitamin D deficiency in African American kidney transplant recipients. Transplantation 85:767–770

Trombetti A, Gerbase MW, Spiliopoulos A, Slosman DO, Nicod LP, Rizzoli R (2000) Bone mineral density in lung-transplant recipients before and after graft: prevention of lumbar spine post-transplantation-accelerated bone loss by pamidronate. J Heart Lung Transplant 19:736–743

Tschopp O, Schmid C, Speich R, Seifert B, Russi EW, Boehler A (2006) Pretransplantation bone disease in patients with primary pulmonary hypertension. Chest 129:1002–1008

Uyar M, Sezer S, Arat Z, Elsurer R, Ozdemir FN, Haberal M (2006) 1,25-dihydroxyvitamin D(3) therapy is protective for renal function and prevents hyperparathyroidism in renal allograft recipients. Transplant Proc 38:2069–2073

Valero M, Loinaz C, Larrodera L, Leon M, Morena E, Hawkins F (1995) Calcitonin and bisphosphonate treatment in bone loss after liver transplantation. Calcif Tissue Int 57:15–19

Valimaki MJ, Kinnunen K, Tahtela R et al (1999a) A prospective study of bone loss and turnover after cardiac transplantation: effect of calcium supplementation with or without calcitonin. Osteoporos Int 10:128–136

Valimaki M, Kinnunen K, Volin L et al (1999b) A prospective study of bone loss and turnover after allogeneic bone marrow transplantation: effect of calcium supplementation with or without calcitonin. Bone Marrow Transplant 23:355–361

Van Cleemput J, Daenen W, Nijs J, Geusens P, Dequeker J, Vanhaecke J (1995) Timing and quantification of bone loss in cardiac transplant recipients. Transpl Int 8:196–200

van den Ham EC, Kooman JP, Christiaans ML, van Hooff JP (2003) The influence of early steroid withdrawal on body composition and bone mineral density in renal transplantation patients. Transpl Int 16:82–87

van Staa TP (2006) The pathogenesis, epidemiology and management of glucocorticoid-induced osteoporosis. Calcif Tissue Int 79:129–137

Van Staa TP, Leufkens HG, Abenhaim L, Zhang B, Cooper C (2000) Use of oral corticosteroids and risk of fractures. J Bone Miner Res 15:993–1000

Vedi J, Greer S, Skingle S et al (1999) Mechanism of bone loss after liver transplantation: a histomorphometric analysis. J Bone Miner Res 14:281–287

Venu M, Martin E, Saeian K, Gawrieh S (2013) High prevalence of vitamin A deficiency and vitamin D deficiency in patients evaluated for liver transplantation. Liver Transpl 19:627–633

Wagner D, Amrein K, Dimai HP et al (2012) Ibandronate and calcitriol reduces fracture risk, reverses bone loss, and normalizes bone turnover after LTX. Transplantation 93:331–336

Walsh SB, Altmann P, Pattison J et al (2009) Effect of pamidronate on bone loss after kidney transplantation: a randomized trial. Am J Kidney Dis 53:856–865

Wang TK, O'Sullivan S, Gamble GD, Ruygrok PN (2013) Bone density in heart or lung transplant recipients – a longitudinal study. Transplant Proc 45:2357–2365

Wang Z, Han Z, Tao J et al (2014) Clinical efficacy and safety of pamidronate therapy on bone mass density in early post-renal transplant period: a meta-analysis of randomized controlled trials. PLoS One 9:e108106

Watson RG, Coulton L, Kanis JA et al (1990) Circulating osteocalcin in primary biliary cirrhosis following liver transplantation and during treatment with cyclosporine. J Hepatol 11:354–358

Wei R, Christakos S (2015) Mechanisms underlying the regulation of innate and adaptive immunity by vitamin D. Nutrients 7:8251–8260

Weinstein R, Jilka R, Parfitt M, Manolagas S (1998) Inhibition of osteoblastogenesis and promotion of apoptosis of osteoblasts and osteocytes by glucocorticoids: potential mechanisms of their deleterious effects on bone. J Clin Invest 102:274–282

Westendorf JJ, Kahler RA, Schroeder TM (2004) Wnt signaling in osteoblasts and bone diseases. Gene 341:19–39

Withold W, Wolf H, Kollbach S, Heyll A, Schneider W, Reinauer H (1996) Monitoring of bone metabolism after bone marrow transplantation by measuring two different markers of bone turnover. Eur J Clin Chem Clin Biochem 34:193–197

Woodle ES, First MR, Pirsch J, Shihab F, Gaber AO, Van Veldhuisen P (2008) A prospective, randomized, double-blind, placebo-controlled multicenter trial comparing early (7 day) corticosteroid cessation versus long-term, low-dose corticosteroid therapy. Ann Surg 248:564–577

Yamamoto S, Suzuki A, Sasaki H et al (2013) Oral alendronate can suppress bone turnover but not fracture in kidney transplantation recipients with hyperparathyroidism and chronic kidney disease. J Bone Miner Metab 31:116–122

Yao S, McCarthy PL, Dunford LM et al (2008) High prevalence of early-onset osteopenia/osteoporosis after allogeneic stem cell transplantation and improvement after bisphosphonate therapy. Bone Marrow Transplant 41:393–398

Yu TM, Lin CL, Chang SN, Sung FC, Huang ST, Kao CH (2014) Osteoporosis and fractures after solid organ transplantation: a nationwide population-based cohort study. Mayo Clin Proc 89:888–895

Yun YS, Kim BJ, Hong TW, Lee CG, Kim MJ (1996) Changes of bone metabolism indices in patients receiving immunosuppressive therapy including low doses of steroids after transplantation. Transplant Proc 28:1561–1564

Zahner M, Teraz F, Pacifici R (1997) T cells mediate a stimulatory effect of cyclosporine A on human osteoclastogenesis while immature osteoclast precursors are directly regulated by glucocorticoids. J Bone Miner Res 12(Suppl 1):198

Zhang AB, Zheng SS, Jia CK, Wang Y (2003) Effect of 1,25-dihydroxyvitamin D3 on preventing allograft from acute rejection following rat orthotopic liver transplantation. World J Gastroenterol 9:1067–1071

Osteoimmunological Aspects of Periodontal Diseases

12

Kristina Bertl, Peter Pietschmann, and Andreas Stavropoulos

Abbreviations

AgP	Aggressive periodontitis
C3	Complement component 3
CP	Chronic periodontitis
IFN	Interferon
Ig	Immunoglobulin
IL	Interleukin
iNOS	Inducible NO-synthases
LPS	Lipopolysaccharide
M-CSF	Macrophage colony-stimulating factor
MMP	Matrix metalloproteinases
NO	Nitric oxide
OPG	Osteoprotegerin
RANK	Receptor activator of NF-kB

K. Bertl, DMD, PhD, MSc (✉)
Department of Periodontology, Faculty of Odontology, University of Malmö, Malmö, Sweden

Division of Oral Surgery, University Clinic of Dentistry, Medical University of Vienna, Vienna, Austria
e-mail: kristina.bertl@mah.se

P. Pietschmann, MD
Department of Pathophysiology and Allergy Research, Center of Pathophysiology, Infectiology and Immunology, Medical University Vienna, Vienna, Austria

A. Stavropoulos, DDS, PhD
Department of Periodontology, Faculty of Odontology, University of Malmö, Malmö, Sweden

P. Pietschmann (ed.), *Principles of Osteoimmunology*,
DOI 10.1007/978-3-319-34238-2_12

RANKL	Receptor activator of nuclear factor "kappa-light-chain-enhancer" of the activated B cell ligand
TGF	Transforming growth factor
Th	T helper cells
TIMP	Tissue inhibitors of metalloproteinase
TLR	Toll-like receptors
TNF	Tumor necrosis factor
Treg	Regulatory T cells

12.1 Osteoimmunological Aspects of Periodontal Diseases

Osteoimmunology describes the cross-talk of cells of the musculoskeletal and the immune system during the pathogenesis of various diseases; among the most prevalent ones is *periodontitis*, a chronic infectious inflammatory disease of the tooth-supporting structures, i.e., the periodontium consisting of the gingiva, alveolar bone, periodontal ligament, and root cementum. Periodontal disease is initiated by oral pathogens that accumulate at the gingival margin of the teeth and thereby trigger a response of the innate and adaptive immunity. In contrast to other osteoimmunological disorders (e.g., osteoporosis, rheumatoid arthritis), the immune system plays a two-sided role in the pathogenesis of periodontitis: it controls the infection and protects the organism from bacterial invasion but also propagates the destruction of the soft and hard tissues surrounding the tooth. Periodontitis, if left untreated, finally results in tooth loss.

The present chapter discusses, in a concise manner, the interplay of pathogens, mediators, and enzymes, and host cells involved in the pathogenesis of periodontitis. Further, a short overview on some of the potential "osteoimmunological targets" for the treatment of periodontitis is included.

12.1.1 Anatomy of the Periodontium

The term "*periodontium*" is derived from the Greek terms "peri-" (around) and "-odous" (tooth); it consists of four main components, which differ in their histological structure but act functionally as a single unit:

- Gingiva
- Periodontal ligament
- Alveolar bone
- Root cementum

The gingiva is via the junctional epithelium attached to the tooth surface and thereby forms a barrier against external factors (e.g., bacterial invasion). The other three components of the periodontium are forming the "attachment apparatus"; the

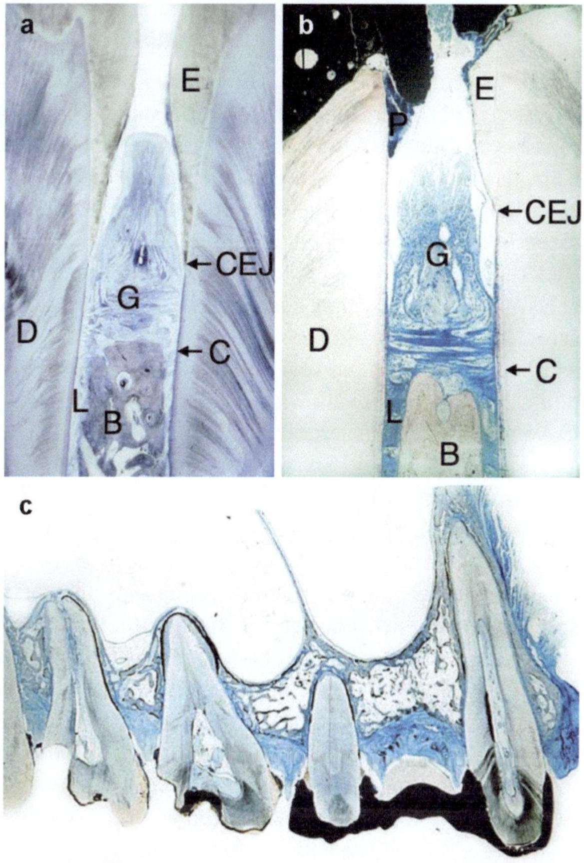

Fig, 12.1 Thin ground section of the periodontium of two adjacent teeth, stained with toluidine blue, in health (**a**) and disease (**b**). *E* enamel, *CEJ* cementoenamel junction, *P* plaque (due to overhanging crown margins), *G* gingival fibers, *D* dentin, *L* periodontal ligament, *B* alveolar bone, *C* root cementum. (**c**) Thin ground section of the lateral teeth of the upper jaw (By courtesy of late Prof. Dr. mult. K. Donath and Prof. DDr. Ulm)

periodontal ligament consists of a rich vasculofibrous connective tissue, with collagen fibers incorporated in the root cementum and in the alveolar bone, and provides tooth support and tooth function (Fig. 12.1a c).

12.1.2 Periodontitis: Disease Initiation

The oral cavity provides a habitat for about *750 different microbial species* and is one of the most complicated microbial ecosystems in the human body (Jenkinson and Lamont 2005). It provides huge amounts of nutrients through the saliva, gingival crevicular fluid, and food remnants, particularly those containing sugar. The first steps of dental plaque and *biofilm formation* occur within minutes after tooth brushing. A pellicle, consisting mainly of salivary proteins, attaches first at surface pits and fissures of the tooth and at the smooth tooth surfaces of the interdental area and the gingival margin, which are not affected much by the constant movement of the cheek and tongue. Thereafter, early bacterial colonizers such as *Streptococci* species,

Actinomyces species, *Capnocytophaga* species, and *Veillonella* species are able to attach to salivary pellicle receptors (Foster and Kolenbrander 2004; Listgarten 1994; Ritz 1967; Teles et al. 2013). The primary bacterial attachment may still be reversible, but if not interrupted, the bacteria start to build up a biofilm by multiplying, forming multilayers, and secreting extracellular matrix. This extracellular matrix helps the bacteria to stick together, form the biofilm architecture, and provide each other nutrients. Further, the early colonizers provide binding sites for the next colonizers (e.g., *Fusobacterium nucleatum*). *F. nucleatum*, which is an obligate anaerobic bacterium, plays a highly relevant role in biofilm maturation. *F. nucleatum* co-aggregates with multiple other bacteria, including early as well as late colonizers, and aerobic as well as anaerobic species and is present in both periodontally healthy sites and in diseased sites, however, in larger amounts in diseased sites. The co-aggregation with *F. nucleatum* allows late colonizers, which are mainly gram-negative facultative and obligate anaerobic bacteria (e.g., *Porphyromonas gingivalis*, *Tannerella forsythia*, *Treponema denticola*, *Aggregatibacter actinomycetemcomitans*), to survive in an aerated environment due to oxygen gradients in the biofilm structure (Bradshaw et al. 1998; Holt and Ebersole 2005).

Biofilm formation at the gingival margin causes in the beginning gingivitis, which is a reversible condition affecting only the soft tissues (Löe et al. 1965). In susceptible patients, however, increasing biofilm accumulation provokes progression to periodontitis, which involves periodontal attachment and bone loss, mostly irreversible. Recently the *"polymicrobial synergy and dysbiosis"* model was proposed for periodontal disease pathogenesis (Hajishengallis et al. 2012; Hajishengallis and Lamont 2012). Herein, single pathogens, so-called keystone pathogens, are able to initiate a shift of the entire microbial colonization from homeostasis to dysbiosis, i.e., the imbalance of a microbial flora in an ecosystem associated with the presence of a disease, thereby causing a strong immune response. Thus far, *P. gingivalis* is the best-documented keystone pathogen. Its presence, although not necessarily in high amounts, elevates the pathogenicity of the entire oral microbial community, e.g., by an increased expression of various virulence factors (Hajishengallis 2014; Hajishengallis and Lamont 2012). An experimental periodontitis model with mice demonstrated that presence of *P. gingivalis* in less than 0.01 % of the total bacterial amount was still enough for periodontal disease initiation (Hajishengallis et al. 2011). However, after this initiating step, the major portion of periodontal tissue degradation is not caused by the biofilm; in fact, it is the host immune response that propagates the loss of periodontal structures.

12.1.3 Periodontitis: Prevalence and Risk Factors

Periodontitis represents besides caries the major reason for tooth loss and is vastly prevalent in the population. In Germany, the *prevalence of periodontitis* according to the case definitions developed by the Centers for Disease Control and Prevention and the American Academy of Periodontology (Eke et al. 2012; Page and Eke 2007) was 71 and 87 % in age cohorts of 35–44 and 65–74 years, respectively; 17 and 42 %, respectively, suffered from severe forms of periodontitis, i.e., advanced loss of the periodontal structures and bone (Holtfreter et al. 2010). *Risk factors* and *risk indicators* for periodontal disease may be both nonmodifiable and modifiable; examples of nonmodifiable factors include age, male gender, ethnicity, and genetic factors, while modifiable factors include smoking, alcohol consumption, low educational level, poorly controlled diabetes, obesity and metabolic syndrome, osteoporosis, and stress and its coping mechanisms (Genco and Borgnakke 2013). Page and Kornman (1997) illustrated the interplay between the bacterial attack, the immune response, and the risk factors in the pathogenesis of periodontal disease as shown in Fig. 12.2.

12.1.4 Periodontal Disease Entities

The current classification system of periodontal diseases was established in 1999 at the International Workshop for the Classification of Periodontal Diseases and Conditions (Armitage 1999):

1. Gingival diseases
2. *Chronic periodontitis*
3. *Aggressive periodontitis*
4. Periodontitis as a manifestation of systemic diseases
5. Necrotizing periodontal diseases
6. Abscesses of the periodontium
7. Periodontitis associated with endodontic lesions
8. Developmental or acquired deformities and conditions

This chapter deals with osteoimmunological aspects associated with alveolar bone loss occurring in patients with aggressive or chronic periodontitis. *Aggressive*

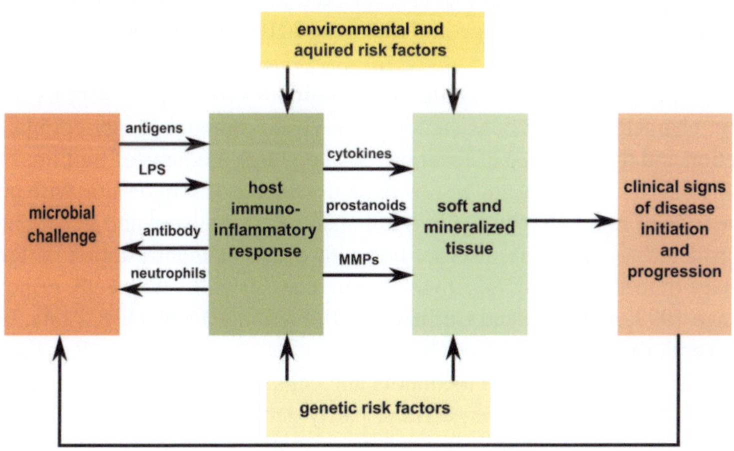

Fig. 12.2 Pathogenesis model of periodontal disease. (Modified from Page and Kornman 1997)

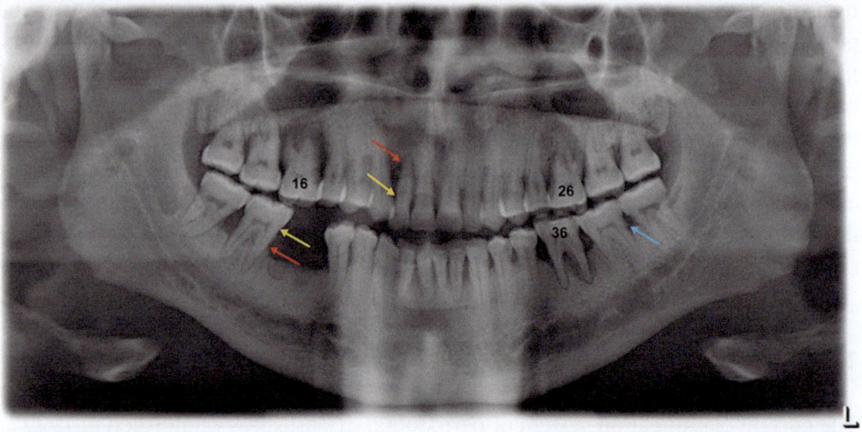

Fig. 12.3 Panoramic radiograph of a 24-year-old man with severe generalized aggressive periodontitis. *Yellow arrows* indicate the regular height of the alveolar bone in a healthy patient, while the *red arrows* indicate the bone level of this patient. The *light blue arrow* is indicating calculus, which is a rare finding in aggressive periodontitis. Severe bone loss is present around the first molar teeth (tooth number 16, 26, and 36; tooth 46 is already missing)

periodontitis (AgP) is characterized by rapid attachment loss, which is most often inconsistent with the amount of microbial deposits. AgP patients are systemically healthy, i.e., no specific immune compromising condition, but familial aggregation is characteristic. AgP appears clinically as localized and generalized AgP, currently considered to represent two different disease entities. Localized AgP is primarily restricted to first molars and incisors and has a circumpubertal onset, and patients show a high serum antibody response against the infecting agents. In contrast, generalized AgP affects individuals around 30 years of age, but patients may be older; further, generalized AgP patients present a poor serum antibody response against the infecting agents and a pronounced episodic destruction of periodontal attachment (Armitage 1999; Armitage and Cullinan 2010; Tonetti and Mombelli 1999) (Fig. 12.3).

Chronic periodontitis (CP) is the most common form of periodontitis in adults and it is characterized by consistency between the amount of periodontal tissue destruction and the presence of local factors, i.e., large amounts of biofilm. Further, a variable microbial pattern and a slow to moderate rate of progression with possible burst periods of rapid attachment loss are observed. Localized and generalized CP are considered as the same disease, but with different manifestation in terms of percentage of teeth affected, i.e., up to or more than 30 % affected teeth, respectively (Armitage 1999; Armitage and Cullinan 2010; Flemmig 1999) (Fig. 12.4a, b).

In general, the biochemical and histological characteristics, as well as the pathological processes and involved mediators and cells in the lesions, are rather similar between AgP and CP. However, certain, partly genetic, but yet not fully clarified, differences are causing the existing variations in the onset, magnitude, extent, and progression rate of tissue destruction (Kulkarni and Kinane 2014).

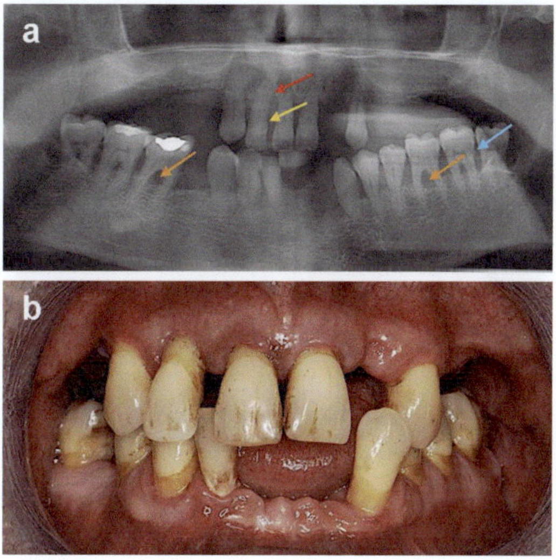

Fig. 12.4 Panoramic radiograph (**a**) and clinical picture (**b**) of a 52-year-old woman with severe generalized chronic periodontitis. The *yellow arrow* indicates the regular height of the alveolar bone in a healthy patient, while the *red arrows* indicate the bone level of this patient. The *orange arrows* are indicating the lost bone between the roots of the lower molars, i.e., furcation involvement, and the light *blue arrow* is indicating calculus

12.1.5 Periodontal Disease Progression

In summary, the first defense mechanism against the biofilm is represented by the *innate immune system*; while at later stages, the *adaptive immune system* becomes more relevant. After the initial step of biofilm accumulation at the tooth surface, biofilm components are perceived as "danger signals" by cells of the *innate immune system*; i.e., leukocytes and resident cells (e.g., gingival epithelial cells, periodontal ligament cells) recognize microbial lipopolysaccharide (LPS), DNA, or peptidoglycans via toll-like receptors (TLR). TLR are indispensable for the host to recognize the bacterial attack and their activation induces an intracellular signal pathway, expression of transcription factors, and subsequently release of pro-inflammatory cytokines and chemokines; altogether, an immune response focused on the infecting agents. TLR-2 and TLR-4 are primarily responsible for sensing periodontal pathogens (e.g., *A. actinomycetemcomitans* or *P. gingivalis*) (Kikkert et al. 2007). Peptidoglycans and atypical LPS from *P. gingivalis* activate TLR-2, LPS from gram-negative bacteria in general activate TLR-4, while *A. actinomycetemcomitans* activates both receptors (Akira and Takeda 2004; Ford et al. 2010). As a response to biofilm accumulation, a massive influx of neutrophils occurs in the tissues and, through the epithelium, in the periodontal sulcus. Neutrophils, except from phagocytosing bacteria, express a variety of pro-inflammatory cytokines, resulting in additional recruitment of various immune cells, including macrophages and T cells, thus propagating inflammation. Among the T cells, mainly $CD4^{+}$ T cells (T helper cells; Th) are involved in periodontal disease pathogenesis; more specifically the cells of the innate immune system activate Th1 and Th17 cells, which results in a strong pro-inflammatory immune response. Later on Th2 cells promote the establishment of a lesion with predominantly B and plasma cells. Additionally, regulatory T cells (Treg) are activated and exert an anti-inflammatory role.

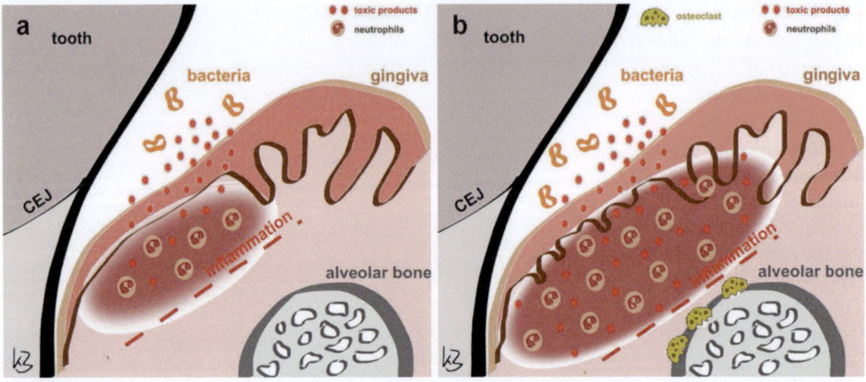

Fig. 12.5 (**a**) The inflammatory reaction is restricted to the gingiva; (**b**) the inflammatory process reaches the proximity of the alveolar bone and induces osteoclastogenesis and alveolar bone loss. (Modified by Graves et al. 2011; Schröder and Lindhe, 1981). *CEJ* cemento enamel junction

As already mentioned, the immune response initially causes a local inflammatory reaction, resulting in tissue destruction restricted to the gingiva, i.e., gingivitis. Yet, if the bacterial invasion cannot be controlled in susceptible hosts, the inflammatory reaction proceeds from the gingival margin apically towards the alveolar bone; i.e., the inflammatory response causes detachment of the junctional epithelium and its conversion into an ulcerated pocket epithelium, loss of the connective tissue attachment to the root, and loss of alveolar bone, i.e., periodontitis. An inflammatory process in close proximity to bone regularly supports differentiation and activation of osteoclast precursors, which results in uncoupling of the physiological balanced bone formation and bone resorption, in favor of the latter (Fig. 12.5a, b).

If this, often non-self-resolving, inflammatory process is left untreated, alveolar bone is continuously lost until tooth exfoliation. Hence, the immune response initiated by the bacterial attack is primary responsible for the extent, rate, and severity of host tissue degradation (Bartold et al. 2010; Brook 2003; Ford et al. 2010; Stabholz et al. 2010).

12.1.6 Mediators and Enzymes Involved in the Pathogenesis of Periodontal Disease

12.1.6.1 RANKL/RANK/OPG System

The receptor activator of nuclear factor "kappa-light-chain-enhancer" of the activated B cell ligand (RANKL)/receptor activator of NF-kB (RANK)/osteoprotegerin (OPG) system (*RANKL/RANK/OPG system)* is regulating the coupling of bone formation and resorption. RANKL, a member of the tumor necrosis factor (TNF) family, induces osteoclastogenesis by binding to its receptor RANK on osteoclast precursors. This step is promoted by macrophage colony-stimulating factor

(M-CSF), which is stimulating RANK expression on osteoclast precursors. The activation of RANK by RANKL is essential for osteoclast formation, differentiation, and activity. The soluble counterpart OPG, expressed by osteoblasts, acts as a decoy receptor binding to RANKL and thereby can inhibit the binding of RANKL to RANK (Boyle et al. 2003).

During periodontal disease, elevated levels of RANKL and/or decreased levels of OPG, resulting in a RANKL/OPG ratio that favors bone resorption, are observed (Bartold et al. 2010; Cochran 2008; Crotti et al. 2003). For example, in a preclinical trial on experimental periodontitis in mice, the expression level of RANKL correlated well with the level of various pro-inflammatory cytokines, i.e., interleukin (IL)-1, TNF-α, and interferon (IFN)-γ, during active bone loss (Garlet et al. 2006). Elevated levels of RANKL are also found in active periodontal lesions in patients, i.e., sites showing progressive alveolar bone loss; in contrast, no remarkable change from the physiological RANKL/OPG ratio was found in samples of gingivitis sites, i.e., sites without alveolar bone loss (Menezes et al. 2008). Osteoblasts, dendritic cells, B and T cells, and resident cells (e.g., periodontal ligament cells, gingival fibroblasts) are considered as important sources for the elevated levels of RANKL in periodontal lesions upon pro-inflammatory stimuli and/or TLR activation (Bar-Shavit 2008; Belibasakis et al. 2007; Kawai et al. 2006).

12.1.6.2 Matrix Metalloproteinases

Matrix metalloproteinases (MMPs) are a family of calcium- and zinc-dependent proteases, which are expressed by immune cells (e.g., neutrophils, macrophages). In healthy periodontal tissues, MMPs control the physiological turnover of extracellular matrix and are regulated by their inhibitors, i.e., *tissue inhibitors of metalloproteinases (TIMPs)*, which are expressed by resident and immune cells (e.g., macrophages, fibroblasts, endothelial cells). During inflammation, presence of pro-inflammatory cytokines deregulates the MMP/TIMP balance, resulting in higher levels of MMPs and/or lower levels of TIMPs. This imbalance leads to increased degradation of periodontal soft tissues (Giannobile 2008; Verstappen and Von den Hoff 2006). Specifically, MMP-8 – being a collagenase – is strongly associated with periodontal disease; progressively increasing levels of MMP-8 are found in sites with increasing severity of disease, i.e., from healthy to gingivitis and to periodontitis. Thus MMP-8 is considered as a strong biomarker for detecting alveolar bone loss (Ebersole et al. 2013; Gursoy et al. 2013; Leppilahti et al. 2014).

12.1.6.3 Pro-inflammatory Interleukins

IL-1 is a strict pro-inflammatory key cytokine expressed by resident (e.g., fibroblasts, epithelial cells) and immune cells (e.g., neutrophils, macrophages). IL-1 stimulates the expression of other pro-inflammatory and chemotactic factors (e.g., IL-6, TNF-α, prostaglandins) and promotes osteoclastogenesis by upregulating RANKL expression. Further, IL-1 enhances matrix degradation by inducing MMPs and impairs the regenerative potential of the tissues by inducing apoptosis of matrix producing cells (Cochran 2008; Garlet et al. 2006; Graves 2008; Graves and Cochran 2003; Wei et al. 2005). Several studies have shown increased levels in gingival crevicular fluid, saliva, and periodontal tissue samples of periodontitis patients (Duarte et al. 2007; Ebersole et al. 2013; Gamonal et al. 2000; Gursoy et al. 2009; Mogi et al. 1999).

IL-2 is a pro-inflammatory cytokine and mainly expressed by Th1 cells. IL-2 affects the differentiation, growth, and activation of T, B, and natural killer cells (Kuziel and Greene 1990). Studies have reported both unaltered and increased levels of IL-2 in chronic periodontitis patients compared to healthy controls (Teles et al. 2009; Tymkiw et al. 2011).

IL-6 is another pro-inflammatory key cytokine, which is expressed by various resident and immune cells (e.g., fibroblasts, neutrophils, macrophages). IL-6 stimulates the expression of other pro-inflammatory factors and promotes – in a similar way as IL-1 – soft and hard tissue degradation by upregulating RANKL and MMP expression and inducing apoptosis of matrix producing cells (Cochran 2008; Garlet et al. 2006; Graves 2008; Graves and Cochran 2003; Wei et al. 2005). However, while IL-1 is a strictly pro-inflammatory cytokine, IL-6 has also a regulatory function; IL-6 was shown to regulate IL-1 and TNF-α by inducing IL-1 receptor antagonist and TNF-α soluble receptor (Irwin and Myrillas 1998). Several studies have shown increased levels of IL-6 in fluid and tissue samples of periodontitis patients (Duarte et al. 2007; Ebersole et al. 2013; Gamonal et al. 2000; Gursoy et al. 2009; Mogi et al. 1999).

IL-8 is a chemokine and secreted by monocytes, macrophages, fibroblasts, keratinocytes, and endothelial cells in the presence of microorganisms and related toxins. Its main function is to attract and activate neutrophils (Bickel 1993). Studies have reported significantly increased levels of IL-8 in periodontally diseased sites (Ertugrul et al. 2013; Gamonal et al. 2000).

IL-12, mainly expressed by immune cells (e.g., dendritic cells, macrophages, neutrophils), is a key mediator of cell-mediated immunity, necessary for the initiation and maintenance of Th1 cell response, including high IFN-γ expression (Park and Scott 2001). The levels of IL-12 in gingival crevicular fluid samples may not always differ significantly between periodontally diseased and healthy patients, but it has been shown that a significant reduction of IL-12 levels occurred after initial treatment in periodontitis patients (Thunell et al. 2010).

IL-17 is a pro-inflammatory cytokine secreted by Th17 cells and by various other cell types (e.g., neutrophils, dendritic cells, periodontal ligament cells). IL-17 increases the levels of MMPs and RANKL and amplifies the pro-inflammatory loop of IL-1–IL-6 and TNF-α (Cardoso et al. 2008; Cheng et al. 2014; Ford et al. 2010; Kotake et al. 1999; Sato et al. 2006). Further, IL-17 plays a role in the defense against pathogens by increasing mobilization and activation of neutrophils (Yu et al. 2007) and by improving the responsiveness of TLR in human gingival epithelial cells (Beklen et al. 2009). Indeed, induced *P. gingivalis* infection resulted in increased alveolar bone loss in IL-17 receptor knockout mice due to an impaired immune response (Yu et al. 2007). *P. gingivalis* is considered to promote a Th17 cell response with the corresponding IL-17 expression, and indeed, increased levels of IL-17 have been described in periodontally diseased subjects (Cheng et al. 2014).

IL-18 is a pro-inflammatory cytokine, expressed by immune cells, osteoblasts, and fibroblasts, and upregulates other pro-inflammatory and chemotactic mediators (e.g., IL-1, TNF-α, IL-8). Significantly higher levels of IL-18 have been shown in periodontitis patients compared to healthy controls (Banu et al. 2015; Ozçaka et al. 2011).

IL-23 is expressed by monocytes, macrophages, and dendritic cells and promotes Th17 cell differentiation. Progressively increased levels of IL-23 in the gingival crevicular fluid from healthy to gingivitis and to periodontitis sites have been reported (Gonzales 2015; Himani et al. 2014).

IL-33 is primarily expressed intracellularly, e.g., in monocytes and endothelial and epithelial cells. It is assumed that IL-33 release after cell necrosis represents an alarming signal for cell damage and, hence, causes a pro-inflammatory immune response and cytokine production (Moussion et al. 2008; Nile et al. 2010). In human gingival tissue biopsies, the levels of IL-33 were increased compared to healthy tissue samples, and in a preclinical trial on experimental periodontitis, IL-33 expression was increased in the presence of bacterial infection and mediated bone loss via the RANKL system (Malcolm et al. 2015).

12.1.6.4 Anti-inflammatory Interleukins

IL-4 is expressed by Th2 cells and at the same time also promotes Th2 cell differentiation; further, IL-4 promotes B cell activation, differentiation, and antibody production (Appay et al. 2008; Cronstein 2007). IL-4 exerts its anti-inflammatory and antiresorptive action by decreasing IFN-γ, MMP, and RANKL expression and by elevating IL-10, TIMPs, and OPG levels (Appay et al. 2008; Garlet et al. 2006; Ihn et al. 2002; Jarnicki and Fallon 2003; Saidenberg-Kermanac'h et al. 2004). Indeed, lower IL-4 levels have been recorded in gingival crevicular fluid of periodontally diseased patients compared to those in healthy controls (Pradeep et al. 2008).

IL-10 is an anti-inflammatory cytokine, which is expressed by various cell types (e.g., T and B cells, macrophages, dendritic cells). IL-10 reduces the activity of pro-inflammatory cytokines (e.g., IL-1–IL-17, IFN-γ) (Jovanovic et al. 1998; Naundorf et al. 2009) and upregulates TIMPs and OPG, thereby reducing soft and mineralized periodontal tissue destruction (Garlet et al. 2004, 2006; Zhang and Teng 2006). IL-10 knockout mice presented a remarkably higher susceptibility against *P. gingivalis* (Sasaki et al. 2004) and a reduced expression of osteoblast and osteocyte markers (Claudino et al. 2010), which indicates that, in addition to its anti-inflammatory and antiresorptive properties, IL-10 appears to also have a direct positive effect on bone formation. In the clinic, patients with a polymorphism reducing IL-10 mRNA transcription showed also reduced TIMP-3 and OPG transcription (Claudino et al. 2008), and therefore IL-10 polymorphism has been suggested as risk factor for chronic periodontitis (Zhong et al. 2012).

IL-13, expressed mainly by Th2 cells, exerts its anti-inflammatory action by downregulating inflammatory cytokine production in monocytes and supports B cell activation and antibody production (Abbas et al. 1996; de Waal Malefyt et al. 1993). During periodontal disease, IL-13 appears to correlate with IL-4 levels but shows in general higher levels (Johnson and Serio 2007).

12.1.6.5 TNF-α

TNF-α is expressed by various resident and immune cells and is, next to IL-1 and IL-6, a pro-inflammatory key cytokine in periodontitis pathogenesis. IL-1, IL-6, and TNF-α seem to somehow complement each other during inflammatory alveolar bone loss, since blocking of all three cytokines is shown to result in a higher inhibition of alveolar bone loss compared to single cytokine blocking (Graves 2008; Zwerina et al. 2004). TNF-α has the potential to induce osteoclastogenesis even in the absence of RANKL and also to diminish bone formation by inhibiting differentiation of osteoblast precursors and proliferation of mature osteoblasts (Bartold et al. 2010; de Vries et al. 2016; Graves et al. 2011; Graves and Cochran 2003; Lacey et al. 2009; Tomomatsu et al. 2009). Progressively increasing gingival crevicular fluid levels of TNF-α from periodontally healthy to gingivitis and to periodontitis patients have been reported (Ertugrul et al. 2013).

12.1.6.6 IFN-γ

IFN-γ is expressed by Th1 cells and natural killer cells and exerts a strong pro-inflammatory action by attracting and enhancing phagocyte activity and by inducing expression of pro-inflammatory cytokines and chemokines. Although it has been shown in vitro that IFN-γ inhibits osteoclastogenesis, due to its strong pro-inflammatory action, i.e., elevation of IL-1, TNF-α, RANKL levels, this direct anti-osteoclastogenic effect is overcome in vivo, resulting altogether in alveolar bone loss (Gao et al. 2007; Garlet et al. 2008; Gemmell and Seymour 1994; Ji et al. 2009). Yet, IFN-γ plays also an important role in the defense mechanism against infections by activating cytotoxic $CD8^+$ T cells and natural killer cells; e.g., it has been shown that IFN-γ knockout mice presented severe defense impairment against *A. actinomycetemcomitans* (Garlet et al. 2008). In preclinical and clinical trials, increasing levels of IFN-γ correlated well with increasing severity of periodontal lesions (Garlet et al. 2003; Honda et al. 2006; Isaza-Guzmán et al. 2015; Teng et al. 2005).

12.1.6.7 TGF-ß

Transforming growth factor beta (TGF-ß) is part of the TGF-ß superfamily that represents a variety of proteins, including the bone morphogenetic proteins. TGF-ß is secreted by a variety of cells (e.g., Treg, macrophages, neutrophils) and can regulate the immune response by reducing pro-inflammatory cytokine (e.g., IL-1, TNF-α) and tissue-degrading enzyme (e.g., MMPs) expression; additionally, TGF-ß is considered as an important factor for wound healing. High levels of TGF-ß have been found in periodontally inflamed tissues, which might be indicative of the constant wound healing response of the host (Mize et al. 2015; Steinsvoll et al. 1999).

12.1.6.8 Complement System

The *complement system* consists of more than 40 proteins and can be activated by three different pathways, which converge with cleavage of the complement component 3 (C3) into C3a and C3b. C3a is attracting leukocytes and C3b binds covalently to target surfaces to make the pathogens more susceptible to phagocytosis by leukocytes, i.e., "complement opsonization." Although the complement system supports the immune system, continuous complement activation can be destructive for the host tissue by indirectly inducing alveolar bone loss (Damgaard et al. 2015). In particular, the membrane attack complex (C5b-9) can lead to activation of phospholipase A2, release of arachidonic acid, and synthesis of prostaglandin E2, which is a very potent osteolytic substance (Klein and Raisz 1970). Further, neutrophils and monocytes express the complement receptor 3, which is strongly activated by *P. gingivalis*; complement receptor 3 activation increases phagocyte recruitment and pro-inflammatory cytokine expression (e.g., IL-1, IL-6, TNF-α) causing increased alveolar bone loss.

12.1.6.9 Lipid Mediators

Arachidonic acid is released by phospholipase A2 from membrane phospholipids and further processed to either pro- or anti-inflammatory mediators. Pro-inflammatory lipid mediators, i.e., *leukotrienes* and *prostaglandins*, induce recruitment of neutrophils (e.g., leukotriene B4) and promote osteoclastogenesis and bone resorption (e.g., prostaglandin E2). Elevated levels of such lipid mediators have been observed in the gingival crevicular fluid of periodontally diseased patients (Offenbacher et al. 1986; Zhong et al. 2007).

The anti-inflammatory lipid mediators, i.e., *lipoxins*, *resolvins*, and *protectins*, are pro-resolving on the inflammation process (Garlet 2010; Graves et al. 2011). Pro-resolution is, in contrast to the passive termination of inflammation, an actively regulated process; e.g., anti-inflammatory lipid mediators were shown to regulate the migration and recruitment of neutrophils and T cells, to attenuate leukotriene B4-dependent pro-inflammatory signals, to reduce the expression of pro-inflammatory cytokines, and to promote T cell apoptosis (Serhan et al. 2008).

12.1.6.10 Nitric Oxide

Nitric oxide (NO), previously named "endothelial-derived relaxing factor," is synthesized during conversion of L-arginine to L-citrulline by NO-synthases. The endothelial and neuronal NO-synthases are constitutively expressed, whereas the inducible NO-synthases (iNOS) are expressed in neutrophils, macrophages, and/or gingival tissue upon pro-inflammatory cytokine expression and/or in the presence of bacterial products. NO has an important role in vascular regulation, platelet aggregation, regulation of mineralized tissue, and the pathogenesis of inflammatory diseases (Alayan et al. 2006; Daghigh et al. 2002; Matejka et al. 1998; Ugar-Cankal and Ozmeric 2006). The role of NO is most likely dual; on the one side, it is part of the defense mechanism and helps to control the bacterial attack, but on the other side, excessive amounts of NO are toxic for various cells (e.g., fibroblasts, epithelial cells) and NO is suspected to increase MMPs and reduce TIMPs (Brennan et al. 2003; Nguyen et al. 1992; Ugar-Cankal and Ozmeric 2006). Indeed, experimental periodontitis in iNOS knockout mice resulted in an elevated amount of neutrophils in the periodontal tissues and more bone, compared to that in wild-type mice (Alayan et al. 2006; Fukada et al. 2008). Further, elevated levels of iNOS have been observed in periodontally diseased tissue (Lappin et al. 2000; Matejka et al. 1998), while reduced amounts of NO_2^-, which is the stable metabolite of NO, have been found in the saliva of periodontally diseased patients (Aurer et al. 2001) (Table 12.1).

Table 12.1 Overview of mediators and enzymes involved in periodontal disease pathogenesis

Mediator/enzyme	Cell source	Main effect
Bone metabolism		
RANKL	Osteoblasts, dendritic cells, B and T cells, periodontal ligament cells, gingival fibroblasts	Induction of osteoclastogenesis
OPG	Osteoblasts	Inhibition of RANKL-RANK binding
Soft tissue metabolism		
MMP	Neutrophils, macrophages	Degradation of periodontal soft tissue
TIMP	Macrophages, fibroblasts, endothelial cells	Inhibition of MMPs
Pro-inflammatory		
IL-1	Various cells of the periodontium and immune system	Pro-inflammatory key cytokine
IL-2	Th1 cells	Activation of T and B cells and natural killer cells
IL-6	Various cells of the periodontium and the immune system	Pro-inflammatory key cytokine
IL-8	Monocytes, macrophages, fibroblasts, keratinocytes, and endothelial cells	Chemotaxis on neutrophils
IL-12	Cells of the immune system	Promotion of a Th1 cell response
IL-17	Th17 cells, neutrophils, dendritic cells, periodontal ligament cells	Pro-inflammatory cytokine and increase in tissue degradation
IL-18	Cells of the immune system, osteoblasts, fibroblasts	Pro-inflammatory cytokine
IL-23	Monocytes, macrophages, and dendritic cells	Promotion of Th17 cell differentiation
IL-33	Intracellular expression in monocytes, and endothelial and epithelial cells	Pro-inflammatory cytokine
TNF-α	Various cells of the periodontium and the immune system	Pro-inflammatory key cytokine
IFN-γ	Th1 cells, natural killer cells	Promotion of phagocyte activity and expression of pro-inflammatory cytokines
Pro-inflammatory lipid mediators	–	Promotion of osteoclastogenesis and recruitment of neutrophils
Anti-inflammatory		
IL-4	Th2 cells	Promotion of Th2 and B cell differentiation and reduction in tissue degradation
IL-10	Cells of the immune system	Anti-inflammatory key cytokine

(continued)

Table 12.1 continued

Mediator/enzyme	Cell source	Main effect
IL-13	Th2 cells	Anti-inflammatory effect on monocytes and promotion of B cell differentiation and antibody production
TGF-ß	Treg, macrophages, neutrophils	Reduction of pro-inflammatory cytokines and tissue-degrading enzymes
Anti-inflammatory lipid mediators	–	Resolution of inflammatory process
Control of bacterial attack		
Complement system	Monocytes/macrophages, fibroblasts, epithelial and endothelial cells, osteoblasts	C3a – attraction of leukocytes; C3b – "complement opsonization"
NO	Neutrophils, macrophages, gingival tissue	Control of the bacterial attack; toxic effect on various cells and promotion of soft tissue degradation

C3 complement component 3, *IFN* interferon, *IL* interleukin, *MMP* matrix metalloproteinases, *NO* nitric oxide, *OPG* osteoprotegerin, *RANK* receptor activator of NF-kB, *RANKL* receptor activator of nuclear factor "kappa-light-chain-enhancer" of the activated B cell ligand, *Th* T helper cells, *TIMP* tissue inhibitors of metalloproteinase, *TGF* transforming growth factor, *TNF* tumor necrosis factor

12.1.7 Cells Involved in the Pathogenesis of Periodontal Disease

12.1.7.1 Gingival Fibroblasts and Periodontal Ligament Cells

Gingival fibroblasts and *periodontal ligament cells* constitute the main cellular component of the gingiva and periodontal ligament, respectively, and are responsible for maintenance of tissue integrity. Both cell types can be considered as multipotent cells, possessing an *osteogenic potential* (Kook et al. 2009; Lee et al. 2009; Mostafa et al. 2011; Rodrigues et al. 2007), and hence may constitute a potential cellular source during bone wound healing. However, this osteogenic potential is diminished in the presence of tissue-degrading pro-inflammatory proteases, i.e., MMPs, during chronic inflammation (Hayami et al. 2008; Joseph et al. 2010).

12.1.7.2 Osteoblasts, Osteoclasts, and Osteocytes

Osteoblasts are derived from mesenchymal cells of the bone marrow, are lining the bone surfaces, and are responsible for bone formation. *Osteoclasts* are multinucleated monocyte/macrophage lineage cells of hematopoietic origin with the capacity to resorb bone and are found close to resorption lacunae (called Howship's lacunae). *Osteocytes* are previous osteoblasts that became entrapped within the bone matrix and reside in lacunae connecting to each other via long cytoplasmic extensions. Osteocytes can regulate the activity of osteoclasts and osteoblasts, resulting in bone loss, through expression of RANKL and sclerostin; specifically, sclerostin expression reduces osteoblast differentiation (Kim et al. 2014, 2015).

Osteoblast and osteoclasts are controlling together bone remodeling and homeostasis, i.e., the lifelong constant coupling of bone formation and resorption (Fuller et al. 1998; Lacey et al. 1998; Quinn et al. 1998; Yasuda et al. 1998). As mentioned earlier, differentiation and activation of osteoclasts are regulated by the RANKL/RANK/OPG system in combination with M-CSF. In periodontitis, osteoclastogenesis is constantly upregulated due to a high RANKL expression by resident and immune cells. Indeed, RANKL levels have been shown to correlate well with the cathepsin K levels; cathepsin K is a protease expressed by osteoclasts and is responsible for degradation of the organic matrix of bone. In addition, LPS from periodontal pathogens seems to impair osteoblastic cell differentiation partially due to TNF-α upregulation (Kadono et al. 1999; Roberts et al. 2008; Tomomatsu et al. 2009; Wang et al. 2010).

12.1.7.3 Neutrophils

Neutrophils (syn. polymorphonuclear leukocytes) differentiate in the bone marrow before entering the blood flow and are the most abundant leukocytes in the blood, accounting for about two thirds of all blood leukocytes. This constant mature state of neutrophils in the blood circulation allows an immediate immune response whenever and wherever necessary; if bacterial invasion occurs, neutrophils can exit the blood flow and migrate into the affected tissue following a chemical gradient, i.e., chemotaxis. In the periodontium, neutrophils enter the tissues constantly from the terminal blood circulation and exit into the gingival crevicular fluid, where they form a "defense wall" against potential invaders.

Neutrophil involvement in the pathogenesis of periodontal disease may be considered as either *hyporesponsive* or as *hyperresponsive*. Hyporesponsive neutrophils (e.g., due to defects in chemotaxis, transendothelial migration, or phagocytosis) weaken the first host defense mechanism and the bacterial attack might overwhelm host immunity. In contrast, an intense response of hyperresponsive neutrophils not only degrades the microbial invaders but also causes collateral tissue damage. A hyperresponsive neutrophil response is characterized by excessive release of toxic products, an elevated oxidative burst, and increased secretion of degrading enzymes (e.g., MMPs) (Gustafsson et al. 2006; Nussbaum and Shapira 2011; Ryder 2010; Shaddox et al. 2010).

Further, neutrophils can interfere directly and indirectly with the bone metabolism, being a source of pro-inflammatory cytokines (e.g., IL-1, TNF-α) that further upregulate the immune response and cause an increased degradation of soft and mineralized periodontal tissue. Neutrophils also promote the adaptive immune response via chemotactic effects on Th17 cells, which in turn release high amounts of the pro-inflammatory cytokine IL-17 (Pelletier et al. 2010). In addition, activated neutrophils can express – but not secrete – on their cell surface RANKL; hence, in close proximity to osteoclast precursors, they can directly activate and promote osteoclastogenesis (Bloemen et al. 2010; Chakravarti et al. 2009).

12.1.7.4 Monocytes

Monocytes are myeloid cells of the hematopoietic system, which migrate from the bone marrow with the blood stream in a quite immature state that enables them to further differentiate according to given requirements. For example, monocytes can enter the tissue and differentiate into *macrophages*, which are typical phagocytosing cells. Macrophages can kill pathogens by producing antimicrobial substances (e.g., myeloperoxidase, reactive oxygen species, reactive nitrogen species) and can ingest, process, and present antigens to T cells. Further, macrophages are contributing to inflammatory tissue damage by secretion of pro-inflammatory cytokines (e.g., IL-1, TNF-α). Another possible differentiation pathway of monocytes – in the presence of M-CSF and RANKL – is into osteoclasts, and thereby they are also contributing to bone resorption (Bar-Shavit 2008; Faust et al. 1999; Massey and Flanagan 1999).

Interestingly, monocytes from periodontally diseased patients may present distinct differences comparing to mononuclear cells from periodontally healthy individuals. Specifically, in vitro studies suggest that a *hyper-reactive phenotype* may exist in periodontitis patients, which upon stimulation with bacteria or LPS releases increased amounts of pro-inflammatory cytokines and generates more oxygen radicals comparing to monocytes from healthy subjects (Gustafsson et al. 2006). Further, discrepancies in *peripheral osteoclastogenesis* have been reported, with monocytes from periodontitis patients spontaneously forming osteoclasts without stimulation with M-CSF and/or RANKL. Hence, it has been suggested that the priming of the osteoclast precursors in periodontally diseased patients may already take place – at least partly – in the peripheral blood (Brunetti et al. 2005; Tjoa et al. 2008).

12.1.7.5 Dendritic Cells

Peripheral *dendritic cells* are professional antigen processing and presenting cells; after antigen incorporation, they prime naive T cells in the lymph nodes (Wolff 1972), and this naive T cell stimulation is promoted in the presence of RANKL (Anderson et al. 1997). In the presence of RANKL and M-CSF, dendritic cells differentiate into mature osteoclasts, while in the absence of RANKL, they differentiate into mature dendritic cells and support the adaptive immunity by processing and presenting antigens. However, once differentiated into mature dendritic cells, they do not posses any longer their osteoclastogenic potential (Liu et al. 2010).

Langerhans cells are dendritic cells located above the basal layer of epithelial cells, e.g., in the skin and oral mucosa. Increased numbers of Langerhans cells have occasionally been observed in the gingiva of periodontitis patients, comparing to what observed in healthy individuals (Ford et al. 2010).

12.1.7.6 Natural Killer Cells

Natural killer cells are a subset of lymphocytes and play an important role in the innate immunity. Natural killer cells express primarily IFN-γ but exhibit additional antimicrobial action by expression of defensins, which are specific antimicrobial peptides. Interestingly, tissue infiltration with activated natural killer cells has been observed in AgP but not CP lesions (Nowak et al. 2013). It has been suggested that this difference is due to the differences among the infecting periodontal pathogens. Specifically, infection with *A. actinomycetemcomitans*, which is characteristically present in localized AgP, induces via dendritic cells activation of natural killer cells and subsequent IFN-γ expression, while infection with *P. gingivalis*, which is a key periodontal pathogen in CP, does not (Chalifour et al. 2004; Gonzales 2015; Kikuchi et al. 2004; Nowak et al. 2013).

12.1.7.7 T Cells

T cells are part of the adaptive immunity, derive from the hematopoietic stem cells in the bone marrow, but migrate and mature in the thymus. They can be subdivided based on the expression of the co-receptors CD8 or CD4. *CD8+ T cells*, also called cytotoxic T cells, control intracellular antigens (e.g., viruses) and may play a role in periodontitis in the presence of coinfection with herpes virus; recognition of herpes virus by the immune system induces a strong $CD8^+$ T cell activation and mobilization (Gonzales 2015; Slots 2010).

CD4+ T cells, also called Th cells, primarily assist other cells of the immune system in their functions; for example, they assist in B cell maturation. $CD4^+$ T cells differentiate further depending on the presence of specific cytokines (Mosmann et al. 1991); while high IL-12 levels favor *Th1 cell* development, high IL-4 levels promote *Th2 cell* differentiation. Th1 and Th2 cells can, in turn, be distinguished by differences in their cytokine expression pattern. Th1 cells secrete mainly IFN-γ, IL-2, IL-12, and TNF-α and initiate a cellular and pro-inflammatory immune response characterized by presence of phagocytosing cells (Gonzales 2015), while Th2 cells appear to have a more protective function by expressing IL-4, IL-5, IL-9, IL-10, and IL-13 and promoting B cell activation, differentiation, and antibody production (Appay et al. 2008; Cronstein 2007; Gonzales 2015; Murphy and Reiner 2002). In addition to Th1 and Th2 cells, *Th17* and *Treg cells* have been detected in periodontally diseased tissues (Cardoso et al. 2008; Gaffen and Hajishengallis 2008; Gonzales 2015; Teng 2006b). Th17 cells differentiate in the presence of IL-23 and express IL-6 and mainly IL-17, which – as already described – exerts a strong pro-inflammatory immune response (Cardoso et al. 2008; Cheng et al. 2014; Ford et al. 2010; Kotake et al. 1999; Sato et al. 2006; Yu et al. 2007). In contrast, *Treg* develops in the presence of TGF-ß and IL-2 and they have a protective role by decreasing periodontal disease progression through expression of IL-10 and TGF-ß (Cardoso et al. 2008; Gonzales 2015).

12.1.7.8 B Cells

B cells belong to the adaptive immunity and they have two main functions: (a) they recognize antigens by using a high-affinity receptor and then process and present them to $CD4^+$ and $CD8^+$ T cells (Gonzales 2015) and (b) they produce and secret antibodies for opsonization of invading pathogens.

In the gingival tissue in periodontitis lesions, B and T cells are the predominant mononuclear cell types, with the number of B cells regularly exceeding that of T cells. B cells exert a dual role in periodontal disease progression; they have a major protective role through opsonization of periodontal pathogens, thus facilitating phagocytosis by neutrophils and macrophages, and through activation of the complement system (Guentsch et al. 2009; Teng 2006a). For example, *A. actinomycetemcomitans* can only be controlled by neutrophils after opsonization by IgG, while other opsonizing factors (e.g., C3b, LPS-binding proteins, other Ig isotypes) appear not effective against it (Guentsch et al. 2009; Teng 2006a). In a longitudinal human trial, patients with active periodontal lesions presented lower serum levels of IgG against *A. actinomycetemcomitans* and *P. gingivalis* compared to patients that were periodontally stable within a regular maintenance program (Rams et al. 2006). However, B cells appear to play also a major role in alveolar bone loss in periodontitis by expression of RANKL after stimulation by periodontal pathogens (Cochran 2008; Han et al. 2009; Han et al. 2006; Kawai et al. 2006); indeed, B cell deletion resulted in remarkably reduced alveolar bone loss (Baker et al. 2009). In addition, autoreactive B cells, that produce antibodies against host tissue components (e.g., collagen, keratin), have also been detected in tissue samples from periodontitis sites (Donati et al. 2009; Koutouzis et al. 2009).

Interestingly, mice severely affected by a combined immunodeficiency resulting in complete deletion of B and T cells experience less bone loss after challenge with *P. gingivalis*, comparing with non-immunodeficient controls, which indicates that lymphocytes may not be indispensable in infection control, but they play a major role in alveolar bone loss (Baker et al. 1999) (Table 12.2).

12.1.8 Periodontal Treatment

12.1.8.1 Standard Treatment

Cause-related periodontal treatment consists of *oral hygiene instructions* aiming at the efficient daily removal of the accumulating biofilm from the teeth surfaces and of the *mechanical removal of the infectious agents* already accumulated on the root surface, i.e., scaling and root planing. The mechanical removal of the biofilm can be combined with the administration of local or systemic use of chemotherapeutic agents (e.g., rinsing solutions based on chlorhexidine, hyaluronan, essential oils; antibiotics). In cases where nonsurgical treatment does not result in disease resolution, periodontal surgery is performed to achieve better access to the root surfaces for efficient mechanical instrumentation. After efficient control of the infection and establishment of inflammation-free tissues, regular maintenance appointments, including mechanical biofilm removal and reinforcement of oral hygiene attitudes, are essential for long-term stability of treatment and for prevention of disease recurrence (Bertl et al. 2015; Deas and Mealey 2010; Zandbergen et al. 2013).

12.1.8.2 "Osteoimmunological Targets" in Periodontal Treatment

Although it is nowadays recognized that the host immune response is the major cause for periodontal tissue destruction, routine periodontal treatment aims primarily to control the bacterial invasion. In periodontitis, unlike in systemic conditions like osteoporosis or rheumatoid arthritis, treatment aiming at *interfering with the host's immune system* (e.g., by applying anti-inflammatory or antiresorptive agents) is associated with two major concerns: firstly, any (severe) side effects of systemically administered medications might not justify the possible – most likely small – additional benefit compared to what achieved with standard periodontal treatment alone and secondly any (partial) blocking of the immune system may result in inefficient immune response against the bacterial attack.

Table 12.2 Overview of the dual role of resident and immune cells in periodontal disease pathogenesis

Cell type	Protective aspect	Destructive aspect
Resident cells		
Gingival fibroblasts	Osteogenic potential	Source of pro-inflammatory cytokines and RANKL
Periodontal ligament cells	Osteogenic potential	Source of pro-inflammatory cytokines and RANKL
Osteoblasts	Bone formation	Source of RANKL
Osteoclasts	–	Bone resorption
Osteocytes	Regulator of bone homeostasis	Source of RANKL and sclerostin
Innate immunity		
Neutrophils	"Defense wall"	Source of pro-inflammatory cytokines and RANKL; causing collateral tissue damage
Monocytes	Differentiation into macrophages	Differentiation into osteoclasts
Macrophages	Phagocytosis of pathogens and antigen presentation to T cells	Source of pro-inflammatory cytokines; causing collateral tissue damage
Dendritic cells	Antigen presentation	Differentiation into osteoclasts
Natural killer cells	Expression of antimicrobial defensins	IFN-γ expression
Adaptive immunity		
Thl cells	Increase in phagocyte activity and in vitro inhibition of osteoclastogenesis via IFN-γ	Expression of IFN-γ, IL-2, IL-12, and TNF-α and promotion of expression of other pro-inflammatory cytokines
Th2 cells	Expression of IL-4, IL-5, IL-9, IL-10, and IL-13 and promotion of B cell activation	Strong Th2 response favors RANKL expressing B cells
Th17 cells	Activation of neutrophils	Expression of IL-6 and IL-17 and promotion of expression of other pro-inflammatory cytokines
Treg cells	Expression of IL-10 and TGF-ß	–
B cells	Production of antibodies	Source of RANKL and possibility of auoantibodies

IFN interferon, *IL* interleukin, *RANKL* receptor activator of nuclear factor "kappa-light-chain-enhancer" of the activated B cell ligand, *Treg* regulatory T cells, *Th* T helper cells, *TGF* transforming growth factor, *TNF* tumor necrosis factor

In the following, a brief overview of major new therapeutic targets considered in periodontitis treatment is discussed (for summary, see Table 12.3).

Doxycycline

Doxycycline delivered in *subantimicrobial doses* (e.g., 20 mg doxycycline twice daily for 3 months) has been shown to downregulate MMP activity and, more recently, also pro-inflammatory cytokine production, i.e., IL-1, IL-17, and TNF-α. A systematic review of the literature and meta-analysis have confirmed that there

Table 12.3 Overview of potential osteoimmunological targets for periodontitis treatment, evaluated in preclinical and/or clinical studies

Target	Agent	Mode of action
Preclinical trials		
RANKL/RANK/OPG system	OPG/OPG analogues	Blockade of RANKL-RANK binding
–	Lipid mediators	Pro-resolving and anti-inflammatory
Sclerostin	Sclerostin antibody	Improved bone formation
Osteoclasts	Bisphosphonates	Reduced osteoclast activity and bone turnover
Cathepsin K	Cathepsin K inhibitors	Reduced bone loss
iNOS	iNOS inhibitors	Reduced NO production
Mitogen-activated protein kinases	Mitogen-activated protein kinases inhibitors	Anti-inflammatory and antiresorptive
–	Melatonin	Antioxidant and free radical scavenger
Angiotensin II receptor	Angiotensin II receptor blocker	Anti-inflammatory
Clinical trials		
MMPs	Doxycycline	Inhibition of MMPs
TNF-α	TNF-α-antagonists	Anti-inflammatory and antiresorptive
–	Statins	Anti-inflammatory and antiresorptive
–	Vitamin D and calcium	Regulation of bone homeostasis and immune system

iNOS inducible NO-synthases, *MMP* matrix metalloproteinases, *NO* nitric oxide, *OPG* osteoprotegerin, *RANK* receptor activator of NF-kB, *RANKL* receptor activator of nuclear factor "kappa-light-chain-enhancer" of the activated B cell ligand, *TNF* tumor necrosis factor

seems to be a significant improvement of clinical periodontal parameters after nonsurgical periodontal treatment with adjunctive delivery of doxycycline in subantimicrobial doses compared to nonsurgical treatment alone (Castro et al. 2016; Sgolastra et al. 2011).

Statins

Statins (e.g., simvastatin, atorvastatin) are widely used to control lipid metabolism by reducing serum cholesterol levels. Additionally, they appear to have anti-inflammatory and antiresorptive properties. Clinical trials have reported a significant positive effect of locally or systemically administered statins as adjunct to nonsurgical periodontal treatment, primarily in terms of posttreatment alveolar bone levels but also regarding the clinical periodontal parameters (Fajardo et al. 2010; Pradeep et al. 2012; Pradeep et al. 2013; Pradeep et al. 2015).

TNF-α Antagonists

TNF-α antagonists are used in the treatment of rheumatoid arthritis. Preclinical trials evaluating the effect of TNF-α antagonists in experimental periodontitis have shown reduced neutrophil infiltration and periodontal inflammation levels (Di Paola et al. 2007; Gonçalves et al. 2014). However, reduced neutrophil infiltration may impair pathogen clearance. For example, increased *A. actinomycetemcomitans* load and inflammatory marker levels have been detected in a TNF-α knockout mouse model; in particular, knockout mice presented reduced activation, migration, and phagocytosing activity of neutrophils and macrophages in periodontal lesions (Garlet et al. 2007). In a case-control study, however, patients with rheumatoid arthritis taking TNF-α-antagonists presented better periodontal indices and lower gingival crevicular fluid levels of TNF-α, compared to patients taking another medication (Mayer et al. 2009).

Lipid Mediators

Pro-resolving *lipid mediators*, i.e., lipoxins, resolvins, and protectins, as adjunct to periodontal therapy, would theoretically reduce the amount of tissue degradation by actively promoting inflammation resolution. Indeed, the extent of tissue breakdown during experimental periodontitis and disease resolution was significantly improved in preclinical trials involving local application of resolvins (Hasturk et al. 2007; Van Dyke et al. 2015).

OPG

Systemic and local administrations of *OPG* or *OPG analogues* in experimental periodontitis models have shown significant alveolar bone loss prevention, due to blocking RANKL binding to RANK and, hence, interfering with the process of osteoclastogenesis (Jin et al. 2007; Lin et al. 2011; Tang et al. 2015; Teng et al. 2000). However, systemic long-term administration of a RANKL inhibitor may entail unwanted systemic side effects on physiologic bone turnover and on the immune system, considering the fact that RANK is not only expressed on osteoclasts and their precursors but also on monocytes/macrophages and dendritic cells (Ferrari-Lacraz and Ferrari 2011; Kong et al. 1999).

Sclerostin Antibody

In a preclinical trial in rats, administration of a *sclerostin antibody* significantly improved bone healing in periodontal defects, by blocking the inhibitory effect of sclerostin on osteogenesis (Taut et al. 2013).

Bisphosphonates

Bisphosphonates reduce osteoclast activity and thereby the amount of bone resorption and are regularly used in the treatment of osteoporosis or tumor-associated osteolysis. Systemic application of bisphosphonates appears not applicable in periodontitis treatment, due to the interference with the physiologic bone turnover, but also because it might cause severe side effects, i.e., bisphosphonate-related osteonecrosis of the jaw; in fact, periodontitis itself appears as risk factor for developing bisphosphonate-related osteonecrosis of the jaw. Yet, it has been demonstrated in preclinical trials of experimental periodontitis models that local administration of bisphosphonates results in reduced alveolar bone destruction and in enhanced alveolar bone regeneration (De Almeida et al. 2015; Furlaneto et al. 2014; Thumbigere-Math et al. 2014).

Vitamin D and Calcium

Vitamin D plays important role in bone homeostasis and also exerts additional immune regulatory properties; for instance, vitamin D deficiency is associated with reduced calcium levels in the bone and bone volume loss. Indeed, a moderate positive effect of *vitamin D* and *calcium* supplement intake has been described in cross-sectional and cohort studies (Garcia et al. 2011; Miley et al. 2009). However, there are currently no randomized controlled clinical trials on vitamin D and calcium supplementation during nonsurgical periodontal treatment.

Cathepsin K Inhibitors

Cathepsin K inhibitors are considered as promising drugs in osteoporosis treatment. Recently, *inhibition of cathepsin K* was shown to reduce the extent of the immune response and of alveolar bone loss in experimental periodontitis in mice (Bartold et al. 2010; Hao et al. 2015).

iNOS Inhibitors

iNOS inhibitors (e.g., mercaptoethyl guanidine, aminoguanidine) are shown in vitro to reduce NO production by human gingival fibroblasts, while, in an experimental periodontitis model, they decreased the level of periodontal inflammation and bone loss (Chang et al. 2014; Daghigh et al. 2002; Di Paola et al. 2004; Lohinai et al. 1998).

Mitogen-Activated Protein Kinase

Mitogen-activated protein kinases are intracellular molecules involved in signal transduction during inflammation and production of pro-inflammatory cytokines. Administration of *mitogen-activated protein kinases inhibitors* in an experimental periodontitis model reduced the inflammatory response and thereby also the alveolar bone loss (Kirkwood et al. 2007; Rogers et al. 2007).

Melatonin

Melatonin, a hormone primary involved in the control of the circadian rhythm, acts also as an antioxidant, immune modulator, and free radical scavenger. It has been demonstrated that the levels of salivary melatonin are decreased in periodontal disease, while systemic melatonin administration during experimental periodontitis reduced the level of pro-inflammatory cytokines and the extent of alveolar bone destruction (Arabacı et al. 2015; Bertl et al. 2013; Kara et al. 2013).

Angiotensin II Receptor Blockers

Angiotensin II receptor blockers (e.g., azilsartan, olmesartan, telmisartan) exert an anti-inflammatory effect by suppressing TNF-α-induced IL-6 gene promoter activity. In an experimental periodontitis model, angiotensin II receptor blocker administration resulted in reduced tissue degradation, including reduced levels of MMPs and RANKL, increased levels of OPG, reduced levels of IL-1, TNF-α, and increased levels IL-10 cytokines (Araújo et al. 2013a, b, 2014) compared to controls (Table 12.3).

12.1.9 Summary

Periodontal disease is initiated by oral pathogens that provoke a response from the immune system, which exerts a two-sided effect: on one side, it controls the infection and protects the organism from bacterial invasion, but on the other side, it causes collateral tissue destruction within the periodontium. This dual effect is reflected in the often two-sided duties of cells and mediators involved in periodontitis pathogenesis. Interestingly, it appears that the intensity of the immune response is not the main relevant factor for periodontal infection control but is decisive for the extent of periodontal tissue destruction. Thus, the stronger the immune response is, the larger the damage caused in the periodontal tissues (Trombone et al. 2009). In contrast, a deficient immune response also leads to increased tissue damage due to failure in controlling the infection. The increased knowledge on the interactions between cells of the immune system and the resident cells in the periodontal tissues and on the involved enzymes and cytokines offers new exciting targets for the treatment of periodontal disease; yet, evaluation of most of these targets is still on the preclinical level (Fig. 12.6).

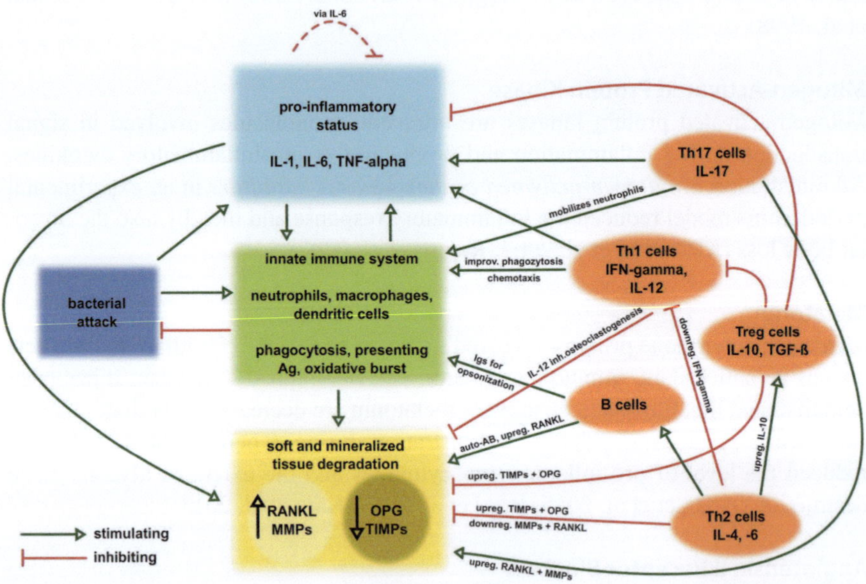

Fig. 12.6 Graphic presentation of the enzymes, mediators, and cells involved in periodontitis pathogenesis, including their interactions and their effects on the soft and mineralized periodontal tissues

References

Abbas AK, Murphy KM, Sher A (1996) Functional diversity of helper T lymphocytes. Nature 383:787–793

Akira S, Takeda K (2004) Toll-like receptor signalling. Nat Rev Immunol 4:499–511

Alayan J, Ivanovski S, Gemmell E, Ford P, Hamlet S, Farah CS (2006) Deficiency of iNOS contributes to Porphyromonas gingivalis-induced tissue damage. Oral Microbiol Immunol 21:360–365

Anderson DM, Maraskovsky E, Billingsley WL et al (1997) A homologue of the TNF receptor and its ligand enhance T-cell growth and dendritic-cell function. Nature 390:175–179

Appay V, van Lier RA, Sallusto F, Roederer M (2008) Phenotype and function of human T lymphocyte subsets: consensus and issues. Cytometry A 73:975–983

Arabacı T, Kermen E, Özkanlar S et al (2015) Therapeutic effects of melatonin on alveolar bone resorption after experimental periodontitis in rats: a biochemical and immunohistochemical study. J Periodontol 86:874–881

Araújo AA, Lopes de Souza G, Souza TO et al (2013a) Olmesartan decreases IL-1β and TNF-α levels; downregulates MMP-2, MMP-9, COX-2, and RANKL; and upregulates OPG in experimental periodontitis. Naunyn Schmiedebergs Arch Pharmacol 386:875–884

Araújo AA, Souza TO, Moura LM et al (2013b) Effect of telmisartan on levels of IL-1, TNF-α, down-regulated COX-2, MMP-2, MMP-9 and RANKL/RANK in an experimental periodontitis model. J Clin Periodontol 40:1104–1111

Araújo AA, Varela H, Brito GA et al (2014) Azilsartan increases levels of IL-10, down-regulates MMP-2, MMP-9, RANKL/RANK, Cathepsin K and up-regulates OPG in an experimental periodontitis model. PLoS One 9, e96750

Armitage GC (1999) Development of a classification system for periodontal diseases and conditions. Ann Periodontol 4:1–6

Armitage GC, Cullinan MP (2010) Comparison of the clinical features of chronic and aggressive periodontitis. Periodontol 2000 53:12–27

Aurer A, Aleksic J, Ivic-Kardum M, Aurer J, Culo F (2001) Nitric oxide synthesis is decreased in periodontitis. J Clin Periodontol 28:565–568

Baker PJ, Dixon M, Evans RT, Dufour L, Johnson E, Roopenian DC (1999) CD4(+) T cells and the proinflammatory cytokines gamma interferon and interleukin-6 contribute to alveolar bone loss in mice. Infect Immun 67:2804–2809

Baker PJ, Boutaugh NR, Tiffany M, Roopenian DC (2009) B cell IgD deletion prevents Alveolar bone loss following murine oral infection. Interdiscip Perspect Infect Dis 2009:864359

Banu S, Jabir NR, Mohan R et al (2015) Correlation of Toll-like receptor 4, interleukin-18, transaminases, and uric acid in patients with chronic periodontitis and healthy adults. J Periodontol 86:431–439

Bar-Shavit Z (2008) Taking a toll on the bones: regulation of bone metabolism by innate immune regulators. Autoimmunity 41:195–203

Bartold PM, Cantley MD, Haynes DR (2010) Mechanisms and control of pathologic bone loss in periodontitis. Periodontol 2000 53:55–69

Beklen A, Sorsa T, Konttinen YT (2009) Toll-like receptors 2 and 5 in human gingival epithelial cells co-operate with T-cell cytokine interleukin-17. Oral Microbiol Immunol 24:38–42

Belibasakis GN, Bostanci N, Hashim A et al (2007) Regulation of RANKL and OPG gene expression in human gingival fibroblasts and periodontal ligament cells by Porphyromonas gingivalis: a putative role of the Arg-gingipains. Microb Pathog 43:46–53

Bertl K, Schoiber A, Haririan H et al (2013) Non-surgical periodontal therapy influences salivary melatonin levels. Clin Oral Investig 17:1219–1225

Bertl K, Bruckmann C, Isberg PE, Klinge B, Gotfredsen K, Stavropoulos A (2015) Hyaluronan in non-surgical and surgical periodontal therapy: a systematic review. J Clin Periodontol 42:236–246

Bickel M (1993) The role of interleukin-8 in inflammation and mechanisms of regulation. J Periodontol 64:456–460

Bloemen V, Schoenmaker T, de Vries TJ, Everts V (2010) Direct cell-cell contact between periodontal ligament fibroblasts and osteoclast precursors synergistically increases the expression of genes related to osteoclastogenesis. J Cell Physiol 222:565–573

Boyle WJ, Simonet WS, Lacey DL (2003) Osteoclast differentiation and activation. Nature 423:337–342

Bradshaw DJ, Marsh PD, Watson GK, Allison C (1998) Role of Fusobacterium nucleatum and coaggregation in anaerobe survival in planktonic and biofilm oral microbial communities during aeration. Infect Immun 66:4729–4732

Brennan PA, Thomas GJ, Langdon JD (2003) The role of nitric oxide in oral diseases. Arch Oral Biol 48:93–100

Brook I (2003) Microbiology and management of periodontal infections. Gen Dent 51:424–428

Brunetti G, Colucci S, Pignataro P et al (2005) T cells support osteoclastogenesis in an in vitro model derived from human periodontitis patients. J Periodontol 76:1675–1680

Cardoso CR, Garlet GP, Moreira AP, Junior WM, Rossi MA, Silva JS (2008) Characterization of CD4+CD25+ natural regulatory T cells in the inflammatory infiltrate of human chronic periodontitis. J Leukoc Biol 84:311–318

Castro ML, Franco GC, Branco-de-Almeida LS et al (2016) Down-Regulation of protease activated receptor 2, interleukin-17 and other pro-inflammatory genes by subantimicrobial doxycycline dose in a rat periodontitis model. J Periodontol 87:203–210

Chakravarti A, Raquil MA, Tessier P, Poubelle PE (2009) Surface RANKL of Toll-like receptor 4-stimulated human neutrophils activates osteoclastic bone resorption. Blood 114:1633–1644

Chalifour A, Jeannin P, Gauchat JF et al (2004) Direct bacterial protein PAMP recognition by human NK cells involves TLRs and triggers alpha-defensin production. Blood 104:1778–1783

Chang PC, Chong LY, Tsai SC, Lim LP (2014) Aminoguanidine inhibits the AGE-RAGE axis to modulate the induction of periodontitis but has limited effects on the progression and recovery of experimental periodontitis: a preliminary study. J Periodontol 85:729–739

Cheng WC, Hughes FJ, Taams LS (2014) The presence, function and regulation of IL-17 and Th17 cells in periodontitis. J Clin Periodontol 41:541–549

Claudino M, Trombone AP, Cardoso CR et al (2008) The broad effects of the functional IL-10 promoter-592 polymorphism: modulation of IL-10, TIMP-3, and OPG expression and their association with periodontal disease outcome. J Leukoc Biol 84:1565–1573

Claudino M, Garlet TP, Cardoso CR et al (2010) Down-regulation of expression of osteoblast and osteocyte markers in periodontal tissues associated with the spontaneous alveolar bone loss of interleukin-10 knockout mice. Eur J Oral Sci 118:19–28

Cochran DL (2008) Inflammation and bone loss in periodontal disease. J Periodontol 79:1569–1576

Cronstein BN (2007) Interleukin-6--a key mediator of systemic and local symptoms in rheumatoid arthritis. Bull NYU Hosp Jt Dis 65(Suppl 1):S11–S15

Crotti T, Smith MD, Hirsch R et al (2003) Receptor activator NF kappaB ligand (RANKL) and osteoprotegerin (OPG) protein expression in periodontitis. J Periodontal Res 38:380–387

Daghigh F, Borghaei RC, Thornton RD, Bee JH (2002) Human gingival fibroblasts produce nitric oxide in response to proinflammatory cytokines. J Periodontol 73:392–400

Damgaard C, Holmstrup P, Van Dyke TE, Nielsen CH (2015) The complement system and its role in the pathogenesis of periodontitis: current concepts. J Periodontal Res 50:283–293

De Almeida J, Ervolino E, Bonfietti LH et al (2015) Adjuvant therapy with sodium alendronate for the treatment of experimental periodontitis in rats. J Periodontol 86:1166–1175

de Vries TJ, Yousovich J, Schoenmaker T, Scheres N, Everts V (2016) Tumor necrosis factor-α antagonist infliximab inhibits osteoclast formation of peripheral blood mononuclear cells but does not affect periodontal ligament fibroblast-mediated osteoclast formation. J Periodontal Res 51:186–195

de Waal Malefyt R, Figdor CG, Huijbens R et al (1993) Effects of IL-13 on phenotype, cytokine production, and cytotoxic function of human monocytes. Comparison with IL-4 and modulation by IFN-gamma or IL-10. J Immunol 151:6370–6381

Deas DE, Mealey BL (2010) Response of chronic and aggressive periodontitis to treatment. Periodontol 2000 53:154–166

Di Paola R, Marzocco S, Mazzon E et al (2004) Effect of aminoguanidine in ligature-induced periodontitis in rats. J Dent Res 83:343–348

Di Paola R, Mazzon E, Muià C et al (2007) Effects of etanercept, a tumour necrosis factor-alpha antagonist, in an experimental model of periodontitis in rats. Br J Pharmacol 150:286–297

Donati M, Liljenberg B, Zitzmann NU, Berglundh T (2009) B-1a cells and plasma cells in periodontitis lesions. J Periodontal Res 44:683–688

Duarte PM, de Oliveira MC, Tambeli CH, Parada CA, Casati MZ, Nociti FH (2007) Overexpression of interleukin-1beta and interleukin-6 may play an important role in periodontal breakdown in type 2 diabetic patients. J Periodontal Res 42:377–381

Ebersole JL, Schuster JL, Stevens J et al (2013) Patterns of salivary analytes provide diagnostic capacity for distinguishing chronic adult periodontitis from health. J Clin Immunol 33:271–279

Eke PI, Page RC, Wei L, Thornton-Evans G, Genco RJ (2012) Update of the case definitions for population-based surveillance of periodontitis. J Periodontol 83:1449–1454

Ertugrul AS, Sahin H, Dikilitas A, Alpaslan N, Bozoglan A (2013) Comparison of CCL28, interleukin-8, interleukin-1β and tumor necrosis factor-alpha in subjects with gingivitis, chronic periodontitis and generalized aggressive periodontitis. J Periodontal Res 48:44–51

Fajardo ME, Rocha ML, Sánchez-Marin FJ, Espinosa-Chávez EJ (2010) Effect of atorvastatin on chronic periodontitis: a randomized pilot study. J Clin Periodontol 37:1016–1022

Faust J, Lacey DL, Hunt P et al (1999) Osteoclast markers accumulate on cells developing from human peripheral blood mononuclear precursors. J Cell Biochem 72:67–80

Ferrari-Lacraz S, Ferrari S (2011) Do RANKL inhibitors (denosumab) affect inflammation and immunity? Osteoporos Int 22:435–46

Flemmig TF (1999) Periodontitis. Ann Periodontol 4:32–38

Ford PJ, Gamonal J, Seymour GJ (2010) Immunological differences and similarities between chronic periodontitis and aggressive periodontitis. Periodontol 2000 53:111–123

Foster JS, Kolenbrander PE (2004) Development of a multispecies oral bacterial community in a saliva-conditioned flow cell. Appl Environ Microbiol 70:4340–4348

Fukada SY, Silva TA, Saconato IF et al (2008) iNOS-derived nitric oxide modulates infection-stimulated bone loss. J Dent Res 87:1155–1159

Fuller K, Wong B, Fox S, Choi Y, Chambers TJ (1998) TRANCE is necessary and sufficient for osteoblast-mediated activation of bone resorption in osteoclasts. J Exp Med 188:997–1001

Furlaneto FA, Nunes NL, Oliveira Filho IL et al (2014) Effects of locally administered tiludronic acid on experimental periodontitis in rats. J Periodontol 85:1291–1301

Gaffen SL, Hajishengallis G (2008) A new inflammatory cytokine on the block: re-thinking periodontal disease and the Th1/Th2 paradigm in the context of Th17 cells and IL-17. J Dent Res 87:817–828

Gamonal J, Acevedo A, Bascones A, Jorge O, Silva A (2000) Levels of interleukin-1 beta, −8, and −10 and RANTES in gingival crevicular fluid and cell populations in adult periodontitis patients and the effect of periodontal treatment. J Periodontol 71:1535–1545

Gao Y, Grassi F, Ryan MR et al (2007) IFN-gamma stimulates osteoclast formation and bone loss in vivo via antigen-driven T cell activation. J Clin Invest 117:122–132

Garcia MN, Hildebolt CF, Miley DD et al (2011) One-year effects of vitamin D and calcium supplementation on chronic periodontitis. J Periodontol 82:25–32

Garlet GP (2010) Destructive and protective roles of cytokines in periodontitis: a re-appraisal from host defense and tissue destruction viewpoints. J Dent Res 89:1349–1363

Garlet GP, Martins WJ, Ferreira BR, Milanezi CM, Silva JS (2003) Patterns of chemokines and chemokine receptors expression in different forms of human periodontal disease. J Periodontal Res 38:210–217

Garlet GP, Martins WJ, Fonseca BA, Ferreira BR, Silva JS (2004) Matrix metalloproteinases, their physiological inhibitors and osteoclast factors are differentially regulated by the cytokine profile in human periodontal disease. J Clin Periodontol 31:671–679

Garlet GP, Cardoso CR, Silva TA et al (2006) Cytokine pattern determines the progression of experimental periodontal disease induced by Actinobacillus actinomycetemcomitans through the modulation of MMPs, RANKL, and their physiological inhibitors. Oral Microbiol Immunol 21:12–20

Garlet GP, Cardoso CR, Campanelli AP et al (2007) The dual role of p55 tumour necrosis factor-alpha receptor in Actinobacillus actinomycetemcomitans-induced experimental periodontitis: host protection and tissue destruction. Clin Exp Immunol 147:128–138

Garlet GP, Cardoso CR, Campanelli AP et al (2008) The essential role of IFN-gamma in the control of lethal Aggregatibacter actinomycetemcomitans infection in mice. Microbes Infect 10:489–496

Gemmell E, Seymour GJ (1994) Cytokines and T cell switching. Crit Rev Oral Biol Med 5:249–279

Genco RJ, Borgnakke WS (2013) Risk factors for periodontal disease. Periodontol 2000 62:59–94

Giannobile WV (2008) Host-response therapeutics for periodontal diseases. J Periodontol 79:1592–1600

Gonçalves DC, Evangelista RC, da Silva RR et al (2014) Infliximab attenuates inflammatory osteolysis in a model of periodontitis in Wistar rats. Exp Biol Med (Maywood) 239:442–453

Gonzales JR (2015) T- and B-cell subsets in periodontitis. Periodontol 2000 69:181–200

Graves D (2008) Cytokines that promote periodontal tissue destruction. J Periodontol 79:1585–1591

Graves DT, Cochran D (2003) The contribution of interleukin-1 and tumor necrosis factor to periodontal tissue destruction. J Periodontol 74:391–401

Graves DT, Li J, Cochran DL (2011) Inflammation and uncoupling as mechanisms of periodontal bone loss. J Dent Res 90:143–153

Guentsch A, Puklo M, Preshaw PM et al (2009) Neutrophils in chronic and aggressive periodontitis in interaction with Porphyromonas gingivalis and Aggregatibacter actinomycetemcomitans. J Periodontal Res 44:368–377

Gursoy UK, Könönen E, Uitto VJ et al (2009) Salivary interleukin-1beta concentration and the presence of multiple pathogens in periodontitis. J Clin Periodontol 36:922–927

Gursoy UK, Könönen E, Huumonen S et al (2013) Salivary type I collagen degradation end-products and related matrix metalloproteinases in periodontitis. J Clin Periodontol 40:18–25

Gustafsson A, Ito H, Asman B, Bergstrom K (2006) Hyper-reactive mononuclear cells and neutrophils in chronic periodontitis. J Clin Periodontol 33:126–129

Hajishengallis G (2014) Immunomicrobial pathogenesis of periodontitis: keystones, pathobionts, and host response. Trends Immunol 35:3–11

Hajishengallis G, Lamont RJ (2012) Beyond the red complex and into more complexity: the polymicrobial synergy and dysbiosis (PSD) model of periodontal disease etiology. Mol Oral Microbiol 27:409–419

Hajishengallis G, Liang S, Payne MA et al (2011) Low-abundance biofilm species orchestrates inflammatory periodontal disease through the commensal microbiota and complement. Cell Host Microbe 10:497–506

Hajishengallis G, Darveau RP, Curtis MA (2012) The keystone-pathogen hypothesis. Nat Rev Microbiol 10:717–725

Han X, Kawai T, Eastcott JW, Taubman MA (2006) Bacterial-responsive B lymphocytes induce periodontal bone resorption. J Immunol 176:625–631

Han X, Lin X, Seliger AR, Eastcott J, Kawai T, Taubman MA (2009) Expression of receptor activator of nuclear factor-kappaB ligand by B cells in response to oral bacteria. Oral Microbiol Immunol 24:190–196

Hao L, Chen J, Zhu Z et al (2015) Odanacatib, a cathepsin K-Specific inhibitor, inhibits inflammation and bone loss caused by periodontal diseases. J Periodontol 86:972–983

Hasturk H, Kantarci A, Goguet-Surmenian E et al (2007) Resolvin E1 regulates inflammation at the cellular and tissue level and restores tissue homeostasis in vivo. J Immunol 179:7021–7029

Hayami T, Kapila YL, Kapila S (2008) MMP-1 (collagenase-1) and MMP-13 (collagenase-3) differentially regulate markers of osteoblastic differentiation in osteogenic cells. Matrix Biol 27:682–692

Himani GS, Prabhuji ML, Karthikeyan BV (2014) Gingival crevicular fluid and interleukin-23 concentration in systemically healthy subjects: their relationship in periodontal health and disease. J Periodontal Res 49:237–245

Holt SC, Ebersole JL (2005) Porphyromonas gingivalis, Treponema denticola, and Tannerella forsythia: the "red complex", a prototype polybacterial pathogenic consortium in periodontitis. Periodontol 2000 38:72–122

Holtfreter B, Kocher T, Hoffmann T, Desvarieux M, Micheelis W (2010) Prevalence of periodontal disease and treatment demands based on a German dental survey (DMS IV). J Clin Periodontol 37:211–219

Honda T, Domon H, Okui T, Kajita K, Amanuma R, Yamazaki K (2006) Balance of inflammatory response in stable gingivitis and progressive periodontitis lesions. Clin Exp Immunol 144:35–40

Ihn H, Yamane K, Asano Y, Kubo M, Tamaki K (2002) IL-4 up-regulates the expression of tissue inhibitor of metalloproteinase-2 in dermal fibroblasts via the p38 mitogen-activated protein kinase dependent pathway. J Immunol 168:1895–1902

Irwin CR, Myrillas TT (1998) The role of IL-6 in the pathogenesis of periodontal disease. Oral Dis 4:43–47

Isaza-Guzmán DM, Cardona-Vélez N, Gaviria-Correa DE, Martínez-Pabón MC, Castaño-Granada MC, Tobón-Arroyave SI (2015) Association study between salivary levels of interferon (IFN)-gamma, interleukin (IL)-17, IL-21, and IL-22 with chronic periodontitis. Arch Oral Biol 60:91–99

Jarnicki AG, Fallon PG (2003) T helper type-2 cytokine responses: potential therapeutic targets. Curr Opin Pharmacol 3:449–455

Jenkinson HF, Lamont RJ (2005) Oral microbial communities in sickness and in health. Trends Microbiol 13:589–595

Ji JD, Park-Min KH, Shen Z et al (2009) Inhibition of RANK expression and osteoclastogenesis by TLRs and IFN-gamma in human osteoclast precursors. J Immunol 183:7223–7233

Jin Q, Cirelli JA, Park CH et al (2007) RANKL inhibition through osteoprotegerin blocks bone loss in experimental periodontitis. J Periodontol 78:1300–1308

Johnson RB, Serio FG (2007) The contribution of interleukin-13 and −15 to the cytokine network within normal and diseased gingiva. J Periodontol 78:691–695

Joseph J, Kapila YL, Hayami T, Kapila S (2010) Disease-associated extracellular matrix suppresses osteoblastic differentiation of human periodontal ligament cells via MMP-1. Calcif Tissue Int 86:154–162

Jovanovic DV, Di Battista JA, Martel-Pelletier J et al (1998) IL-17 stimulates the production and expression of proinflammatory cytokines, IL-beta and TNF-alpha, by human macrophages. J Immunol 160:3513–3521

Kadono H, Kido J, Kataoka M, Yamauchi N, Nagata T (1999) Inhibition of osteoblastic cell differentiation by lipopolysaccharide extract from Porphyromonas gingivalis. Infect Immun 67:2841–2846

Kara A, Akman S, Ozkanlar S et al (2013) Immune modulatory and antioxidant effects of melatonin in experimental periodontitis in rats. Free Radic Biol Med 55:21–26

Kawai T, Matsuyama T, Hosokawa Y et al (2006) B and T lymphocytes are the primary sources of RANKL in the bone resorptive lesion of periodontal disease. Am J Pathol 169:987–998

Kikkert R, Laine ML, Aarden LA, van Winkelhoff AJ (2007) Activation of toll-like receptors 2 and 4 by gram-negative periodontal bacteria. Oral Microbiol Immunol 22:145–151

Kikuchi T, Hahn CL, Tanaka S, Barbour SE, Schenkein HA, Tew JG (2004) Dendritic cells stimulated with Actinobacillus actinomycetemcomitans elicit rapid gamma interferon responses by natural killer cells. Infect Immun 72:5089–5096

Kim JH, Lee DE, Cha JH, Bak EJ, Yoo YJ (2014) Receptor activator of nuclear factor-kB ligand and sclerostin expression in osteocytes of alveolar bone in rats with ligature-induced periodontitis. J Periodontol 85:e370–e378

Kim JH, Lee DE, Woo GH, Cha JH, Bak EJ, Yoo YJ (2015) Osteocytic sclerostin expression in alveolar bone in rats with diabetes mellitus and ligature-induced periodontitis. J Periodontol 86:1005–1011

Kirkwood KL, Li F, Rogers JE et al (2007) A p38alpha selective mitogen-activated protein kinase inhibitor prevents periodontal bone loss. J Pharmacol Exp Ther 320:56–63
Klein DC, Raisz LG (1970) Prostaglandins: stimulation of bone resorption in tissue culture. Endocrinology 86:1436–1440
Kong YY, Yoshida H, Sarosi I et al (1999) OPGL is a key regulator of osteoclastogenesis, lymphocyte development and lymph-node organogenesis. Nature 397:315–323
Kook YA, Lee SK, Son DH et al (2009) Effects of substance P on osteoblastic differentiation and heme oxygenase-1 in human periodontal ligament cells. Cell Biol Int 33:424–428
Kotake S, Udagawa N, Takahashi N et al (1999) IL-17 in synovial fluids from patients with rheumatoid arthritis is a potent stimulator of osteoclastogenesis. J Clin Invest 103:1345–1352
Koutouzis T, Haber D, Shaddox L, Aukhil I, Wallet SM (2009) Autoreactivity of serum immunoglobulin to periodontal tissue components: a pilot study. J Periodontol 80:625–633
Kulkarni C, Kinane DF (2014) Host response in aggressive periodontitis. Periodontol 2000 65:79–91
Kuziel WA, Greene WC (1990) Interleukin-2 and the IL-2 receptor: new insights into structure and function. J Invest Dermatol 94:27S–32S
Lacey DL, Timms E, Tan HL et al (1998) Osteoprotegerin ligand is a cytokine that regulates osteoclast differentiation and activation. Cell 93:165–176
Lacey DC, Simmons PJ, Graves SE, Hamilton JA (2009) Proinflammatory cytokines inhibit osteogenic differentiation from stem cells: implications for bone repair during inflammation. Osteoarthritis Cartilage 17:735–742
Lappin DF, Kjeldsen M, Sander L, Kinane DF (2000) Inducible nitric oxide synthase expression in periodontitis. J Periodontal Res 35:369–373
Lee HJ, Pi SH, Kim Y et al (2009) Effects of nicotine on antioxidant defense enzymes and RANKL expression in human periodontal ligament cells. J Periodontol 80:1281–1288
Leppilahti JM, Hernández-Ríos PA, Gamonal JA et al (2014) Matrix metalloproteinases and myeloperoxidase in gingival crevicular fluid provide site-specific diagnostic value for chronic periodontitis. J Clin Periodontol 41:348–356
Lin X, Han X, Kawai T, Taubman MA (2011) Antibody to RANKL Ameliorates T cell-mediated Periodontal Bone Resorption. Infect Immun 79:911–7
Listgarten MA (1994) The structure of dental plaque. Periodontol 2000 5:52–65
Liu YC, Lerner UH, Teng YT (2010) Cytokine responses against periodontal infection: protective and destructive roles. Periodontol 2000 52:163–206
Loe H, Theilade E, Jensen SB (1965) Experimental gingivitis in man. J Periodontol 36:177–187
Lohinai Z, Benedek P, Feher E et al (1998) Protective effects of mercaptoethylguanidine, a selective inhibitor of inducible nitric oxide synthase, in ligature-induced periodontitis in the rat. Br J Pharmacol 123:353–360
Malcolm J, Awang RA, Oliver-Bell J et al (2015) IL-33 exacerbates periodontal disease through induction of RANKL. J Dent Res 94:968–975
Massey HM, Flanagan AM (1999) Human osteoclasts derive from CD14-positive monocytes. Br J Haematol 106:167–170
Matejka M, Partyka L, Ulm C, Solar P, Sinzinger H (1998) Nitric oxide synthesis is increased in periodontal disease. J Periodontal Res 33:517–518
Mayer Y, Balbir-Gurman A, Machtei EE (2009) Anti-tumor necrosis factor-alpha therapy and periodontal parameters in patients with rheumatoid arthritis. J Periodontol 80:1414–1420
Menezes R, Garlet TP, Letra A et al (2008) Differential patterns of receptor activator of nuclear factor kappa B ligand/osteoprotegerin expression in human periapical granulomas: possible association with progressive or stable nature of the lesions. J Endod 34:932–938
Miley DD, Garcia MN, Hildebolt CF et al (2009) Cross-sectional study of vitamin D and calcium supplementation effects on chronic periodontitis. J Periodontol 80:1433–1439
Mize TW, Sundararaj KP, Leite RS, Huang Y (2015) Increased and correlated expression of connective tissue growth factor and transforming growth factor beta 1 in surgically removed periodontal tissues with chronic periodontitis. J Periodontal Res 50:315–319

Mogi M, Otogoto J, Ota N, Inagaki H, Minami M, Kojima K (1999) Interleukin 1 beta, interleukin 6, beta 2-microglobulin, and transforming growth factor-alpha in gingival crevicular fluid from human periodontal disease. Arch Oral Biol 44:535–539

Mosmann TR, Schumacher JH, Street NF et al (1991) Diversity of cytokine synthesis and function of mouse CD4+ T cells. Immunol Rev 123:209–229

Mostafa NZ, Uludağ H, Varkey M, Dederich DN, Doschak MR, El-Bialy TH (2011) In vitro osteogenic induction of human gingival fibroblasts for bone regeneration. Open Dent J 5:139–145

Moussion C, Ortega N, Girard JP (2008) The IL-1-like cytokine IL-33 is constitutively expressed in the nucleus of endothelial cells and epithelial cells in vivo: a novel 'alarmin'. PLoS One 3, e3331

Murphy KM, Reiner SL (2002) The lineage decisions of helper T cells. Nat Rev Immunol 2:933–944

Naundorf S, Schroder M, Hoflich C, Suman N, Volk HD, Grutz G (2009) IL-10 interferes directly with TCR-induced IFN-gamma but not IL-17 production in memory T cells. Eur J Immunol 39:1066–1077

Nguyen T, Brunson D, Crespi CL, Penman BW, Wishnok JS, Tannenbaum SR (1992) DNA damage and mutation in human cells exposed to nitric oxide in vitro. Proc Natl Acad Sci U S A 89:3030–3034

Nile CJ, Barksby E, Jitprasertwong P, Preshaw PM, Taylor JJ (2010) Expression and regulation of interleukin-33 in human monocytes. Immunology 130:172–180

Nowak M, Krämer B, Haupt M et al (2013) Activation of invariant NK T cells in periodontitis lesions. J Immunol 190:2282–2291

Nussbaum G, Shapira L (2011) How has neutrophil research improved our understanding of periodontal pathogenesis. J Clin Periodontol 38(Suppl 11):49–59

Offenbacher S, Odle BM, Van Dyke TE (1986) The use of crevicular fluid prostaglandin E2 levels as a predictor of periodontal attachment loss. J Periodontal Res 21:101–112

Ozçaka O, Nalbantsoy A, Buduneli N (2011) Interleukin-17 and interleukin-18 levels in saliva and plasma of patients with chronic periodontitis. J Periodontal Res 46:592–598

Page RC, Eke PI (2007) Case definitions for use in population-based surveillance of periodontitis. J Periodontol 78:1387–1399

Page RC, Kornman KS (1997) The pathogenesis of human periodontitis: an introduction. Periodontol 2000 14:9–11

Park AY, Scott P (2001) Il-12: keeping cell-mediated immunity alive. Scand J Immunol 53: 529–532

Pelletier M, Maggi L, Micheletti A et al (2010) Evidence for a cross-talk between human neutrophils and Th17 cells. Blood 115:335–343

Pradeep AR, Roopa Y, Swati PP (2008) Interleukin-4, a T-helper 2 cell cytokine, is associated with the remission of periodontal disease. J Periodontal Res 43:712–716

Pradeep AR, Priyanka N, Kalra N, Naik SB, Singh SP, Martande S (2012) Clinical efficacy of subgingivally delivered 1.2-mg simvastatin in the treatment of individuals with Class II furcation defects: a randomized controlled clinical trial. J Periodontol 83:1472–1479

Pradeep AR, Kumari M, Rao NS, Martande SS, Naik SB (2013) Clinical efficacy of subgingivally delivered 1.2% atorvastatin in chronic periodontitis: a randomized controlled clinical trial. J Periodontol 84:871–879

Pradeep AR, Karvekar S, Nagpal K, Patnaik K, Guruprasad CN, Kumaraswamy KM (2015) Efficacy of locally delivered 1.2% rosuvastatin gel to non-surgical treatment of patients with chronic periodontitis: a randomized, placebo-controlled clinical trial. J Periodontol 86:738–745

Quinn JM, Elliott J, Gillespie MT, Martin TJ (1998) A combination of osteoclast differentiation factor and macrophage-colony stimulating factor is sufficient for both human and mouse osteoclast formation in vitro. Endocrinology 139:4424–4427

Rams TE, Listgarten MA, Slots J (2006) Actinobacillus actinomycetemcomitans and Porphyromonas gingivalis subgingival presence, species-specific serum immunoglobulin G antibody levels, and periodontitis disease recurrence. J Periodontal Res 41:228–234

Ritz HL (1967) Microbial population shifts in developing human dental plaque. Arch Oral Biol 12:1561–1568

Roberts HC, Moseley R, Sloan AJ, Youde SJ, Waddington RJ (2008) Lipopolysaccharide alters decorin and biglycan synthesis in rat alveolar bone osteoblasts: consequences for bone repair during periodontal disease. Eur J Oral Sci 116:207–216

Rodrigues TL, Marchesan JT, Coletta RD et al (2007) Effects of enamel matrix derivative and transforming growth factor-beta1 on human periodontal ligament fibroblasts. J Clin Periodontol 34:514–522

Rogers JE, Li F, Coatney DD et al (2007) A p38 mitogen-activated protein kinase inhibitor arrests active alveolar bone loss in a rat periodontitis model. J Periodontol 78:1992–1998

Ryder MI (2010) Comparison of neutrophil functions in aggressive and chronic periodontitis. Periodontol 2000 53:124–137

Saidenberg-Kermanac'h N, Bessis N, Lemeiter D, de Vernejoul MC, Boissier MC, Cohen-Solal M (2004) Interleukin-4 cellular gene therapy and osteoprotegerin decrease inflammation-associated bone resorption in collagen-induced arthritis. J Clin Immunol 24:370–378

Sasaki H, Okamatsu Y, Kawai T, Kent R, Taubman M, Stashenko P (2004) The interleukin-10 knockout mouse is highly susceptible to Porphyromonas gingivalis-induced alveolar bone loss. J Periodontal Res 39:432–441

Sato K, Suematsu A, Okamoto K et al (2006) Th17 functions as an osteoclastogenic helper T cell subset that links T cell activation and bone destruction. J Exp Med 203:2673–2682

Schröder HE, Lindhe J (1981) Conditions and pathological features of rapidly destructive, experimental periodontitis in dogs. J Periodontol 51:6–19

Serhan CN, Chiang N, Van Dyke TE (2008) Resolving inflammation: dual anti-inflammatory and pro-resolution lipid mediators. Nat Rev Immunol 8:349–361

Sgolastra F, Petrucci A, Gatto R, Giannoni M, Monaco A (2011) Long-term efficacy of subantimicrobial-dose doxycycline as an adjunctive treatment to scaling and root planing: a systematic review and meta-analysis. J Periodontol 82:1570–1581

Shaddox L, Wiedey J, Bimstein E et al (2010) Hyper-responsive phenotype in localized aggressive periodontitis. J Dent Res 89:143–148

Slots J (2010) Human viruses in periodontitis. Periodontol 2000 53:89–110

Stabholz A, Soskolne WA, Shapira L (2010) Genetic and environmental risk factors for chronic periodontitis and aggressive periodontitis. Periodontol 2000 53:138–153

Steinsvoll S, Halstensen TS, Schenck K (1999) Extensive expression of TGF-beta1 in chronically-inflamed periodontal tissue. J Clin Periodontol 26:366–373

Tang H, Mattheos N, Yao Y, Jia Y, Ma L, Gong P (2015) In vivo osteoprotegerin gene therapy preventing bone loss induced by periodontitis. J Periodontal Res 50:434–443

Taut AD, Jin Q, Chung JH et al (2013) Sclerostin antibody stimulates bone regeneration after experimental periodontitis. J Bone Miner Res 28:2347–2356

Teles RP, Likhari V, Socransky SS, Haffajee AD (2009) Salivary cytokine levels in subjects with chronic periodontitis and in periodontally healthy individuals: a cross-sectional study. J Periodontal Res 44:411–417

Teles R, Teles F, Frias-Lopez J, Paster B, Haffajee A (2013) Lessons learned and unlearned in periodontal microbiology. Periodontol 2000 62:95–162

Teng YT (2006a) Protective and destructive immunity in the periodontium: Part 1--innate and humoral immunity and the periodontium. J Dent Res 85:198–208

Teng YT (2006b) Protective and destructive immunity in the periodontium: Part 2--T-cell-mediated immunity in the periodontium. J Dent Res 85:209–219

Teng YT, Nguyen H, Gao X et al (2000) Functional human T-cell immunity and osteoprotegerin ligand control alveolar bone destruction in periodontal infection. J Clin Invest 106:R59–R67

Teng YT, Mahamed D, Singh B (2005) Gamma interferon positively modulates Actinobacillus actinomycetemcomitans-specific RANKL+ CD4+ Th-cell-mediated alveolar bone destruction in vivo. Infect Immun 73:3453–3461

Thumbigere-Math V, Michalowicz BS, Hodges JS et al (2014) Periodontal disease as a risk factor for bisphosphonate-related osteonecrosis of the jaw. J Periodontol 85:226–233

Thunell DH, Tymkiw KD, Johnson GK et al (2010) A multiplex immunoassay demonstrates reductions in gingival crevicular fluid cytokines following initial periodontal therapy. J Periodontal Res 45:148–152

Tjoa ST, de Vries TJ, Schoenmaker T, Kelder A, Loos BG, Everts V (2008) Formation of osteoclast-like cells from peripheral blood of periodontitis patients occurs without supplementation of macrophage colony-stimulating factor. J Clin Periodontol 35:568–575

Tomomatsu N, Aoki K, Alles N et al (2009) LPS-induced inhibition of osteogenesis is TNF-alpha dependent in a murine tooth extraction model. J Bone Miner Res 24:1770–1781

Tonetti MS, Mombelli A (1999) Early-onset periodontitis. Ann Periodontol 4:39–53

Trombone AP, Ferreira SBJ, Raimundo FM et al (2009) Experimental periodontitis in mice selected for maximal or minimal inflammatory reactions: increased inflammatory immune responsiveness drives increased alveolar bone loss without enhancing the control of periodontal infection. J Periodontal Res 44:443–451

Tymkiw KD, Thunell DH, Johnson GK et al (2011) Influence of smoking on gingival crevicular fluid cytokines in severe chronic periodontitis. J Clin Periodontol 38:219–228

Ugar-Cankal D, Ozmeric N (2006) A multifaceted molecule, nitric oxide in oral and periodontal diseases. Clin Chim Acta 366:90–100

Van Dyke TE, Hasturk H, Kantarci A et al (2015) Proresolving nanomedicines activate bone regeneration in periodontitis. J Dent Res 94:148–156

Verstappen J, Von den Hoff JW (2006) Tissue inhibitors of metalloproteinases (TIMPs): their biological functions and involvement in oral disease. J Dent Res 85:1074–1084

Wang YH, Jiang J, Zhu Q et al (2010) Porphyromonas gingivalis lipids inhibit osteoblastic differentiation and function. Infect Immun 78:3726–3735

Wei S, Kitaura H, Zhou P, Ross FP, Teitelbaum SL (2005) IL-1 mediates TNF-induced osteoclastogenesis. J Clin Invest 115:282–290

Wolff K (1972) The langerhans cell. Curr Probl Dermatol 4:79–145

Yasuda H, Shima N, Nakagawa N et al (1998) Osteoclast differentiation factor is a ligand for osteoprotegerin/osteoclastogenesis-inhibitory factor and is identical to TRANCE/RANKL. Proc Natl Acad Sci U S A 95:3597–3602

Yu JJ, Ruddy MJ, Wong GC et al (2007) An essential role for IL-17 in preventing pathogen-initiated bone destruction: recruitment of neutrophils to inflamed bone requires IL-17 receptor-dependent signals. Blood 109:3794–3802

Zandbergen D, Slot DE, Cobb CM, Van der Weijden FA (2013) The clinical effect of scaling and root planing and the concomitant administration of systemic amoxicillin and metronidazole: a systematic review. J Periodontol 84:332–351

Zhang X, Teng YT (2006) Interleukin-10 inhibits gram-negative-microbe-specific human receptor activator of NF-kappaB ligand-positive CD4+−Th1-cell-associated alveolar bone loss in vivo. Infect Immun 74:4927–4931

Zhong Y, Slade GD, Beck JD, Offenbacher S (2007) Gingival crevicular fluid interleukin-1beta, prostaglandin E2 and periodontal status in a community population. J Clin Periodontol 34:285–293

Zhong Q, Ding C, Wang M, Sun Y, Xu Y (2012) Interleukin-10 gene polymorphisms and chronic/aggressive periodontitis susceptibility: a meta-analysis based on 14 case–control studies. Cytokine 60:47–54

Zwerina J, Hayer S, Tohidast-Akrad M et al (2004) Single and combined inhibition of tumor necrosis factor, interleukin-1, and RANKL pathways in tumor necrosis factor-induced arthritis: effects on synovial inflammation, bone erosion, and cartilage destruction. Arthritis Rheum 50:277–290

Index

P. Pietschmann (ed.), *Principles of Osteoimmunology*, DOI 10.1007/978-3-319-34238-2

MIX
Papier aus verantwortungsvollen Quellen
Paper from responsible sources
FSC® C105338

If you have any concerns about our products,
you can contact us on
ProductSafety@springernature.com

In case Publisher is established outside the EU,
the EU authorized representative is:
Springer Nature Customer Service Center GmbH
Europaplatz 3, 69115 Heidelberg, Germany

Printed by Libri Plureos GmbH
in Hamburg, Germany